T0140124

Advances in Intelligent Systems and Computing

Volume 1051

The series "Advances in Intelligent Systems and Computing" contains publications on theory, applications, and design methods of Intelligent Systems and Intelligent Computing. Virtually all disciplines such as engineering, natural sciences, computer and information science, ICT, economics, business, e-commerce, environment, healthcare, life science are covered. The list of topics spans all the areas of modern intelligent systems and computing such as: computational intelligence, soft computing including neural networks, fuzzy systems, evolutionary computing and the fusion of these paradigms, social intelligence, ambient intelligence, computational neuroscience, artificial life, virtual worlds and society, cognitive science and systems, Perception and Vision, DNA and immune based systems, self-organizing and adaptive systems, e-Learning and teaching, human-centered and human-centric computing, recommender systems, intelligent control, robotics and mechatronics including human-machine teaming, knowledge-based paradigms, learning paradigms, machine ethics, intelligent data analysis, knowledge management, intelligent agents, intelligent decision making and support, intelligent network security, trust management, interactive entertainment, Web intelligence and multimedia.

The publications within "Advances in Intelligent Systems and Computing" are primarily proceedings of important conferences, symposia and congresses. They cover significant recent developments in the field, both of a foundational and applicable character. An important characteristic feature of the series is the short publication time and world-wide distribution. This permits a rapid and broad dissemination of research results.

**** Indexing: The books of this series are submitted to ISI Proceedings, EI-Compendex, DBLP, SCOPUS, Google Scholar and Springerlink ****

More information about this series at http://www.springer.com/series/11156

Jerzy Świątek · Leszek Borzemski ·
Zofia Wilimowska
Editors

Information Systems Architecture and Technology: Proceedings of 40th Anniversary International Conference on Information Systems Architecture and Technology – ISAT 2019

Part II

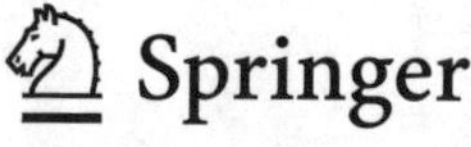

Editors
Jerzy Świątek
Faculty of Computer Science
and Management
Wrocław University of Science
and Technology
Wrocław, Poland

Leszek Borzemski
Faculty of Computer Science
and Management
Wrocław University of Science
and Technology
Wrocław, Poland

Zofia Wilimowska
University of Applied Sciences in Nysa
Nysa, Poland

ISSN 2194-5357 ISSN 2194-5365 (electronic)
Advances in Intelligent Systems and Computing
ISBN 978-3-030-30603-8 ISBN 978-3-030-30604-5 (eBook)
https://doi.org/10.1007/978-3-030-30604-5

This Springer imprint is published by the registered company Springer Nature Switzerland AG
The registered company address is: Gewerbestrasse 11, 6330 Cham, Switzerland

Preface

We are pleased to present before you the proceedings of the 2019 40th Anniversary International Conference Information Systems Architecture and Technology (ISAT), or ISAT 2019 for short, held on September 15–17, 2019 in Wrocław, Poland. The conference was organized by the Department of Computer Science, Faculty of Computer Science and Management, Wrocław University of Science and Technology, Poland, and the University of Applied Sciences in Nysa, Poland.

The International Conference on Information Systems Architecture and Technology has been organized by the Wrocław University of Science and Technology from the eighties of the last century. Most of the events took place in Szklarska Poręba and Karpacz charming small towns in the Karkonosze Mountains, Lower Silesia in the southwestern part of Poland. This year 2019, we celebrate the 40th anniversary of the conference in Wrocław—the capital of Lower Silesia, a city with a thousand-year history. A beautiful and modern city that is developing dynamically and is a meeting point for people from all over the world. It is worth noting that Wrocław is currently one of the most important centers for the development of modern software and information systems in Poland.

The past four decades have also been a period of dynamic development of computer science, which we can recall when reviewing conference materials from these years—their shape and content were always created with current achievements of national and international IT.

The purpose of the ISAT is to discuss a state-of-art of information systems concepts and applications as well as architectures and technologies supporting contemporary information systems. The aim is also to consider an impact of knowledge, information, computing and communication technologies on managing of the organization scope of functionality as well as on enterprise information systems design, implementation, and maintenance processes taking into account various methodological, technological, and technical aspects. It is also devoted to information systems concepts and applications supporting the exchange of goods and services by using different business models and exploiting opportunities offered by Internet-based electronic business and commerce solutions.

ISAT is a forum for specific disciplinary research, as well as on multi-disciplinary studies to present original contributions and to discuss different subjects of today's information systems planning, designing, development, and implementation.

The event is addressed to the scientific community, people involved in a variety of topics related to information, management, computer and communication systems, and people involved in the development of business information systems and business computer applications. ISAT is also devoted as a forum for the presentation of scientific contributions prepared by MSc. and Ph.D. students. Business, Commercial, and Industry participants are welcome.

This year, we received 141 papers from 20 countries. The papers included in the three proceedings volumes have been subject to a thoroughgoing review process by highly qualified peer reviewers. The final acceptance rate was 60%. Program Chairs selected 85 best papers for oral presentation and publication in the 40th International Conference Information Systems Architecture and Technology 2019 proceedings.

The papers have been clustered into three volumes:

Part I—discoursing about essential topics of information technology including, but not limited to, Computer Systems Security, Computer Network Architectures, Distributed Computer Systems, Quality of Service, Cloud Computing and High-Performance Computing, Human–Computer Interface, Multimedia Systems, Big Data, Knowledge Discovery and Data Mining, Software Engineering, E-Business Systems, Web Design, Optimization and Performance, Internet of Things, Mobile Systems, and Applications.

Part II—addressing topics including, but not limited to, Pattern Recognition and Image Processing Algorithms, Production Planning and Management Systems, Big Data Analysis, Knowledge Discovery, and Knowledge-Based Decision Support and Artificial Intelligence Methods and Algorithms.

Part III—is gain to address very hot topics in the field of today's various computer-based applications—is devoted to information systems concepts and applications supporting the managerial decisions by using different business models and exploiting opportunities offered by IT systems. It is dealing with topics including, but not limited to, Knowledge-Based Management, Modeling of Financial and Investment Decisions, Modeling of Managerial Decisions, Production and Organization Management, Project Management, Risk Management, Small Business Management, Software Tools for Production, Theories, and Models of Innovation.

We would like to thank the Program Committee Members and Reviewers, essential for reviewing the papers to ensure a high standard of the ISAT 2019 conference, and the proceedings. We thank the authors, presenters, and participants of ISAT 2019 without them the conference could not have taken place. Finally, we thank the organizing team for the efforts this and previous years in bringing the conference to a successful conclusion.

We hope that ISAT conference is a good scientific contribution to the development of information technology not only in the region but also internationally. It happens, among others, thanks to cooperation with Springer Publishing House, where the AISC series is issued from 2015. We want to thank Springer's people who deal directly with the publishing process, from publishing contracts to the delivering of printed books. Thank you for your cooperation.

September 2019

Leszek Borzemski
Jerzy Świątek
Zofia Wilimowska

Organization

ISAT 2019 Conference Organization

General Chair

Leszek Borzemski, Poland

Program Co-chairs

Leszek Borzemski, Poland
Jerzy Świątek, Poland
Zofia Wilimowska, Poland

Local Organizing Committee

Leszek Borzemski (Chair)
Zofia Wilimowska (Co-chair)
Jerzy Świątek (Co-chair)
Mariusz Fraś (Conference Secretary, Website Support)
Arkadiusz Górski (Technical Editor)
Anna Kamińska (Technical Secretary)
Ziemowit Nowak (Technical Support)
Kamil Nowak (Website Coordinator)
Danuta Seretna-Sałamaj (Technical Secretary)

International Program Committee

Leszek Borzemski (Chair), Poland
Jerzy Świątek (Co-chair), Poland
Zofia Wilimowska (Co-chair), Poland
Witold Abramowicz, Poland
Dhiya Al-Jumeily, UK
Iosif Androulidakis, Greece
Patricia Anthony, New Zealand
Zbigniew Banaszak, Poland
Elena N. Benderskaya, Russia
Janos Botzheim, Japan
Djallel E. Boubiche, Algeria
Patrice Boursier, France
Anna Burduk, Poland
Andrii Buriachenko, Ukraine

Udo Buscher, Germany
Wojciech Cellary, Poland
Haruna Chiroma, Malaysia
Edward Chlebus, Poland
Gloria Cerasela Crisan, Romania
Marilia Curado, Portugal
Czesław Daniłowicz, Poland
Zhaohong Deng, China
Małgorzata Dolińska, Poland
Ewa Dudek-Dyduch, Poland
Milan Edl, Czech Republic
El-Sayed M. El-Alfy, Saudi Arabia
Peter Frankovsky, Slovakia
Naoki Fukuta, Japan
Bogdan Gabryś, UK
Piotr Gawkowski, Poland
Manuel Graña, Spain
Katsuhiro Honda, Japan
Marian Hopej, Poland
Zbigniew Huzar, Poland
Natthakan Iam-On, Thailand
Biju Issac, UK
Arun Iyengar, USA
Jürgen Jasperneite, Germany
Janusz Kacprzyk, Poland
Henryk Kaproń, Poland
Yury Y. Korolev, Belarus
Yannis L. Karnavas, Greece
Ryszard Knosala, Poland
Zdzisław Kowalczuk, Poland
Lumír Kulhanek, Czech Republic
Binod Kumar, India
Jan Kwiatkowski, Poland
Antonio Latorre, Spain
Radim Lenort, Czech Republic
Gang Li, Australia
José M. Merigó Lindahl, Chile
Jose M. Luna, Spain
Emilio Luque, Spain
Sofian Maabout, France
Lech Madeyski, Poland
Zygmunt Mazur, Poland
Elżbieta Mączyńska, Poland
Pedro Medeiros, Portugal
Toshiro Minami, Japan
Marian Molasy, Poland
Zbigniew Nahorski, Poland
Kazumi Nakamatsu, Japan
Peter Nielsen, Denmark
Tadashi Nomoto, Japan
Cezary Orłowski, Poland
Sandeep Pachpande, India
Michele Pagano, Italy
George A. Papakostas, Greece
Zdzisław Papir, Poland
Marek Pawlak, Poland
Jan Platoš, Czech Republic
Tomasz Popławski, Poland
Edward Radosinski, Poland
Wolfgang Renz, Germany
Dolores I. Rexachs, Spain
José S. Reyes, Spain
Leszek Rutkowski, Poland
Sebastian Saniuk, Poland
Joanna Santiago, Portugal
Habib Shah, Malaysia
J. N. Shah, India
Jeng Shyang, Taiwan
Anna Sikora, Spain
Marcin Sikorski, Poland
Małgorzata Sterna, Poland
Janusz Stokłosa, Poland
Remo Suppi, Spain
Edward Szczerbicki, Australia
Eugeniusz Toczyłowski, Poland
Elpida Tzafestas, Greece
José R. Villar, Spain
Bay Vo, Vietnam
Hongzhi Wang, China
Leon S. I. Wang, Taiwan
Junzo Watada, Japan
Eduardo A. Durazo Watanabe, India
Jan Werewka, Poland
Thomas Wielicki, USA
Bernd Wolfinger, Germany
Józef Woźniak, Poland
Roman Wyrzykowski, Poland
Yue Xiao-Guang, Hong Kong
Jaroslav Zendulka, Czech Republic
Bernard Ženko, Slovenia

ISAT 2019 Reviewers

Hamid Al-Asadi, Iraq
S. Balakrishnan, India
Zbigniew Banaszak, Poland
Agnieszka Bieńkowska, Poland
Grzegorz Bocewicz, Poland
Leszek Borzemski, Poland
Janos Botzheim, Hungary
Krzysztof Brzostowski, Poland
Anna Burduk, Poland
Wojciech Cellary, Poland
Haruna Chiroma, Malaysia
Grzegorz Chodak, Poland
Piotr Chwastyk, Poland
Gloria Cerasela Crisan, Romania
Anna Czarnecka, Poland
Mariusz Czekała, Poland
Yousef Daradkeh, Saudi Arabia
Grzegorz Debita, Poland
Anna Jolanta Dobrowolska, Poland
Jarosław Drapała, Poland
Maciej Drwal, Poland
Tadeusz Dudycz, Poland
Grzegorz Filcek, Poland
Mariusz Fraś, Poland
Piotr Gawkowski, Poland
Dariusz Gąsior, Poland
Arkadiusz Górski, Poland
Jerzy Grobelny, Poland
Krzysztof Grochla, Poland
Houda Hakim Guermazi, Tunisia
Biju Issac, UK
Jerzy Józefczyk, Poland
Ireneusz Jóźwiak, Poland
Krzysztof Juszczyszyn, Poland
Tetiana Viktorivna Kalashnikova, Ukraine
Jan Kałuski, Poland
Anna Maria Kamińska, Poland
Radosław Katarzyniak, Poland
Agata Klaus-Rosińska, Poland
Grzegorz Kołaczek, Poland
Iryna Koshkalda, Ukraine
Zdzisław Kowalczuk, Poland
Dorota Kuchta, Poland
Binod Kumar, India
Jan Kwiatkowski, Poland
Wojciech Lorkiewicz, Poland
Marek Lubicz, Poland
Zbigniew Malara, Poland
Mariusz Mazurkiewicz, Poland
Vojtěch Merunka, Czech Republic
Rafał Michalski, Poland
Bożena Mielczarek, Poland
Peter Nielsen, Denmark
Ziemowit Nowak, Poland
Donat Orski, Poland
Michele Pagano, Italy
Jonghyun Park, Korea
Agnieszka Parkitna, Poland
Marek Pawlak, Poland
Dolores Rexachs, Spain
Paweł Rola, Poland
Stefano Rovetta, Italy
Abdel-Badeeh Salem, Egypt
Joanna Santiago, Portugal
Danuta Seretna-Sałamaj, Poland
Anna Sikora, Spain
Marcin Sikorski, Poland
Jan Skonieczny, Poland
Malgorzata Sterna, Poland
Janusz Stokłosa, Poland
Grażyna Suchacka, Poland
Joanna Szczepańska, Poland
Edward Szczerbicki, Australia
Jerzy Świątek, Poland
Kamila Urbańska, Poland
Jan Werewka, Poland
Zofia Wilimowska, Poland
Marek Wilimowski, Poland
Bernd Wolfinger, Germany
Józef Woźniak, Poland
Krzysztof Zatwarnicki, Poland
Jaroslav Zendulka, Czech Republic
Chunbiao Zhu, People's Republic of China

ISAT 2019 Keynote Speaker

Professor Cecilia Zanni-Merk, Normandie Université, INSA Rouen, LITIS, Saint-Etienne-du-Rouvray, France

Topic: **On the Need of an Explainable Artificial Intelligence**

Contents

Systems Analysis and Modeling

Knowledge Discovery and Data Mining

Production Planning and Management System

Synthesis of No-Wait Cyclic Schedules for Cascade-Like Systems of Repetitive Processes with Fixed Periods

Robert Wójcik[1(✉)], Grzegorz Bocewicz[2], and Zbigniew Banaszak[2]

[1] Department of Computer Engineering, Faculty of Electronics, Wrocław University of Science and Technology, Wrocław, Poland
robert.wojcik@pwr.edu.pl

[2] Department of Computer Science and Management, Koszalin University of Technology, Koszalin, Poland
{grzegorz.bocewicz, zbigniew.banaszak}@ie.tu.koszalin.pl

Abstract. The paper deals with the problem of synthesis of cyclic schedules for cascade-like topology repetitive systems that share resources using the mutual exclusion rule. Such processes are represented by, for example, repetitive transport tasks occurring in AGV systems, in which the carriages share sectors of routes belonging to neighboring manufacturing cells or cyclic operations carried out in urban transport systems such as train, subway, in which individual lines use shared hubs (i.e. inter change stations, cross-platforms, etc.) combining mesh-like communication networks. In this context, the problem in question concerns determining the operation times of cascade-like repetitive processes and their start times, which guarantee the existence of a no-wait cyclic schedule of processes. An extended variant of the cascade-like system is considered, in which, in comparison to the classical "chain" model with two operations carried out as part of the process, each cyclic component process contains additional operations. The purpose of this work is to provide necessary and sufficient conditions for process operations and their start times to ensure that there is a cyclic schedule free of waiting for resources, and then apply the developed conditions in the synthesis procedure of the cascade-like system parameters.

Keywords: Cascade-like topology · Repetitive processes · Cycle time · No-wait cyclic schedules · Synthesis of parameters · Constraint satisfaction problem

1 Introduction

In many modern transport systems [7, 22, 23] (e.g. metro, railway, bus lines) and automated production systems [10, 17, 24] (e.g. AMSs, AGVs) there are repetitive tasks executed concurrently, each of which has the form of a sequence of operations, carried out using system resources (e.g. machines, buffers, sectors of transport networks). Particular tasks, often identified with certain processes implemented in the

J. Świątek et al. (Eds.): ISAT 2019, AISC 1051, pp. 3–15, 2020.
https://doi.org/10.1007/978-3-030-30604-5_1

system, are defined by means of various parameters [3, 7, 16, 24], which describe, among others: resource and operation sets, way of flow of tasks (processes) through the system (e.g. flow-shop, job-shop), restrictions on the sequence of operations (e.g. in the form of directed graphs of operations), operation times (i.e. availability of resources), as well as additional restrictions [16, 21, 32] resulting from the specific use of resources by processes and system activities (e.g. sharing resources in the mutual exclusion mode, blocking the resources while waiting for other resources, no waiting for the resources).

In order to ensure the effective operation of the systems, appropriate schedules of the flow of tasks (processes) that will guarantee their correct implementation as well as the optimization of selected criteria are necessary. The cyclic scheduling problems considered concern mainly minimization of cycle time [1, 12, 27], minimization of waiting times for resources, as well as maximization of the level of resource utilization in flow-shop [1, 16, 19] or job-shop systems [12, 13, 24]. Inclusion of additional restrictions in the form of resource blocking (no-store, no-buffer restrictions) in classical cyclic scheduling problems [1, 2, 13, 27] while waiting for availability of other resources necessary to perform another operation, and/or no waiting the initiated tasks for the execution of subsequent operations after the end of the previous ones [5, 6, 19, 23] leads to decision and optimization problems, which are difficult computationally [18, 21] (NP-hard). Formulated problems often do not have any solutions acceptable due to the occurrence of deadlock state of processes and the need to wait for shared resources.

In production systems, the no-store constraint [1, 16, 27] means no buffer storage of the semi-products between the machines or the limitation of their capacity. As a consequence, it can lead to blocking machines while waiting for products to perform subsequent operations (machine downtime), and a system deadlock state [9, 20, 23], which will stop the execution of tasks. In turn, in transport systems, the no-buffer restriction is related to the lack of parallelization of sectors belonging to the routes and blocking of sectors by transport devices (e.g. trains, AGVs) [10, 11, 22, 23] while waiting for entry to subsequent sectors.

In the case of a no-wait constraint, the initiated tasks (processes) can not wait for the next operations to start, however, depending on the interpretation [5, 6, 23, 29, 30] this may apply to all operations or only selected operations, e.g. on machines right after transport operations. Often, no-wait and no-buffer restrictions are adopted simultaneously, which makes the problem of cyclic scheduling even more difficult, as the deadlock problem needs to be solved [9, 15, 20].

Most deterministic cyclic scheduling problems are decision NP-complete analysis problems or NP-hard optimization problems [18, 21], which can be reduced to the answer to the questions: is there a cyclic schedule that meets the limitations of the problem described by a given set of parameters or is there an acceptable schedule that minimizes the period of completing the tasks? Problems of this type use different models and are solved using various methods. Due to the high computational complexity of the problems, heuristic and metaheuristic methods are mainly used [2, 6, 19, 26, 27], which allow rapid determination of solutions, but they do not guarantee finding an optimal solution, or even an acceptable solution, because in the case of limitations no-buffer and no-wait types, such solutions may not exist for systems with given parameters (e.g. fixed structure and operation times). The most frequently used are: graph models, in particular including directed graphs [12, 23, 27]; models using mathematical programming [7, 26, 27, 32], Petri nets [7, 14, 28], max-plus algebra [7], and constraint programming models [10, 11]. Among many methods the most important are: the critical path method [12, 24, 27], constraint programming method [10, 11], methods using different variants of Tabu search metaheuristics [12, 26, 27], methods using genetic algorithms [2], as well as variants of branch and bound method [2, 32]. Complementing these methods is a computer simulation that allows rapid prototyping and analysis of solutions [24, 25].

Known cyclic scheduling methods allow to solve problems of analysis consisting in the construction of acceptable cyclic schedules, meeting the constraints for various variants of flow-shop and job-shop systems, having fixed parameters [24, 32]. In this work, the reverse problem is considered, consisting in determining (synthesis) system parameters of cyclic processes (repetitive tasks) with a given configuration, which guarantee the existence of a cyclic schedule that meets the problem (e.g. no blocking, no waiting for resources, executing tasks with fixed periods). In the cyclic scheduling literature, there are few approaches to solving the problem of synthesizing system parameters, aimed at the system meeting specific properties. The existing, few analytical approaches concern only the synthesis of simple systems composed of several processes sharing one resource [4, 29].

In this context, the main achievement of the work is the method of synthesis of operation times and determination of start times of concurrent processes with cascade-like (chain-like) topology [4, 8, 33], see Fig. 1, for which there is a no-wait cyclic schedule.

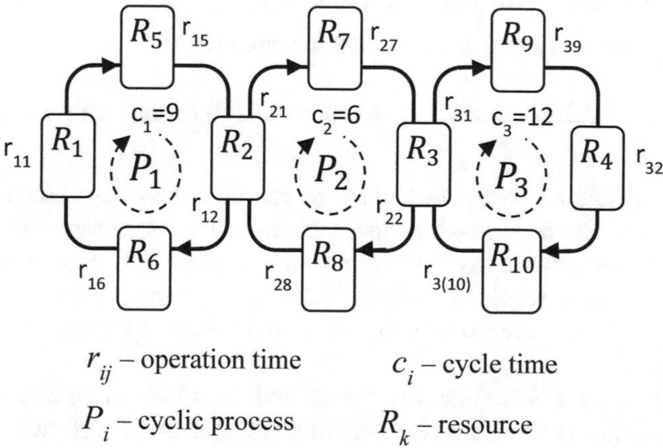

Fig. 1. An example of the extended cascade-like system.

1.1 System Description

The system of processes with a cascade-like structure has been defined in a classical form in the works [4, 8, 31, 33]. It consists of the sequences of connected two-process subsystems (2-P), each of which contains two cyclic processes sharing one resource (Fig. 2) [3, 14]. The individual 2-P subsystems (P_i, P_j) connect using a shared resource R, which is used by processes in the mutual exclusion mode, creating a system with a chain-like topology, i.e. chain-like system (CS). The system defined in this work, designated as the CS-E system, is an extension of the classical CS system [4, 8, 31], because it takes into account more than two operations within each cyclic process (Fig. 1) included in the cascade. In practice, the CS-E system can represent a transport system consisting of AGVs that move cyclically along routes composed of sectors of the production system. In this case, the trolleys are cyclic processes and the sectors belonging to the trolleys routes represent shared resources, which can be blocked by the processes.

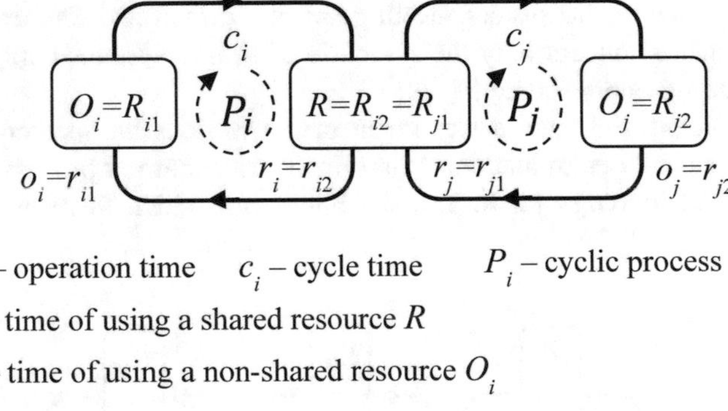

Fig. 2. A basic system of two processes (2-P) sharing one resource.

Despite the resources being blocked by the processes under consideration, the CS-E system is a structurally deadlock-free one, as there will never be a state in which a cycle of process requests exists such that none of the requests can be realized. The CS-E system with no process waiting for beginning of any operation will reach a steady state [8, 31, 33] with a cycle time equal to the least common multiple (lcm) of cycle times of component processes.

In this work, the CS-E system is considered, in which cycle times of repetitive processes forming a cascade are set. The times of operations performed by processes are sought and the start times of processes for which there exist no-wait cyclic schedules. The developed method of synthesis of no-wait cyclic schedules enables determination of appropriate operation times for each 2-P subsystem of the cascade. It uses the known necessary and sufficient conditions [29–31] for the existence of no-wait cyclic schedules for a basic 2-P system composed of two cyclic processes sharing one resource (Fig. 2), each of which performs two operations (one with the shared resource and the other with non-shared, local resource). In this paper the existing conditions have been implemented for the 2-P component subsystems, creating a new type of cascade-like system CS-E, in which each process can perform more than two operations. The obtained constraints were used to carry out the task of no-wait cyclic schedules synthesis.

1.2 Content of the Sections

In Sect. 2 a model of the extended cascade-like system (CS-E system) of concurrent repetitive processes is presented. Section 3 defines a problem of determining the operation and start times of the processes that guarantee the existence of no-wait cyclic schedules for CS-E systems with fixed periods of the processes. The conditions for no-wait execution of the simple 2-P subsystem are transformed into the new necessary and sufficient conditions for no-wait execution of the CS-E system. Section 4 contains a procedure enabling the synthesis of the CS-E system parameters for which no-wait cyclic schedules exist. Finally, the conclusions are presented in Sect. 5.

2 Model of the Extended Cascade-Like System

The cascade-like system behaviour (Fig. 1) depends on dynamics of subsystems composed of two processes sharing a resource [3, 4]. Hence, a CS-E system of repetitive processes $(P_1, \ldots, P_i, \ldots, P_n)$, $(i = 1, 2, \ldots, n)$, can be seen as composed of 2-P (Fig. 2) subsystems (P_i, P_j) sharing the resource $R = R_{i2} = R_{j1}$, i.e. $((P_1, P_2)$, $(P_2, P_3), \ldots, (P_i, P_j), \ldots, (P_{n-1}, P_n))$, where $(i = 1, \ldots, n-1; j = i+1)$. Each cyclic process $P_i (i = 1, 2, \ldots, n)$ (Fig. 3), executes periodically a sequence of operations using resources $Z_i = (R_{i1}, R_{i1(1)}, \ldots, R_{i1(a(i))}, R_{i2}, R_{i2(1)}, \ldots, R_{i2(b(i))})$, where $R_{i2} = R_{j1} = R_{(i+1)1}$, for $(i = 1, \ldots, n-1), R_{i1}, R_{i2}, R_{ij(k)} \in RE$, $a(i)$ – number of resources between R_{i1} and R_{i2}, $b(i)$ – number of resources between R_{i2} and R_{i1}, and $RE = \{R_1, \ldots, R_k, \ldots, R_{lr}\}$ is a set of resources, each one of unit capacity (i.e. only one process can use the resource at any time), lr – a number of resources. In particular, there can be no resources between R_{i1} and R_{i2}, as well as between R_{i2} and R_{i1}, which corresponds to the case of the classical cascade-like system [4, 8, 9], Fig. 2. The operations times are given by a sequence $ZT_i = (r_{i1}, r_{i1(1)}, \ldots, r_{i1(a(i))}, r_{i2}, r_{i2(1)}, \ldots, r_{i2(b(i))})$, and a cycle time of the process P_i is equal to $c_i = r_{i1} + r_{i1(1)} + \ldots + r_{i1(a(i))} + r_{i2} + r_{i2(1)} + \ldots + r_{i2(b(i))}$, where r_{i1}, r_{i2}, $r_{ij(k)} \in N$ are defined in the uniform time units (N - set of natural numbers). For instance the CS-E system shown in Fig. 1 consists of ten resources and three cyclic processes. The resources R_2, R_3 are shared ones, since each one is used by two processes, and the resources R_1, R_4, R_5, R_6, R_7, R_8, R_9, R_{10}, are non-shared resources because each one is exclusively used by only one process. The processes P_1, P_2, P_3 are executing operations using resources given by the sequences: $Z_1 = (R_{11}, R_{11(1)}, R_{12}, R_{12(1)}) = (R_1, R_5, R_2, R_6)$, $Z_2 = (R_{21}, R_{21(1)}, R_{22}, R_{22(1)}) = (R_2, R_7, R_3, R_8)$, $Z_3 = (R_{31}, R_{31(1)}, R_{32}, R_{32(1)}) = (R_3, R_9, R_4, R_{10})$. Process cycle times are equal to: $c_1 = r_{11} + r_{15} + r_{12} + r_{16} = 9$, $c_2 = r_{21} + r_{27} + r_{22} + r_{28} = 6$, $c_3 = r_{31} + r_{39} + r_{32} + r_{3(10)} = 12$. The CS-E system is composed of two 2-P subsystems of cyclic processes sharing a resource: $S_2 = (P_1, P_2)$ – sharing resource R_2, and

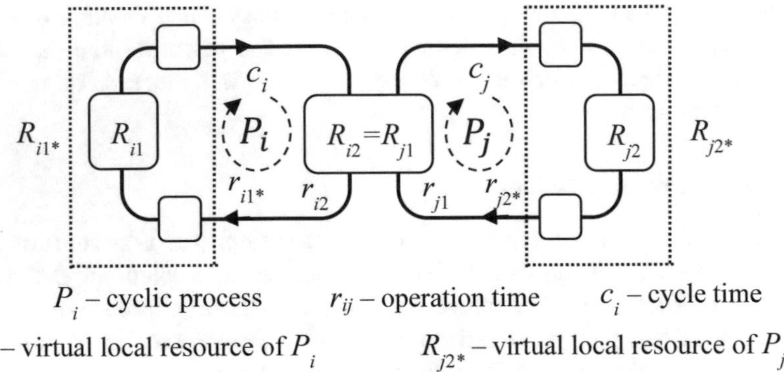

Fig. 3. An extended system of two processes sharing a resource.

$S_3 = (P_2, P_3)$ – sharing resource R_3 (where an index i of the S_i denotes a number of the shared resource R_i).

It can be noticed that is possible to analyse behaviour of each (P_i, P_j) subsystem of two processes sharing a resource, where $(i = 1, 2, \ldots, n-1; j = i+1)$, separately to find the operation times and start times of the processes for which no-wait cyclic schedules exist. It was shown, that dynamics of 2-P systems (P_i, P_j) depends on the operation times, periods, and start times of the processes [3, 4, 29, 31]. The necessary and sufficient conditions for existence of no-wait cyclic schedule for a basic 2-P system (Fig. 2) can be used to find parameters (i.e. operation and start times of the processes) of the extended 2-P system with several local resources (see Fig. 3), that guarantee existence of its no-wait cyclic schedule. If we assume that a start time of using a shared resource by P_i is such that $a_i(0) = 0$, then using the formulas presented in [29–31] it is possible to calculate (if exist) all start times $a_j(0) \in [0, c_i)$ of using the shared resource by process P_j, that belong to a no-wait cyclic schedule of the extended 2-P subsystem (Fig. 3). We can repeat the computations for all (P_i, P_j) subsystems, where $(i = 1, 2, \ldots, n-1;\ j = i+1)$, to obtain operation and start times of the processes of cascade-like (CS-E) system (Fig. 1).

In the following section we will consider a problem of CS-E system parameters synthesis for which no-wait cyclic schedules exist.

3 Problem Formulation

Let us consider a CS-E system of cyclic process $\left((P_1, P_2), \ldots, \left(P_i, P_j\right), \ldots, (P_{n-1}, P_n)\right)$, where $(i = 1, \ldots, n-1;\ j = i+1)$ and its component 2-P subsystem $S_R = (P_i, P_j)$, that consists of two cyclic processes sharing an R resource (Fig. 1). We assume that P_i and P_j execute periodically a sequence of operations using resources $Z_i = (R_{i1}, R_{i2}) = (O_i, R)$, and $Z_j = \left(R_{j1}, R_{j2}\right) = \left(R, O_j\right)$, respectively, where $O_i = R_{i1}$, $O_j = R_{j2}$ denote the non-shared resources used by the processes. Suppose that the processes start from the operations using the shared resource $R = R_{i2} = R_{j1}$. The operations times are given by a sequence $ZT_i = (r_{i1}, r_{i2}) = (o_i, r_i)$, and $ZT_j = \left(r_{j1}, r_{j2}\right) = \left(r_j, o_j\right)$, where $o_i = r_{i1}, r_i = r_{i2}, r_j = r_{j1}, o_j = r_{j2}$, and $r_i, o_i, r_j, o_j \in N$.A cycle times of P_i and P_j are defined as $c_i = r_{i1} + r_{i2} = o_i + r_i$, and $c_j = r_{j1} + r_{j2} = r_j + o_j$. It is possible to design recurrent modulus equations defining times at which P_j can request access to the shared resource R in relation to the allocation time of this resource to the process P_i [29–31]. These times will be denoted as $t_{ij}(l)$, for subsequent $l = 0, 1, \ldots$, and will be used as local start times of the process P_j in relation to a start time of process P_i in the time interval of length c_i (Fig. 4).

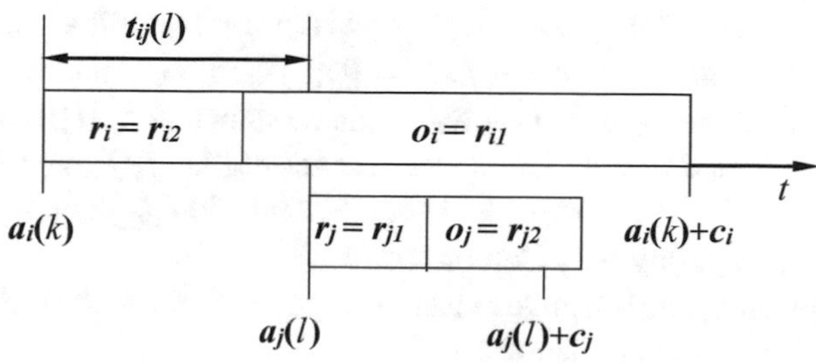

Fig. 4. A local start time $t_{ij}(l)$ of process P_j in the time interval of length c_i.

Assuming that the cycle time c_i of each process P_i is known, the problem is to determine for each subsystem $(P_i, P_j), (i = 1, \ldots, n-1;\ j = i+1)$, that is part of the cascade-like system CS-E, process operation times for which there are local process start times, guaranteeing the existence of a no-wait cyclic schedule for the CS-E system.

3.1 Start Times of Processes in the Basic 2-P System

Let us consider a basic 2-P system of processes (P_i, P_j) sharing a resource (Fig. 2) and assume that $a_i(l), a_j(l) \in N \cup \{0\}, l = 0, 1, 2, \ldots$, denote times at which P_i, P_j receive access to the shared resource R. Suppose that a start time $a_i(0)$ of the process P_i is equal to $a_i(0) = 0$, and a start- time of the process P_j is such that $a_j(0) \geq 0$. In case when the processes cannot wait for the shared resource R they can be seen as executed independently of each other (no sharing of resources). Hence, subsequent allocation times $a_i(l)$, $a_j(l)$ can be calculated according to the equation: $a_j(l + 1) = a_j(l) + c_j = a_j(0) + l \cdot c_j$, where ($l = 0, 1, 2, \ldots$), see Fig. 4.

Let $t_{ij}(l) \in N \cup \{0\}$ define a local start time $t_{ij}(l) \in [0, c_i)$ (Fig. 4) of process P_j. It can be noticed that in order to avoid waiting for the shared resource by P_j in each time interval of length c_i, that begins at $a_i(k)$, there must be $r_i \leq t_{ij}(l)$, for $l = 0, 1, 2, \ldots.$, and to avoid waiting by process P_i, the following should hold: $t_{ij}(l) \leq c_i - r_j$ (Fig. 4). From our previous works [29–31] it follows that local start times $t_{ij}(l) \in [0, c_i)$ of process P_j, defined in relation to resource allocation times of process P_i, with cycle time c_i, are given by the following formula:

$$t_{ij}(l) = a_j(l) \bmod c_i = f_{ij}(l)D_{ij} + y_{ij}(l) \ \& \ l = 0, 1, 2, \ldots \tag{1}$$

where:

$a_j(0), j \in \{1, 2, \ldots, n\}$ - is given; gcd - the greatest common divisor;
$D_{ij} = D_{ji} = gcd(c_i, c_j) \ \& \ c_i = D_{ij}m_{ij} \ \& \ c_j = D_{ji}m_{ji} \ \& \ gcd(m_{ij}, m_{ji}) = 1 \ \& \ m_{ij}, m_{ji} \in N$;
$f_{ij}(l) = [f_{ij}(0) + l \times m_{ji}] \bmod m_{ij} \ \& \ y_{ij}(l) = y_{ij}(0)$;
$t_{ij}(0) = a_j(0) \bmod c_i \ \& \ f_{ij}(0) = t_{ij}(0) \operatorname{div} D_{ij} \ \& \ y_{ij}(0) = t_{ij}(0) \bmod D_{ij}$;
$0 \le f_{ij}(0) < m_{ij} \ \& \ 0 \le y_{ij}(0) < D_{ij}$, i.e. $f_{ij}(0) \in \{0, \ldots, m_{ij} - 1\}, y_{ij}(0) \in \{0, \ldots, D_{ij} - 1\}$.

Let $W_{ij} = \{0, \ldots, m_{ij} - 1\}$. From $f_{ij}(l) = [f_{ij}(0) + l \times m_{ji}] \bmod m_{ij}$ (1) it follows that a range of $f_{ij}(l)$ is such that $f_{ij}(l) \in W_{ij}$. It can be shown [29, 31], that $f_{ij}(l)$ achieves periodically, with period m_{ij}, all values from the set W_{ij}, i.e. $f_{ij}(l) = f_{ij}(l + k \times m_{ij})$ for $k = 0, 1, 2, \ldots$, and $l = 0, \ldots, m_{ij} - 1$. Therefore, function $f_{ij}(l)$ defines permutation $f_{ij} = (f_{ij}(0), f_{ij}(1), \ldots, f_{ij}(m_{ij} - 1))$ on the set W_{ij}.

Hence, a no-wait cyclic schedule exists for a given (P_i, P_j) system if there are local start times $t_{ij}(l)$ (1) such that relations:

$$r_i \le t_{ij}(l) \le c_i - r_j \tag{2}$$

hold for all $l \in W_{ij}$, where $W_{ij} = \{0, \ldots, m_{ij} - 1\}$. If we take into account notation of operation times used for CS-E system (see Figs. 2 and 4), where $r_i = r_{i2}$ and $r_j = r_{j1}$, and relation (1) for values of $f_{ij}(l) = 0$ and $f_{ij}(l) = m_{ij} - 1$, the following formula holds:

$$r_{i2} \le t_{ij}(l) \le c_i - r_{j1} \quad \textit{iff} \quad r_{i2} \le y_{ij}(0) \le D_{ij} - r_{j1} \tag{3}$$

We will use these relations to calculate operation and start times of the processes for which no-wait cyclic schedules exists for the CS-E system.

3.2 System CS-E Parameters that Meet the No-Wait Constraints

It can be noticed that any value $y_{ij}(0)$ (3) exists if the interval $[r_{i2}, D_{ij} - r_{j1}]$ is not empty. i.e. when $r_{i2} \le D_{ij} - r_{j1}$. Taking into account relations $1 \le r_{i2}$ and $1 \le r_{j1}$ the necessary and sufficient condition for existence of a no-wait cyclic schedule for a basic system (P_i, P_j), where $j = i + 1$, can be given:

$$2 \le r_{i2} + r_{j1} \le D_{ij}, \ i.e. \ 2 \le r_{i2} + r_{(i+1)1} \le D_{i(i+1)} \tag{4}$$

Hence, the following theorem holds.

Theorem 1. A no-wait cyclic schedules can be constructed for a basic 2-P system (P_i, P_j) (Fig. 2) with fixed cycle times c_i, c_j of the processes, if and only if there exist the operation times $r_{i1}, r_{i2}, r_{j1}, r_{j2} \in N$ and local start times $t_{ij}(0) \in [0, c_i)$, such that $t_{ij}(0) = f_{ij}(0)D_{ij} + y_{ij}(0)$ (1), and $f_{ij}(0) \in \{0, \ldots, m_{ij} - 1\}$ (1), and $y_{ij}(0) \in \{r_{i2}, \ldots, D_{ij} - r_{j1}\}$ (3), and $2 \le r_{i2} + r_{j1} \le D_{ij}$ (4), where $c_i = r_{i1} + r_{i2}$, and $c_j = r_{j1} + r_{j2}$, and $D_{ij} = gcd(c_i, c_j)$.

In the basic 2-P system (P_i, P_j) each process P_i is executing periodically sequence of two operations using resources $Z_i = (R_{i1}, R_{i2})$. On the other hand, each process belonging to the cascade-like system CS-E can perform additional operations using specific resources, i.e. $Z_i = (R_{i1}, R_{i1(1)}, \ldots, R_{i1(a(i))}, R_{i2}, R_{i2(1)}, \ldots, R_{i2(b(i))})$, see Fig. 1. Let ($P_i$, P_j) be an extended 2-P system (Fig. 3) that is part of the CS-E system. In order to determine the parameters of the CS-E system, guaranteeing the existence of a no-wait schedule, we will apply for each extended subsystem (P_i, P_j) necessary and sufficient conditions defined in the form of Theorem 1 for the basic system. For the P_i process belonging to the extended system (P_i, P_j) we define the virtual resource R_{i1*}, which includes all resources used by the P_i process without the R_{i2} resource, i.e. $R_{i1*} = (R_{i2(1)}, \ldots, R_{i2(b(i))}), R_{i1}, R_{i1(1)}, \ldots, R_{i1(a(i))})$, see Fig. 3. The resource R_{i1*} corresponds to the resource R_{i1} in the basic system shown in Fig. 2, and the time of the virtual operation using this resource is equal to the sum of operation times using its component resources, i.e. $r_{i1*} = (r_{i2(1)} + \ldots + r_{i2(b(i))} + r_{i1} + r_{i1(1)} + \ldots + r_{i1(a(i))})$. Similarly, for the P_j process we define the virtual resource R_{j2*}, which includes all resources used by the P_j process without the R_{j1} resource, i.e. $R_{j2*} = (R_{j1(1)}, \ldots, R_{j1(a(j))}, R_{j2}, R_{j2(1)}, \ldots, R_{j2(b(j))})$, see Fig. 3. The resource R_{j2*} corresponds to the resource R_{j2} in the basic system shown in Fig. 2. Then, the necessary and sufficient conditions for the existence of no-wait cyclic schedules of the CS-E system, which consists of extended subsystems (P_i, P_j), $i = 1,\ldots, n-1; j = i + 1$, and using virtual resources R_{i1*} and R_{j2*} (Fig. 3) defines the following theorem.

Theorem 2. A no-wait cyclic schedules can be constructed for the CS-E system composed of 2-P subsystems ((P_1, P_2), …, (P_i, P_j), …, (P_{n-1}, P_n)) with fixed cycle times $c_1, c_2, \ldots, c_i, c_j, \ldots, c_{n-1}, c_n$ of the processes, if and only if for each subsystem (P_i, P_j), $i = 1,\ldots, n-1; j = i + 1$, there exist the operation times $r_{i1*}, r_{i2}, r_{j1}, r_{j2*} \in N$ and local start times $t_{ij}(0) \in [0, c_i)$, such that $t_{ij}(0) = f_{ij}(0)D_{ij} + y_{ij}(0)$ (1), and $f_{ij}(0) \in \{0,\ldots,m_{ij}-1\}$ (1), $y_{ij}(0) \in \{r_{i2},\ldots,D_{ij}-r_{j1}\}$ (3), and $2 \in r_{i2} + r_{j1} \leq D_{ij}$ (4), where: $c_i = r_{i1*} + r_{i2}, c_j = r_{j1} + r_{j2*}$, and $r_{i1*} = \left(r_{i2(1)} + \ldots + r_{i2(b(i))} + r_{i1} + r_{i1(1)} + \ldots + r_{i1(a(i))}\right)$, and $r_{j2*} = \left(r_{j1(1)} + \ldots + r_{j1(a(j))} + r_{j2} + r_{j2(1)} + \ldots + r_{j2(b(j))}\right)$, and $r_{kl}, r_{pq(s)} \in N$, and $D_{ij} = gcd(c_i, c_j)$.

If it is possible to construct a no-wait cyclic schedule for the CS-E system, then its cycle time is equal to $T = lcm(c_1,\ldots,c_i,\ldots,c_n)$.

Using conditions given by Theorem 2 we can consider a problem of determining parameters of the extended cascade-like system, i.e. operation and start times of the processes for which no-wait cyclic schedules exists.

4 Synthesis of No-Wait Cyclic Schedule Parameters

Let us consider a CS-E system (Fig. 1) of processes $(P_1,\ldots,P_i,P_{i+1},\ldots,P_n)$ with set cycle times $(c_1,\ldots,c_i,c_{i+1},\ldots,c_n)$. The CS-E system can be seen as composition $((P_1,P_2),\ldots(P_1,P_{I+1}),\ldots,(P_{n-1},P_n))$ of 2-P extended subsystems, where $i = 1, \ldots, n-1$. Our goal is to synthesize process operation times and their start times that satisfy the conditions of Theorem 2. We calculate the vector $D = (D_{12},\ldots, D_{i(i+1)},\ldots,D_{(n-1)n})$ such that $D_{i(i+1)} = gcd(c_i,c_{i+1})$ and then determine the permissible operating times $r_{i2}, r_{(i+1)1} \in N$ meeting the constraints (4). Let $s_{i(i+1)} = r_{i2} + r_{(i+1)1}$, where $s_{i(i+1)} \in \{2,\ldots,D_{i(i+1)}\}$. If we assume that $r_{i2} = d$, then it must be $(r_{i2}, r_{(i+1)1}) \in SR_{i(i+1)}$, and $SR_{i(i+1)} = \{(d, s_{i(i+1)} - d) : d \in \{1,\ldots,D_{i(i+1)} - 1\}$ & $s_{i(i+1)} \in \{2,\ldots,D_{i(i+1)}\}$ & $_{i(i+1)} - d \geq 1\}$. Hence, the vector $csr = ((r_{12}, r_{21}),\ldots,(r_{i2}, r_{(i+1)1}),\ldots,(r_{(n-1)2}, r_{n1}))$, which takes into account all subsystems (P_i, P_{i+1}), for $i = 1, \ldots, n-1$, and meets restrictions 4, belongs to the Cartesian product $SR = SR_{12}\,\mathrm{x}\ldots\mathrm{x}SR_{i(i+1)}\mathrm{x}\ldots\mathrm{x}SR_{(n-1)n})$, i.e. $csr \in SR$. We may notice that the number of elements of the $SR_{i(i+1)}$ set is the sum of the items in the $\{1,\ldots,D_{i(i+1)} - 1\}$, i.e. $|SR_{i(i+1)}| = (1 + 2 + \ldots + (D_{i(i+1)} - 1))$. Therefore, $|SR| = |SR_{12}| \cdot \ldots \cdot |SR_{i(i+1}| \cdot \ldots \cdot |SR_{(n-1)n})|$ is the number of all possible vectors csr that satisfy constraints (4). However, as acceptable process operation times, we can only accept values of the operation times given by $csr = ((r_{12}, r_{21}),\ldots,(r_{i2}, r_{(i+1)1}),\ldots,(r_{(n-1)2}, r_{n1}))$ for which it is possible to find operation times $r_{kl}, r_{pq(s)} \in N$, such that constraints $c_i = r_{i1*} + r_{i2}$, and $r_{i1*} = (r_{i2(1)} + \ldots + r_{i2(b(i))} + r_{i1} + r_{i1(1)} + \ldots + r_{i1(a(i))})$ hold, for $i = 1,\ldots,n$. After setting the operation times of the processes, we can define the permissible values of local process start times belonging to no-wait cyclic schedules using formula $t_{ij}(0) = f_{ij}(0)D_{ij} + y_{ij}(0)$ (1), and $f_{ij}(0) \in \{0,\ldots,m_{ij} - 1\}$ (1), $y_{ij}(0) \in [r_{i2}, D_{ij} - r_{j1}]$ (3). Example 1 illustrates the presented method of synthesizing parameters of CS-E systems.

Example 1. As shown in Fig. 1, the CS-E system considered consists of $n = 3$ processes P_1, P_2, P_3 that are executing operations using resources given by the sequences: $Z_1 = (R_{11}, R_{11(1)}, R_{12}, R_{12(1)}) = (R_1, R_5, R_2, R_6)$, $Z_2 = (R_{21}, R_{21(1)}, R_{22}, R_{22(1)}) = (R_2, R_7, R_3, R_8)$, $Z_3 = (R_{31}, R_{31(1)}, R_{32}, R_{32(1)}) = (R_3, R_9, R_4, R_{10})$. Process cycle times are equal to: $c_1 = r_{11} + r_{15} + r_{12} + r_{16} = 9$, $c_2 = r_{21} + r_{27} + r_{22} + r_{28} = 6$, $c_3 = r_{31} + r_{39} + r_{32} + r_{3(10)} = 12$. The CS-E system is composed of two 2-P extended subsystems $((P_1,P_2), (P_2,P_3))$, i.e. (P_1,P_2) – sharing resource R_2, and (P_2,P_3) – sharing resource R_3. The following relations hold (1): $D_{12} = D_{21} = gcd(c_1, c_2) = 3, m_{12} = 3, m_{21} = 2$, and $c_1 = D_{12}m_{12}, c_2 = D_{21}m_{21}; D_{23} = D_{32} = gcd(c_2, c_3) = 6, m_{23} = 1, m_{32} = 2$, and $c_2 = D_{23}m_{23}, c_3 = D_{32}m_{32}$. Hence, $s_{12} = r_{12} + \mathrm{r}_{21}$, and $s_{12} \in \{2,\ldots,D_{12}\} = \{2,3\}$, as well as $s_{23} = r_{22} + \mathrm{r}_{31}$, and $s_{23} \in \{2,\ldots,D_{23}\} = \{2,3,4,5,6\}$. The vector of acceptable operation times has a form: $csr = ((r_{12}, \mathrm{r}_{21}), (\mathrm{r}_{22}, \mathrm{r}_{31}))$, $csr \in SR$, where $SR = SR_{12}$ $\mathrm{SR}_{23} = \{(d, s_{12}\text{-}d)\text{: } d \in \{1, 2\}$ & $s_{12} \in \{2, 3\}$ & $s_{12} - d \geq 1\} \times \{(d, s_{23}\text{-}d)\text{: } d$ $\{1, 2, 3, 4, 5\}$ & $s_{23} \in \{2, 3, 4, 5, 6\}$ & $s_{23} - d \geq 1\} = \{(1, 1); (1, 2), \underline{(2, 1)}\} \times \{ (1, 1); (1, 2), (2, 1); (1, 3), (2, 2), \underline{(3, 1)}; (1, 4), (2, 3), (3, 2), (4, 1); (1\ 5), (2, 4), (3, 3), (4, 2), (5, 1)\}$. The operation times of the no-wait CS-E system must satisfy the following constraints: $((r_{12}, r_{21}), (r_{22}, r_{31})) \in SR$, and $c_1 = r_{11^*} + r_{12} = 9$ and $r_{11^*} = (r_{16} + r_{11} + r_{15})$, and $c_2 = r_{21^*} + r_{22} = 6$, and $r_{21^*} = (r_{28} + r_{21} + r_{27})$, and $c_3 = r_{31^*} + r_{32} = 12$, and $r_{31^*} = (r_{3(10)} + r_{31} + r_{39})$. Let us select a vector $((r_{12}, r_{21}), (r_{22}, r_{31}))$ from the SR set, e.g. $((r_{12}, r_{21}), (r_{22}, r_{31})) = ((2, 1), (3, 1))$ - selected elements were underlined. Hence, the remaining operation times must meet the constraints: $c_1 - r_{12} = r_{11^*} = (r_{16} + r_{11} + r_{15}) = 7$, and $c_2 - r_{22} = r_{21*} = (r_{28} + r_{21} + r_{27}) = (r_{28} + 1 + r_{27}) = 3$, and $c_3 = r_{31^*} + r_{32} = (r_{3(10)} + r_{31} + r_{39}) + r_{32} = (r_{3(10)} + 1 + r_{39}) + r_{32} = 12$, where $(r_{16}, r_{11}, r_{15}), (r_{28}, r_{27}), (r_{3(10)}, r_{39}) \in N$. We may notice that the only solution is $r_{28} = r_{27} = 1$, and for the remaining operation times, an example solution was found, assuming the lowest possible values $(r_{16} = r_{15}) = (r_{3(10)} = r_{39}) = 1$. Hence, we obtain $r_{11} = 7 - 2 = 5$, and $r_{32} = 12 - 3 = 9$. Finally, the possible operation times of the CS-E system for which there are no-wait cyclic schedules are as follows: $ZT_1 = (r_{11}, r_{15}, r_{12}, r_{16}) = (5, 1, 2, 1)$, $ZT_2 = (r_{21}, r_{27}, r_{22}, r_{28}) = (1, 1, 3, 1)$, $ZT_3 = (r_{31}, r_{39}, r_{32}, r_{3(10)}) = (1, 1, 9, 1)$. Assuming that in subsystem (P_1,P_2), processes are started from an operation using the R_2 resource, and in subsystem (P_2,P_3), from operation using the R_3 resource, we can calculate the permissible values of local process start times belonging to cyclic schedules of the CS-E system composed of subsystems $((P_1,P_2), (P_2,P_3))$, by using formulas: $t_{12}(0) = f_{12}(0)D_{12} + y_{12}(0)$ (1), and $f_{12}(0) \in \{0, \ldots, m_{12} - 1\}$ (1), $y_{12}(0) \in [r_{12}, D_{12} - r_{21}]$ (3), as well as $t_{23}(0) = f_{23}(0)D_{23} + y_{23}(0)$ (1), and $f_{23}(0) \in \{0, \ldots, m_{23} - 1\}$ (1), $y_{23}(0) \in [r_{22}, D_{23} - r_{31}]$ (3). Since, $D_{12} = D_{21} = gcd(c_1,c_2) = 3$, $m_{12} = 3$, $m_{21} = 2$, and $c_1 = D_{12}m_{12}$, $c_2 = D_{21}m_{21}$; $D_{23} = D_{32} = gcd(c_2,c_3) = 6$, $m_{23} = 1$, $m_{32} = 2$, and $c_2 = D_{23}m_{23}$, $c_3 = D_{32}m_{32}$, we obtain: $t_{12}(0) = f_{12}(0) \times 3 + y_{12}(0)$ (1), and $f_{12}(0) \in \{0, 1, 2\}$ (1), $y_{12}(0) \in [2, 2]$ (3), and also $t_{23}(0) = f_{23}(0) \times 6 + y_{23}(0)$ (1), and $f_{23}(0) \in \{0\}$ (1), $y_{23}(0) \in [3, 5]$ (3). In Fig. 5 we see an example of a no-wait cyclic schedule for $t_{12}(0) = 2$ and $t_{23}(0) = 3$. Its cycle time is $T = lcm(c_1, c_2, c_3) = lcm(9, 6, 12) = 36$.

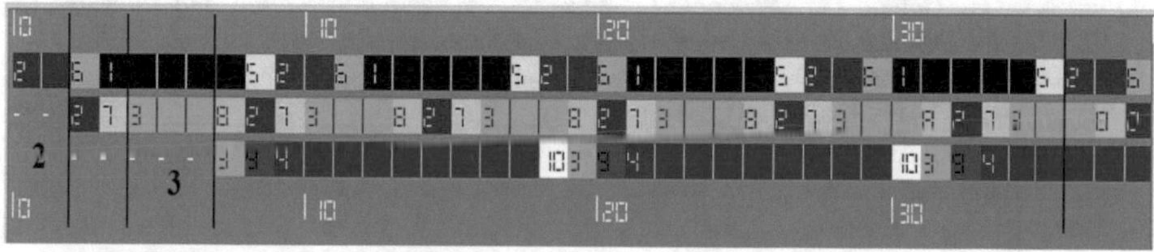

Fig. 5. Example of a no-wait cyclic schedule for the system (P_1, P_2, P_3) shown in Fig. 1; where (2, 6, 1, 5) denote numbers of the resources used by the process P_1, (2, 7, 3, 8) – by the process P_2, and (3, 9, 4, 10) by the process P_3.

5 Conclusions

The problem of synthesis of parameters of repetitive process systems with cascade-like structure, i.e. operation times and process start times, which guarantee the existence of no-wait cyclic schedules, was considered. A new class of systems was defined, the so-called extended cascade-like systems CS-E, in which each process can perform additional operations in comparison to classical systems with "chain-like" structure, in which cyclic processes perform two operations each. Using the known conditions of the existence of no-wait cyclic schedules in the basic two process system, the necessary and sufficient conditions for the existence of no-wait cyclic schedules in CS-E systems, which were applied in the developed procedure of synthesis of parameters of the CS-E system with set periods of cyclic processes, were formulated in a declarative manner. The method developed makes it possible to determine whether there is an acceptable solution in time comparable to the determination of the *gcd* of process cycle times and can be used to construct no-wait cyclic schedules for repetitive process systems with mesh-like structure.

References

1. Abadi, I.N.K., Hall, N.G., Sriskandarajah, C.: Minimizing cycle time in a blocking flowshop. Oper. Res. **48**, 177–180 (2000)
2. AitZai, A., Benmedjdoub, B., Boudhar, M.: A branch and bound and parallel genetic algorithm for the job shop scheduling problem with blocking. Int. J. Oper. Res. **14**(3), 343–365 (2012)
3. Alpan, G., Jafari, M.A.: Dynamic analysis of timed Petri nets: a case of two processes and a shared resource. IEEE Trans. Robot. Autom. **13**(3), 338–346 (1997)
4. Alpan, G., Jafari, M.A.: Synthesis of sequential controller in the presence of conflicts and free choices. IEEE Trans. Robot. Autom. **14**(3), 488–492 (1998)
5. Allahverdi, A.: A survey of scheduling problems with no-wait in process. Eur. J. Oper. Res. **255**(3), 665–686 (2016)
6. Aschauer, A., Roetzer, F., Steinboeck, A., Kugi, A.: An efficient algorithm for scheduling a flexible job shop with blocking and no-wait constraints. IFAC-PapersOnLine **50**(1), 12490–12495 (2017). 20th IFAC World Congress
7. Baccelli, F., Cohen, G., Olsder, G.J., Quadrat, J.-P.: Synchronization and Linearity (An Algebra for Discrete State Systems). Wiley, Chichester (1992)
8. Banaszak, Z., Jędrzejek, K.: On self-synchronization of cascade-like coupled processes. Appl. Math. Comput. Sci. **3**(4), 39–60 (1993)
9. Banaszak, Z., Krogh, B.: Deadlock avoidance in flexible manufacturing systems with concurrently competing process flows. IEEE Trans. Robot. Autom. **6**(6), 724–734 (1990)
10. Bocewicz, G., Nielsen, I., Banaszak, Z.: Automated guided vehicles fleet match-up scheduling with production flow constraints. Eng. Appl. Artif. Intell. **30**, 49–62 (2014)
11. Bocewicz, G., Wójcik, R., Banaszak, Z., Pawlewski, P.: Multimodal processes rescheduling: cyclic steady states space approach. Math. Probl. Eng. (2013). http://dx.doi.org/10.1155/2013/407096
12. Bożejko, W., Pempera, J., Wodecki, M.: Minimal cycle time determination and golf neighborhood generation for the cyclic flexible job shop problem. Bull. Pol. Acad. Sci. Tech. Sci. **66**(3), 333–344 (2018)
13. Brucker, P., Kampmeyer, T.: Cyclic job shop scheduling problems with blocking. Ann. Oper. Res. **159**(1), 161–181 (2008)
14. Gaujal, B., Jafari, M., Baykal, G.-M., Alpan, G.: Allocation sequences of two processes sharing a resource. IEEE Trans. Robot. Autom. **11**(5), 748–753 (1995)
15. Gold, E.: Deadlock prediction: easy and difficult cases. SIAM J. Comput. **7**, 320–336 (1978)
16. Hall, N.G., Sriskandarajah, C.: A survey of machine scheduling problems with blocking and no-wait in process. Oper. Res. **44**(3), 510–525 (1996)
17. Krenczyk, D., Kalinowski, K., Grabowik, C.: Integration production planning and scheduling systems for determination of transitional phases in repetitive production. In: Hybrid Artificial Intelligent Systems. LNCS, vol. 7209, pp. 274–283 (2012)
18. Kamoun, H., Sriskandarajah, C.: The complexity of scheduling jobs in repetitive manufacturing systems. Eur. J. Oper. Res. **70**, 350–364 (1993)
19. Kumar, S., Bagchi, T.P., Sriskandarajah, C.: Lot streaming and scheduling heuristics for m-machine no-wait flowshops. Comput. Ind. Eng. **38**, 149–172 (2000)

20. Lawley, M., Reveliotis, S.: Deadlock avoidance for sequential resource allocation systems: hard and easy cases. Int. J. Flex. Manuf. Syst. **13**(4), 385–404 (2001)
21. Levner, E., Kats, V., Alcaide, D., Pablo, L., Cheng, T.C.E.: Complexity of cyclic scheduling problems: a state-of-the-art survey. Comput. Ind. Eng. **59**(2), 352–361 (2010)
22. Liu, S.-Q., Kozan, E.: Scheduling trains with priorities: a no-wait blocking parallel-machine job-shop scheduling model. Transp. Sci. **45**(2), 175–198 (2011)
23. Louaqad, S., Kamach, O., Iguider, A.: Scheduling for job shop problems with transportation and blocking no-wait constraints. J. Theor. Appl. Inf. Technol. **96**(10), 2782–2792 (2018)
24. Pinedo, M.: Planning and Scheduling in Manufacturing and Services. Springer, New York (2005)
25. Polak, M., Majdzik, P., Banaszak, Z., Wójcik, R.: The performance evaluation tool for automated prototyping of concurrent cyclic processes. Fundamenta Informaticae **60**(1–4), 269–289 (2004)
26. Schuster, ChJ: No-wait job shop scheduling: tabu search and complexity of subproblems. Math. Methods Oper. Res. **63**(3), 473–491 (2006)
27. Smutnicki, Cz.: Minimizing cycle time in the manufacturing system based on the flow of various jobs. IFAC Proc. **42**(4), 1137–1142 (2009). 13th IFAC Symposium on Information Control Problems in Manufacturing
28. Song, J.-S., Lee, T.-E.: Petri net modeling and scheduling for cyclic job shops with blocking. Comput. Ind. Eng. **34**(2), 281–295 (1998)
29. Wójcik, R.: Constraint programming approach to designing conflict-free schedules for repetitive manufacturing processes. In: Cunha, P.F., Maropoulos, P.G. (eds.) Digital Enterprise Technology. Perspectives and Future Challenges, pp. 267–274. Springer, New York (2007)
30. Wójcik, R.: Designing a no-wait cyclic schedule for a class of concurrent repetitive production processes. IFAC-PapersOnLine **51**(11), 1305–1310 (2018). Designing a no-wait cyclic schedule for a class of concurrent repetitive production processes
31. Wójcik, R.: Towards strong stability of concurrent repetitive processes sharing resources. Syst. Sci. **27**(2), 37–47 (2001)
32. Von Kampmeyer, T.: Cyclic scheduling problems. Ph.D. dissertation, Mathematik/Informatik, Universität Osnabrück (2006)
33. Zaremba, M.B., Jędrzejek, K.J., Banaszak, Z.A.: Design of steady-state behavior of concurrent repetitive processes: an algebraic approach. IEEE Trans. Syst. Man Cybern. Part A Syst. Hum. **28**(2), 199–212 (1998)

Robust Competence Allocation for Multi-project Scheduling

Grzegorz Bocewicz[1](✉), Jarosław Wikarek[2], Paweł Sitek[2], and Zbigniew Banaszak[1]

[1] Faculty of Electronics and Computer Science, Koszalin University of Technology, Koszalin, Poland
bocewicz@ie.tu.koszalin.pl, zbiniew.banaszak@tu.koszalin.pl

[2] Institute of Management and Control Systems, Kielce University of Technology, Kielce, Poland
{j.wikarek, sitek}@tu.kielce.pl

Abstract. To ensure proper functioning of an organization (e.g. production system) in settings in which there exists a risk of disruptions caused by unexpected employee absenteeism and/or a changing demand for employees with specific qualifications, one must move away from traditional approaches oriented towards the determination of procedures for generating workable baseline schedules that minimize the project makespan in a deterministic environment. The planner must take into account time uncertainties caused by an unexpected deviation of activity durations and resource availability uncertainties related to the fact that execution of tasks in a project depends on the availability of resources. The aim of this study was to develop a method "determining" a competence structure of employees, allowing for realizing proactive planning robust to unexpected staff absenteeism. To this end, we proposed a procedure for designing a competence framework that returns competence structures robust to disruptions caused by employee absenteeism. The procedure, derived from a declarative model, can be used to determine structures that allow to substitute redundant competences, thus ensuring robustness of the generated schedules to the types of absenteeism that are known *a priori*. The possibilities of practical application of the presented approach are illustrated with examples.

Keywords: Competence allocation · Robust planning · Employee competences

1 Introduction

During the production process, there may occur various types of unpredictable events, or disruptions, related to the availability of resources (e.g. failures of machines, robots), orders (e.g. change of completion date), estimation of operation completion times, changes in the duration of operations (e.g. an employee's absence or ill health), machine failures, or differences between actual and planned operation completion times. Conventional planning-support approaches for production flows assume that production is a stationary and deterministic process. Unfortunately, failure to account

J. Świątek et al. (Eds.): ISAT 2019, AISC 1051, pp. 16–30, 2020.
https://doi.org/10.1007/978-3-030-30604-5_2

for disruptions, which are a natural part of any process found in industrial practice, reduces the quality of the plans constructed in this way. One of the approaches that allow to mitigate the negative impact of disruptions on production flow is proactive, or robust, scheduling. Schedules designed under this approach show lower susceptibility to disruptions associated with the availability of resources required for the execution of planned operations, fluctuations in the duration of operations, etc. In other words, the objective of robust scheduling is to counteract the instability of baseline schedules, taking into account the variability and uncertainty of system parameters. Among the many techniques used in this area, such as redundancy-based techniques or contingent scheduling, one that is particularly worth mentioning is sensitivity analysis, which focuses on the impact of changes in parameters, e.g. operation times, on the quality of the baseline schedule.

While the existing literature offers numerous publications on robust scheduling, including studies in the area of redundancy-based techniques, contingent scheduling and sensitivity analysis [2, 5, 8, 10], there is still a scarcity of reports addressing the issues of planning resource structures that can guarantee the achievement of business objectives in a dynamically changing setting, i.e. structures robust to disruptions.

This scarcity is, among others, the reason why there are no computationally efficient methods of planning resource structures that can ensure that orders are completed in a manner robust to selected disruptions within the time specified by the baseline schedule. One of the reasons for this state of affairs is the NP-hardness of this class of problems. Planning of robust resource structures is henceforth understood as the design and/or selection of specific resource characteristics, such as employee competences, that can guarantee the completion of planned orders in the presence of certain types of disruptions. The results of preliminary research aimed at developing a method for designing resource structures, e.g. competence structures, robust to employee absences, confirm the attractiveness of the approaches derived from the declarative modeling paradigm [15, 16]. Projects are often implemented under risk and uncertainty. In this context, the problem of allocation of competences and, hence, employees (viewed from the perspective of their competences) to the individual activities (operations, workstations) of the task under execution is a task assignment problem. Problems of this type are found in many areas of science and business, such as goods distribution, production management, project planning, etc. They can all be reduced to assigning a known set of operations to a given set of executing agents (e.g. employees, vehicles, warehouses). The next section presents the state of the art of operations assignment planning in conditions connected with unforeseen employee absenteeism. In Sect. 3, we propose a reference model that allows to search for structures robust to selected types of absences of individual and multiple employees and present a procedure for assessing and synthesizing competence structures robust to disruption. Section 4 reports computational experiments performed in the GUROBI environment, which illustrate the possibilities of applying the proposed method. In Sect. 5, we provide conclusions and indicate directions of further research.

2 Background

In recent years, the issue of robust scheduling has often been discussed in the context of the Resource Constrained Project Scheduling Problem (RCPSP). The keen interest in project scheduling is associated with an increasing number of production orders executed as part of Make-To-Order (MTO) and Engineer-To-Order (ETO) projects. In the MTO production approach, customized products are manufactured for individual recipients, whose requirements are variable and can be unpredictable. Each such production order is treated as a separate project that has to be carried out in consultation with the customer. The process of planning the assignment of employees to workstations can be divided into three stages: (1) strategic planning of the personnel structure aimed at ensuring an adequate staff capacity (e.g. making sure that the required personnel competences are available when needed) that can match the planned production capacity, (2) tactical scheduling/structure planning focused on the allocation of specific operations to employees (operation/task/job assignment), and (3) operational planning of allocation of current operations to available employees [1]. The literature describes many decision support methods and models for competence assessment, identification of competence gaps, prototyping changes to the competence structure, etc. [2, 4, 11].

A factor that determines the quality of the generated work schedules and operations assignments is their robustness to disruptions caused by uncertainty of demand, uncertainty of arrival connected with unpredictable work-load [6, 7, 9, 17, 19] (prolonged machine maintenance time, unknown number of patients in a hospital, etc.), and uncertainty of capacity related to employees' health problems, machine failures, etc. Common approaches to improving the robustness of operations assignments use either reactive scheduling (in which the existing schedule is modified to accommodate the disruption identified) or proactive scheduling (in which robust personnel rosters and schedules are constructed taking into account different types of possible disruptions) [14].

One common approach to improving the robustness of operations assignments is to introduce time buffers or capacity buffers. Capacity buffers (understood as surplus resources), also referred to as reserve personnel, are commonly used in services, such as passenger transport, school services, hospital services, etc. [18], where normal service provision is often disrupted by employee sickness or technical failures [12, 18]. One instance of an approach which assumes that a system should necessarily have extra (financial, material, or human) resources is the solution presented in [2], which allows to find a competence structure that will minimize the risk (caused by a specific type of disruption) of non-performance of scheduled operations.

3 Robust Multi-project Scheduling

Motivating example beginning this section provides introduction to the problem of robust competence allocation. The problem considered boils down to answering the question: What competence structure will guarantee a personnel assignment that can ensure robust to unexpected staff absenteeism known *a priori* timely execution of planned orders? Then, a reference model that allows to search for structures robust to selected types of absences of individual and multiple employees provides a formal framework allowing one to formulate above mentioned problem of robust competence allocation in terms of Constraint Satisfaction Problem. Finally its implementation in a constraint programming, i.e. GUROBI, environment allows to analyze and design (minimum) competence structures robust to assumed employee absences.

3.1 Illustrative Example

A company uses a cyclic multi-item batch flow production system to complete two production orders: $\{J_1, J_2\}$ – Fig. 1. Each order comprises a set of operations Z_i: $J_1 = \{Z_1, \ldots, Z_5\}$, $J_2 = \{Z_6, \ldots, Z_{11}\}$, executed in a given technological order, operation times l_i, and an operations schedule determined by the critical path, as shown in Fig. 2. Order placement operations J_1 are executed along the route marked in blue, and their duration times are: 3 h for Z_1, 4 h for Z_2, 5 h for Z_3, 2 h for Z_4, and 2 h for Z_5. Order placement operations J_2 are executed along the route marked in orange, and their duration times are: 2 h for Z_6, 3 h for Z_7, 2 h for Z_8, 1 h for Z_9, 4 h for Z_{10}, and 2 h for Z_{11}. The baseline order execution schedule, developed under the assumption that production is not disrupted by any adverse events (machine failures or absences of machine operator), assumes that order completion time is 10 h.

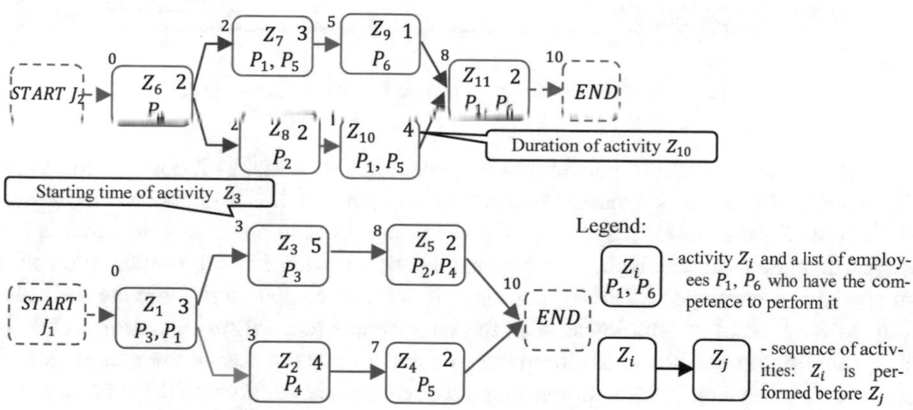

Fig. 1. Structure of production orders J_1, J_2

A set of 11 workstations/machines $\{Z_1, \ldots, Z_{11}\}$ is operated by 6 employees. Table 1 shows options of assignment of employees to workstations. It is assumed that ongoing operations cannot be interrupted, and the employee who performs them can be released only after their completion. Moreover, it is assumed that employees are engaged in the execution of given jobs for no less than 2 h and no more than 8 h.

Table 1. Competence structure of a staff of six employees $P_1, \ldots, P_6$

G	Z_1	Z_2	Z_3	Z_4	Z_5	Z_6	Z_7	Z_8	Z_9	Z_{10}	Z_{11}
P_1	1	0	0	0	0	0	1	0	0	1	1
P_2	0	0	0	0	1	0	0	1	0	0	0
P_3	1	0	1	0	0	0	0	0	0	0	0
P_4	0	1	0	0	0	1	0	0	0	0	0
P_5	0	0	0	1	1	0	1	0	0	1	0
P_6	0	0	0	0	0	0	0	0	1	0	1

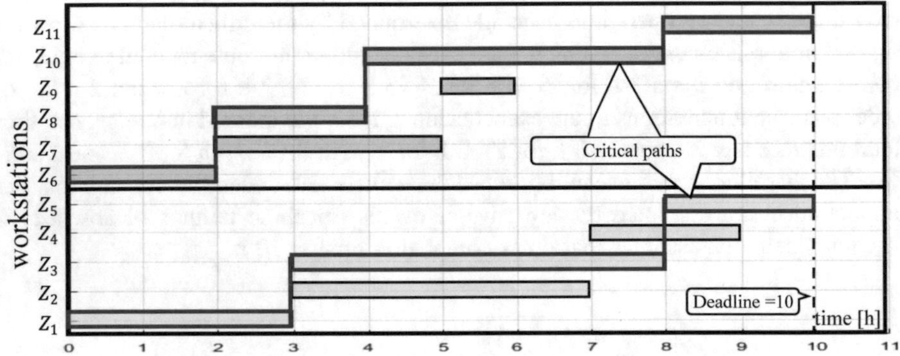

Fig. 2. Order completion schedule for orders from Fig. 1

It is easy to see that the personnel assignment in Fig. 3 makes it possible to follow the schedule from Fig. 1 whereas that in Fig. 4a, which takes into account the possibility of an absence of employee P_1, allows to complete the order in a time that is 1 h longer than the assumed 10 h. The personnel assignment in Fig. 4b (which allows for an absence of employee P3) does not make it possible to plan and complete the order (due to the lack of an employee with the competence to perform operation Z_3). It is clear that the competence structure given in Table 1 does not ensure the completion of planned orders when it is assumed that production may be disrupted by an absence (known at order execution start time) of any (one) employee during the entire order execution horizon. In the context of these deliberations, the following question arises: Does there exist a competence structure for staff $\{P_1, \ldots, P_6\}$ that allows to generate a personnel assignment – and if so what form should it have to guarantee timely (10 h) execution of planned orders $\{J_1, J_2\}$ in the event of an absence of any given employee? It is easy to see that this question is an instance of the more general question: What competence structure of how large a staff will guarantee a personnel assignment that can ensure timely execution of planned orders in the event of collective absenteeism known *a priori*?

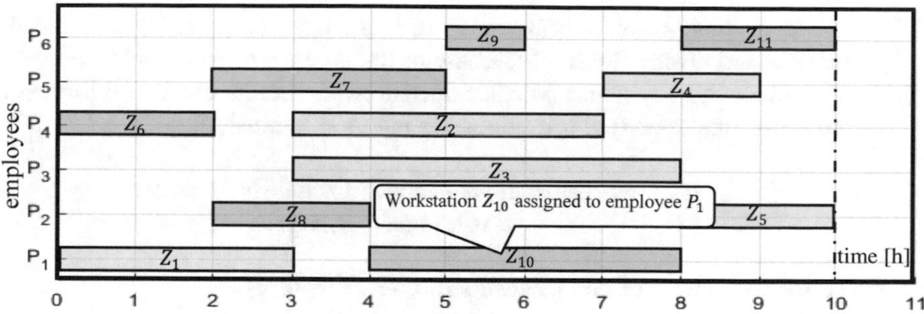

Fig. 3. Allocation of employees to workstations that allows to implement the schedule from Fig. 2

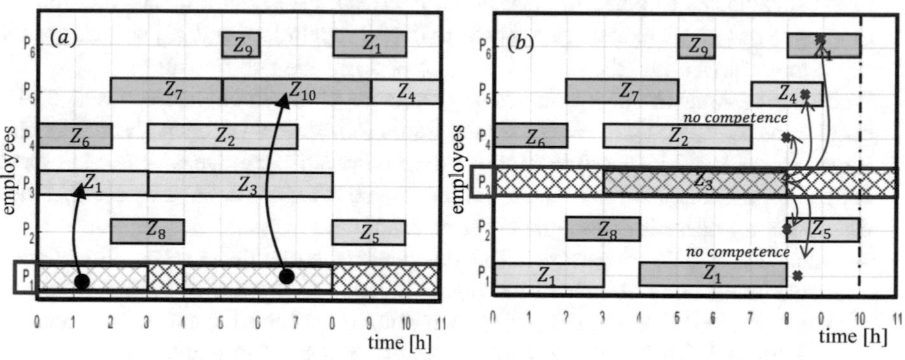

Fig. 4. Personnel assignment for a situation of (a) an absence of P_1, (b) an absence of P_3

3.2 A Reference Model

Limiting further discussion to the case of competence structures disturbed by collective absenteeism (ω simultaneously absent employees) and assuming, at the same time, that these disruptions are known *a priori* already at the stage of planning the allocation of operations, e.g. at the beginning of the day's shift, we adopted the declarative model.

Sets:

Z_i: set of operations, indexed by $i = 1, \ldots, n$

P_k: set of employees, indexed by $k = 1, \ldots, m$

Parameters:

ω: number of employees absent at the same time ($\omega \in \mathbb{N}$), $\omega < m$

U_ω: set of combinations of ω elements from set $\{1, \ldots, m\}$: $U_\omega = \{\{u_1, .., u_i, \ldots, u_\omega\} | u_i \in \{1, \ldots, m\}\}$, combinations in set U_ω describe absent employees

l_i: number of hours of i-th operation Z_i

s_k^j: minimum number of working hours of k-th employee ($s_k \in \mathbb{N}$), the so-called lower working time limit, in the case of the absence of j-th employee

z_k^j: maximum number of working hours of k-th employee ($z_k \in \mathbb{N}$), the so-called upper working time limit, in the case of the absence of j-th employee

$w_{a,b}$: a parameter that specifies whether operations Z_a and Z_b can be performed by the same employee (the operations are mutually exclusive):

$$w_{a,b} = \begin{cases} 1 & \text{when operations } Z_a \text{ and } Z_b \text{ exclude one another} \\ 0 & \text{in the remaining cases} \end{cases}$$

R^* expected robustness of competence structure, $R^* \in [0, 1]$

Decision variables:

$G = (g_{k,i} | k = 1 \ldots m; i = 1 \ldots n)$: competence structure, where $g_{k,i}$ stands for employees' competences to perform operations; $g_{k,i} \in \{0, 1\}$, $g_{k,i} = 0$ means that the k-th employee has no competences to perform the i-th operation, and $g_{k,i} = 1$ means that the k-th employee has the competences to perform the i-th operation.

R_ω: measure of robustness of competence structure G to simultaneous absence of ω employees $R_\omega \in [0, 1]$. $R_\omega = 0$ – stands for no robustness, i.e. each absence defined by set U_ω makes it impossible to allocate operations to employees; $R_\omega = 1$ – stands for full robustness, i.e. for each case of employee absence (defined by set U_ω) there does exist an operations assignment. For example:

- value $R_\omega = 0.25$ means that the competence structure ensures allocation of operations in one-quarter of the possible cases of absence of ω employees,
- value $R_\omega = 0.5$ means that the competence structure ensures allocation of operations in half of the possible cases of absence of ω employees,

G^Θ: an auxiliary variable representing the competence structure for the absence of ω employees described by set $\Theta \in U_\omega$, $G^\Theta = (g_{k,i}^\Theta | k = 1 \ldots m; i = 1 \ldots n)$, where: $g_{k,i}^\Theta \in \{0, 1\}$ stands for employees' competences to perform operations in the case of absence of ω employees described by set Θ. It is assume that $g_{k,i}^\Theta = 0$ for each absent employees: $k \in \Theta$.

$X^\Theta = (x_{k,i}^\Theta | k = 1 \ldots m; i = 1 \ldots n)$: – allocation of operations in the event of an absence of ω employees described by set $\Theta \in U_\omega$, where $x_{k,i}^\Theta \in \{0, 1\}$:

$$x_{k,i}^\Theta = \begin{cases} 1 & \text{when operation } Z_i \text{ has been assigned to employee } P_k \\ 0 & \text{in the remaining cases} \end{cases}$$

c^Θ: an auxiliary variable specifying the fulfillment of the allocation conditions X^Θ. $c^\Theta = 1$ in case then operation is assigned to exactly one employee (3) and the workload of employee is less than the lower time limit and not exceeds the upper time limit (4), (5); $c^\Theta = 0$ – otherwise. The value of variable $c^\Theta \in \{0, 1\}$ depends on auxiliary sub-variables: $c_{1,i}^\Theta$, $c_{2,k}^\Theta$, $c_{3,k}^\Theta$ which specify whether constraints (3)–(5) are satisfied.

Constraints:

1. Construction of a competence structure for a situation when the ω employees are absent from his scheduled duty hours (the sets Θ of absent employees are collected in U_ω):

$$\mathrm{g}_{\mathrm{k,i}}^{\Theta} = \begin{cases} \mathrm{g}_{\mathrm{k,i}} & \text{when } \mathrm{k} \notin \Theta \\ 0 & \text{when } \mathrm{k} \in \Theta \end{cases}, \tag{1}$$

2. Operations can only be performed by employees who have appropriate competences

$$x_{k,i}^{\Theta} = 0, \quad \text{when } g_{k,i}^{\Theta} = 0, \text{for } k = 1\ldots m; i = 1\ldots n; \ \forall \Theta \in U_\omega. \tag{2}$$

3. Operation Z_i is assigned to exactly one employee:

$$\left(\sum\nolimits_{k=1}^{m} x_{k,i}^{\Theta} = 1\right) \Leftrightarrow \left(c_{1,i}^{\Theta} = 1\right), \text{for } i = 1\ldots n; \ \forall \Theta \in U_\omega. \tag{3}$$

4. Workload of the k-th employee should be no less than the lower time limit s_k^{Θ}:

$$\left(\sum\nolimits_{i=1}^{n} x_{k,i}^{\Theta} \cdot l_i \geq s_k^{\Theta}\right) \Leftrightarrow \left(c_{2,k}^{\Theta} = 1\right), \text{ for } k = 1\ldots m; \ \forall \Theta \in U_\omega. \tag{4}$$

5. Workload of the k-th employee should not exceed the upper working time limit z_k^j:

$$\left(\sum\nolimits_{i=1}^{n} x_{k,i}^{\Theta} \cdot l_i \leq z_k^{\Theta}\right) \Leftrightarrow \left(c_{3,k}^{\Theta} = 1\right), \text{for } k = 1\ldots m; \ \forall \Theta \in U_\omega. \tag{5}$$

6. Execution of mutually exclusive operations:

$$x_{k,a}^{\Theta} + x_{k,b}^{\Theta} \leq 1, \quad \text{when } w_{a,b} = 0, \text{for } k = 1\ldots m; \ \forall \Theta \in U_\omega. \tag{6}$$

7. Robustness of the competence structure:

$$R_\omega = \frac{LP}{|U_\omega|} \geq R^*, \tag{7}$$

$$LP = \sum\nolimits_{\Theta \in U_\omega} c^{\Theta}, \tag{8}$$

$$c^{\Theta} = \prod\nolimits_{i=1}^{n} c_{1,i}^{\Theta} \prod\nolimits_{k=1}^{m} c_{2,k}^{\Theta} \prod\nolimits_{k=1}^{m} c_{3,k}^{\Theta}. \tag{9}$$

where: LP – number of sets $\Theta \in U_\omega$ (simultaneous absence of employees described by set Θ) for which there exists an operations assignment X^{Θ}.

The concepts of competence structure and operations assignment are represented in the model by decision variables G, X^{Θ} and R_ω. An operations assignment X^{Θ} which satisfies constraints (2)–(6) is referred to as an admissible assignment in the situation of an absence of employees from set Θ. In this context, the question considered previously can be narrowed down to the following: Does there exist a competence structure G that can ensure robustness $R_\omega \geq R^*$ in the case when ω employees are absent?

3.3 Problem of Robust Competence Allocation

The structure of the proposed model, which comprises a set of decision variables and a set of constraints that relate those variables to one another, in a natural way allows to formulate the problem in hand as a Constraint Satisfaction Problem (CSP) and implement it in a constraint programming environment [3]:

$$PS = ((\mathcal{V}, \mathcal{D}), \mathcal{C}), \tag{10}$$

where:

$\mathcal{V} = \{G, X^{\Theta \in U_\omega}, R_\omega\}$ – a set of decision variables which includes: competence structure G, the operations assignments $X^{\Theta \in U_\omega}$ and robustness R_ω.
$\mathcal{D}$ – a finite set of decision variable domains $\{G, X^{\Theta \in U_\omega}, R_\omega\}$,
$\mathcal{C}$ – a set of constraints specifying the relationships between the competence structure and its robustness (constraints 1–9).

To solve problem PS (10), one has to find such values (determined by domains $\mathcal{D}$) of decision variables G (personnel competence structure), X^Θ (operations assignment) and R_ω (robustness to absenteeism of ω employees), for which all the constraints given in set $\mathcal{C}$ are satisfied. In other words, what is sought is a solution that guarantees a given level R^* of robustness R_ω.

In the general case, a CSP defined in this way can be formulated as an optimization problem COP whose goal is to determine the minimum competence structure G_{OPT} (e.g. one that requires the minimum number of changes to be made to the baseline (original) competence structure: $minimize: \sum_{k=1}^{m} \sum_{i=1}^{n} g_{k,i}$). CSP (10) allows to analyze and design (minimum) competence structures robust to simultaneous absences of ω employees. In the general case, when other types of disruptions can be considered, robustness of a competence structure can be defined as a function of the parameter that characterizes the type of disruption. Thus understood, robustness can be evaluated using various measures. For example, the robustness of a competence structure can be expressed as:

- a measure of robustness to the simultaneous absence of ω employees (7): number of absences for which there exists an operations assignment that guarantees timely completion of orders relative to all possible cases of absenteeism,

Generally, the competence assignment problem can be extended to cover other cases that ensure robustness of the planned resource structure, such as

- a measure of robustness to an employee's loss of qualifications (competences): number of cases of lost qualifications for which there exists an operations assignment that guarantees timely completion of orders relative to all possible cases of loss of qualifications,

- a measure of robustness to changes in the operations structure (changes in the number of operations, technological routes, etc.): number of changes in the operations structure for which there exists an operations assignment that guarantees timely completion of orders relative to all possible cases of change in the operations structure.

4 Computational Experiments

Given is the production system from Fig. 1, in which orders are executed by a staff of six employees $\{P_1,\ldots,P_6\}$. Orders are processed according to the schedule from Fig. 2. In the schedule, operations executed in the same time are mutually exclusive. This means that operations that require the same competences in the same time interval, must be performed by different employees. The goal is to find an answer to the following question:

1. Does there exist a competence structure G that can ensure full robustness ($R^* = 1$) in the event of an absence of one employee ($\omega = 1, U_1 = \{\{P_1\},\{P_2\},\{P_3\},\{P_4\},\{P_5\},\{P_6\}\}$)?

   We assume, that G is partially known (see Table 1) and we can add only the new competences (change values of $g_{k,i}$ from 0 to 1). To answer this question, one must solve CSP (10), which assumes the competence structure G from Table 1 and the parameters of the model shown in Fig. 1. The problem was implemented in the GUROBI environment (Intel i7-4770, 8 GB RAM). The first admissible solution was obtained in less than 0.1 s. The structure for which $R^* = 1$ is shown in Table 2. Only if these competences are obtained, can the company achieve full robustness of its competence structure to the absence of any given staff member – absence of each case from the set U_1, Fig. 5.
2. Does there exist a competence structure G that can ensure full robustness ($R^* = 1$) in the event of an absence of any two employees ($\omega = 2$, $U_2 = \{\{P_1,P_2\},\{P_1,P_3\},\ldots,\{P_5,P_6\}\}$.)?
   The answer to this question, obtained in 1 s, is given in Table 3. Acquisition of 17 new competences will guarantee full ($R^* = 1$) robustness of the competence structure to the absence of any two staff members (absences from the set U_2). An example of a personnel assignment for a situation when employees P_2 and P_4 are absent from work is shown in Fig. 6.
3. Does there exist a competence structure G that can ensure full robustness

   ($R^* = 1$) in the event of an absence of any three employees ($\omega = 3$, $U_3 = \{\{P_1,P_2,P_3\},\{P_1,P_2,P_4\},\ldots,\{P_4,P_5,P_6\}\}$)?

   The answer to this question, obtained in 5 s, is shown in Table 4. Acquisition of 28 new competences guarantees full ($R^* = 1$) robustness of the competence structure to the absence of any three staff members (absences from the set U_3). This guarantee, however, comes at the price of a longer project completion time (11 h). An example of a personnel assignment for a situation when employees P_1, P_4 and P_5 are absent from work is given in Fig. 7.

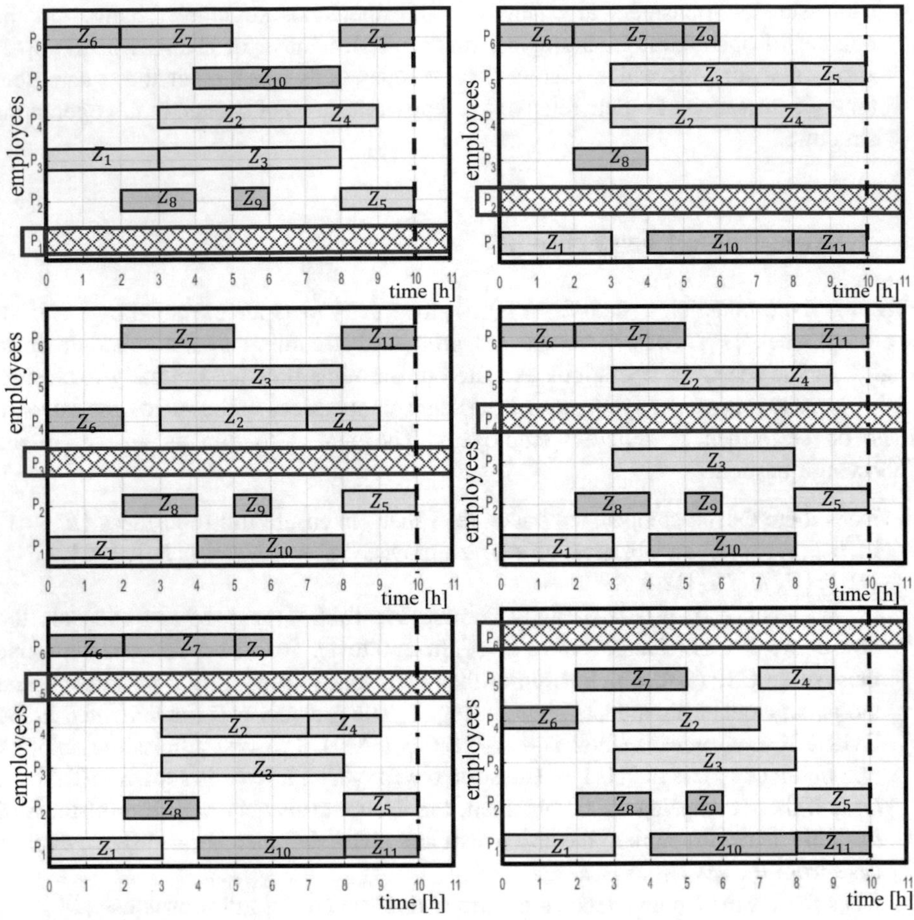

Fig. 5. Personnel assignment options for situations involving the absence of single employees

Table 2. A competence structure of a staff of six employees $P_1, \ldots, P_6$ that guarantees full ($R^* = 1$) robustness to the absence of any chosen staff member - 7 new competences

G	Z_1	Z_2	Z_3	Z_4	Z_5	Z_6	Z_7	Z_8	Z_9	Z_{10}	Z_{11}
P_1	1	0	0	0	0	0	1	0	0	1	1
P_2	0	0	0	0	1	0	0	1	**1**	0	0
P_3	1	0	1	0	0	0	0	**1**	0	0	0
P_4	0	1	0	**1**	0	1	0	0	0	0	0
P_5	0	**1**	**1**	1	1	0	1	0	0	1	0
P_6	0	0	0	0	0	**1**	**1**	0	1	0	1

Table 3. A personnel competence structure that guarantees full ($R^* = 1$) robustness to the absence of any two staff members - 17 new competences

G	Z_1	Z_2	Z_3	Z_4	Z_5	Z_6	Z_7	Z_8	Z_9	Z_{10}	Z_{11}
P_1	1	**1**	**1**	0	0	0	1	**1**	**1**	1	1
P_2	**1**	0	**1**	0	1	0	0	1	0	**1**	**1**
P_3	1	0	1	0	**1**	0	0	0	0	0	0
P_4	0	1	0	**1**	0	1	0	0	0	0	0
P_5	0	0	0	1	1	**1**	1	**1**	**1**	1	0
P_6	0	**1**	0	**1**	0	**1**	**1**	0	1	0	1

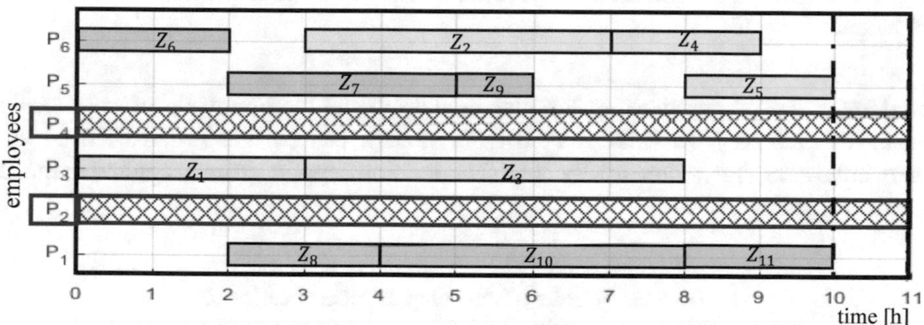

Fig. 6. An example of a personnel assignment for a situation when employees P_2 and P_4 are both absent from work

Table 4. A personnel competence structure that guarantees full ($R^* = 1$) robustness to the absence of any three staff members ($\omega = 3$) - 28 new competences

G	Z_1	Z_2	Z_3	Z_4	Z_5	Z_6	Z_7	Z_8	Z_9	Z_{10}	Z_{11}
P_1	1	**1**	**1**	0	**1**	**1**	1	**1**	**1**	1	1
P_2	0	**1**	0	**1**	1	**1**	**1**	1	0	**1**	**1**
P_3	1	0	1	0	**1**	0	0	**1**	0	**1**	0
P_4	**1**	1	**1**	**1**	0	1	**1**	0	**1**	0	**1**
P_5	0	**1**	0	1	1	**1**	1	**1**	**1**	1	0
P_6	**1**	0	**1**	**1**	0	0	0	0	1	0	1

The method was verified in a series of experiments involving different numbers of employees (6–10) and different numbers of operations (10–100). The experiments were carried out to determine the time needed to synthesize a competence structure robust ($R^* = 1$) to the absence of ($\omega = 1, 2, 3$) employees. The results are shown in Table 5. It is easy to notice that in cases in which the size of the structure does not exceed 10 employees and 50 operations, a competence structure robust to the absence of any three employees can be found in less than 25 min. In our future research, we plan to

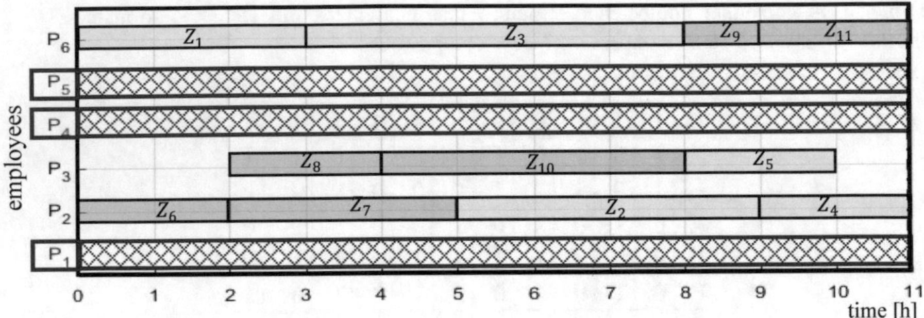

Fig. 7. An example of a personnel assignment for a situation when employees P_1, P_4 and P_5 are all absent from work

implement the proposed computational module in the environments of optimization packages such as IBM ILOG CPLEX, OzMozart, etc. so that it can be used as a software overlay for commercially available decision support systems applied in human resources management.

Table 5. Results of the computational experiment

Number of absent employees ω	Total number of employees n	Number of operations m	Calculation time [s]
1	6	10	0.12
1	6	50	0.15
1	6	100	0.21
1	10	10	0.13
1	10	50	0.24
1	10	100	0.46
2	6	10	1.01
2	6	50	6.58
2	6	100	40.97
2	10	10	5.98
2	10	50	67.64
2	10	100	181.03
3	6	10	5.01
3	6	50	206.26
3	6	100	756.634
3	10	10	8.65
3	10	50	1345.45
3	10	100	>2700

5 Conclusions

The proposed method of developing robust resource structures allows to plan the allocation of production operations (that require specific employee competences) to resources (employees with the given competences) in settings in which there exists a risk of disruptions caused by employee absenteeism. According to this method, to build a competence structure robust to unforeseen disruptions, an organization must determine what additional (redundant) competences contractors need to have to compensate for the competences lost as a result of employee absenteeism. The proposed measure of robustness of competence structures enables interactive, on-line synthesis of structures with a given level of robustness, in particular robustness to absences of one or many employees (a team of contractors).

Being NP-hard, the problems under consideration must be solved using heuristic methods, e.g. ones that implement the declarative programming paradigm. The constraint satisfaction model CSP (10) and the method of synthesizing resource structures robust to employee absenteeism based on this model uses a redundant number of decision variables representing competence-depleted structures (structures that arise as a consequence of occurrence of various disruption scenarios). The redundancy of the set of decision variables, on the one hand, increases the computational complexity of the problem, but, on the other, allows to find solutions at a given level of robustness (especially in the case of fully robust competence structures). The experiments have shown that the method can be effectively used to solve problems in organizational units of up to 10 employees and 50 operations (effectively means here that a robust competence structure can be synthesized on-line in under 1.500 s). It may be possible to increase the scale of the problems solved, by using hybrid methods [13] dedicated to models that use scarce data structures (in the model under consideration, the competence structure contains mostly "0" values). We wish to devote our future research to the implementation of this type of techniques.

References

1. Abernathy, W., Baloff, N., Hershey, J.: A three-stage man power planning and scheduling model - a service sector example. Oper. Res. **21**, 693–711 (1973)
2. Antosz, K.: Maintenance – identification and analysis of the competency gap. Eksploatacja i Niezawodnosc – Maintenance Reliab. **20**(3), 484–494 (2018)
3. Banaszak, Z., Bocewicz, G.: Declarative modeling for production order portfolio scheduling. Found. Manage. **6**(3), 7–24 (2014)
4. Bombiak, E.: Human resources risk as an aspect of human resources management in turbulent environments. In: Pînzaru, F., Zbuchea, A., Brătianu, C., Vătămănescu, E.M., Mitan, A. (eds.) Shift! Major Challenges of Today's Economy, pp. 121–132. Tritonic Publishing House, Bucharest (2017)
5. Chaari, T., Chaabane, S., Aissani, N., Trentesaux, D.: Scheduling under uncertainty: survey and research directions. In: 2014 International Conference on Advanced Logistics and Transport (ICALT), Hammamet, pp. 229–234 (2014). https://doi.org/10.1109/icadlt.2014.6866316

6. Hadi, M., Shadrokh, S.: A robust scheduling for the multi-mode project scheduling problem with a given deadline under uncertainty of activity duration. Int. J. Prod. Res. **57**(10), 3138–3167 (2019). https://doi.org/10.1080/00207543.2018.1552371
7. Hazir, O., Haouari, M., Erel, E.: Robust scheduling and robustness measures for the discrete time/cost trade-off problem. Eur. J. Oper. Res. **207**(1), 633–643 (2010)
8. Herroelen, W., Leus, R.: Project scheduling under uncertainty: survey and research potentials. Eur. J. Oper. Res. **165**, 289–306 (2005). https://doi.org/10.1016/j.ejor.2004.04.002
9. Klimek, M., Łebkowski, P.: Robustness of schedules for project scheduling problem with cash flow optimisation. Bull. Pol. Acad. Sci. Tech. Sci. **61**(4), 1005–1015 (2013). https://doi.org/10.2478/bpasts-2013-0108
10. Klimek, M., Łebkowski, P.: Robust Buffer allocation for scheduling of a project with predefined milestones. Decis. Making Manuf. Serv. **3**(1–2), 49–72 (2009). https://doi.org/10.7494/dmms.2009.3.2.49
11. Korytkowski, P.: Competences-based performance model of multi-skilled workers with learning and forgetting. Expert Syst. Appl. **77**, 226–235 (2017)
12. Potthoff, D., Huisman, D., Desaulniers, G.: Column generation with dynamic duty selection for railway crew rescheduling. Transp. Sci. **44**(4), 493–505 (2010)
13. Wikarek, J.: Implementation aspects of hybrid solution framework. In: Recent Advances in Automation, Robotics and Measuring Techniques. Advances in Intelligent Systems and Computing, vol. 267, pp. 317–328 (2014). https://doi.org/10.1007/978-3-319-05353-0_31
14. Stijn, V.V.: Proactive-reactive procedures for robust project scheduling. Ph.D. theses, no. 247, Faculteit Economische En Toegepaste, Economische Wetenschappen, Katholieke Universiteit Leuven (2006)
15. Szwarc, E., Bocewicz, G., Bach-Dąbrowska, I.: Planning a teacher staff competence structure robust to unexpected personnel absence. In: Manufacturing Modelling, Management and Control (MIM), Berlin (2019, in print)
16. Szwarc, E., Bach-Dąbrowska, I., Bocewicz, G.: Competence management in teacher assignment planning. In: Damaševičius, R., Vasiljevienė, G. (eds.) Information and Software Technologies, ICIST 2018. CCIS, vol. 920, pp. 449–460 (2018)
17. Tam, B., Ehrgott, M., Ryan, D.M., Zakeri, G.: A comparison of stochastic programming and bi-objective optimisation approaches to robust airline crew scheduling. OR Spectrum **33**(1), 49–75 (2011)
18. Topaloglu, S., Selim, H.: Nurse scheduling using fuzzy modelling approach. Fuzzy Sets Syst. **161**(11), 1543–1563 (2010). https://doi.org/10.1016/j.fss.2009.10.003
19. Van den Bergh, J., Beliën, J., De Bruecker, P., Demeulemeester, E., De Boeck, L.: Personnel scheduling: a literature review. Eur. J. Oper. Res. **226**, 367–385 (2013)

Calculation of Labour Input in Multivariant Production with Use of Simulation

Joanna Kochańska, Dagmara Górnicka, and Anna Burduk(✉)

Faculty of Mechanical Engineering, Wroclaw University of Science and Technology, Wybrzeże Wyspiańskiego 27, 50-370 Wrocław, Poland
{joanna.kochanska, dagmara.gornicka, anna.burduk}@pwr.edu.pl

Abstract. In the paper, the problem of multivariant production management is considered. Individual approach of manufacturing companies to their customers is a very desirable feature, but it is also linked with additional challenges in the field of organisation and management. This paper presents a solution implemented in a company, consisting in calculation of labour input based on simulation of production processes. As a tool for increasing the control of multivariant production, the "labour intensity calculator" is suggested. On the grounds of measurement data from representative processes and the production documentation, using simulations of similar processes, the program estimates manufacturing data for the products with preset configurations of parameters. The suggested solution can make an aid for analysis and control of execution stages of variable projects (production parameters, processes effectiveness), with no necessity to take separate measurements for each considered case, as well as for identification and solution of the appearing problems.

Keywords: Labour intensity calculator · Simulation program · Multivariant production · Human resource management · Lean manufacturing (LM)

1 Introduction

Multivariant production enables companies to react flexibly to individual customers' needs. By offering a wider product line, the company can expect a larger group of customers and higher levels of their satisfaction with the purchases. Unfortunately, production of numerous variants of products involves several challenges, not only in the field of technology, but also in the areas of organisation and management [1]. Deciding on the multivariant production, companies want to retain control over runs of the processes in order to analyse the production capabilities correctly and thus to manage them in a proper way. The aim of the study was to build a "labour intensity calculator" and testing it as a simulation tool for analysis and control of multivariant production stages, as well as for identifying and solving problems within its limits. Nowadays, computer models and simulations are widely applied for analysis [2–4] and forecasting [5] of process courses and calculation of process costs [6]. They are also used for solving variable problems [7, 8], including verification of results of potential improvements [9–12].

J. Świątek et al. (Eds.): ISAT 2019, AISC 1051, pp. 31–40, 2020.
https://doi.org/10.1007/978-3-030-30604-5_3

2 Calculation of Labour Input

Human labour input in the production area can be defined as utilisation of human capital (composed of knowledge, education, skills, capabilities, health state, energy, motivation to work etc. [13]) for execution of the company functions. Correct management of human labour input makes it possible to use the available labour resources in order to attain production goals and to reach possibly best results. In the 21st century, it is just an individual that makes the basic component of the work process, and thus the most important factor of the organisation [14], so its success is strictly conditioned by utilisation of the human labour potential.

Labour input can be expressed as the time expended on manufacture. Therefore, measurements of real production times make it possible to obtain reliable results. Labour inputs in manufacturing processes include the following groups of factors:

- preparation of production stations,
- internal transport of materials,
- preparation of tools,
- retooling of the machines,
- setting-up and replacement of tooling,
- execution of production activities,
- execution of auxiliary activities,
- operation of the machines,
- checking of correctness of the executed activities,
- maintaining cleanness of the production station.

Time measurements of the activities belonging to the above groups for a product with preset parameters make it possible to evaluate labour intensity of the manufacturing process. Calculation of labour inputs makes it possible to evaluate labour costs in order to take them into account in the product price and, on the other hand, to evaluate profitability of work distribution by analysis of the present state.

A number of methods for measurements of working time are distinguished, including analytical measuring, analytical calculation and analytical comparison methods [15]. These methods differ from each other in the way of obtaining the results and thus in time-consumption of the measurements, reliability and exactness of these results.

Correct evaluation of labour intensity makes a basis for suitable management and rewarding employees, which, according to the idea of soft management with human resources shared by the Harvard school, is translated into their commitment and, in a longer perspective, into increase of the organisation effectiveness [16].

3 Labour Intensity Calculator

In the case of multivariant production, performing exact measurements for the processes of all the available variants would require a very long time. Therefore, it is suggested to evaluate labour intensity for similar products on the grounds of measurements for a few products from the given group and the production documentation containing information about the product structure and the technological process.

3.1 Development of the Calculator

Development of the program for the labour intensity calculator is schematically illustrated in Fig. 1.

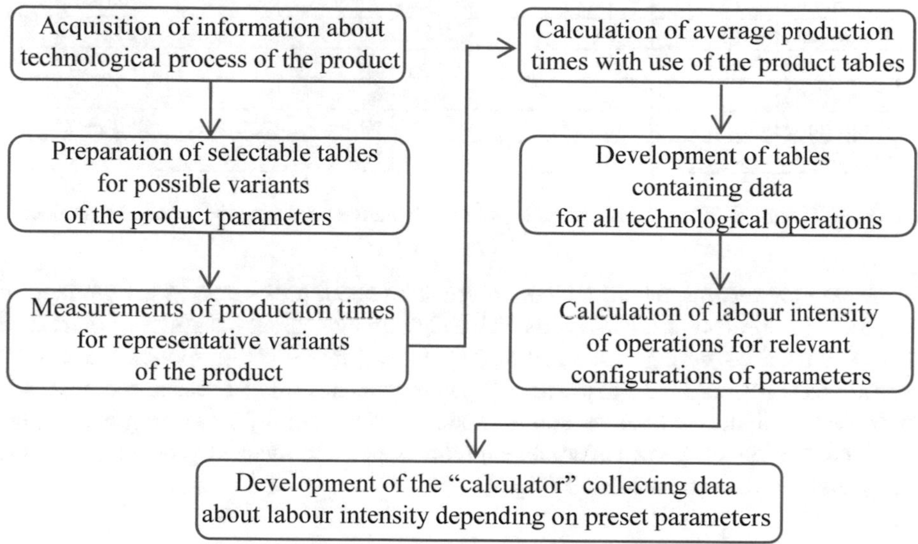

Fig. 1. Layout of development of the "labour intensity calculator"

The development process of the "calculator" begins from acquiring input data about the technological process and any parameters characterising the product together with their available variants (Fig. 2).

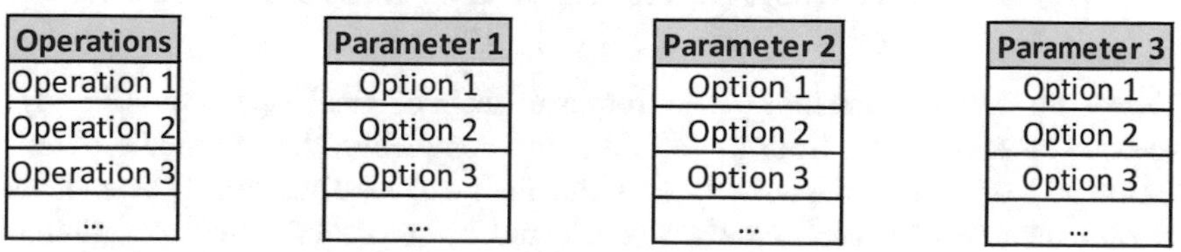

Fig. 2. Tables of data concerning the technological process and product parameters

At the next stage, measurements are taken of production activity times within each operation for the representative variants of products (Fig. 3).

Operation	Cut	Time [h]	Unit	Quantity [unit/FP]	Time [h/unit	Parameter 1	Parameter 2	Parameter 3	...
Operation 1	Cut 1		product		=...	Option 1	Option 2	Option 3	...
Operation 1	Cut 1		product		=...	Option 1	Option 2	Option 3	...
Operation 1	Cut 1		product		=...	Option 1	Option 2	Option 3	...
Operation 1	Cut 2		cm		=...	Option 3	Option 3	Option 2	...
Operation 1	Cut 2		cm		=...	Option 3	Option 3	Option 2	...
Operation 1	Cut 2		cm		=...	Option 3	Option 3	Option 2	...
Operation 1	...		...		=...	...	...	...	...
Operation 2	Cut 1		lot		=...	Option 2	Option 3	Option 2	...
Operation 2	...		...		=...	...	...	...	...

Fig. 3. Fragment of measurement table for representative products (FP = finished product)

Measuring records should include information about parameters of the product for that the measurements were taken, as well as information about quantities of individual activities required during manufacture (“Quantity [unit/FP]”), introducing a unit of measurement (“Unit”) for that the time dependence is determined. Because of that, it is possible to evaluate the time per unit of product (“Time [h/unit]”). On the ground of the obtained data, using pivot tables, average times per one finished product for all the analysed activities are calculated (Fig. 4).

Parameter 1	Option 1
Parameter 2	Option 2
Parameter 3	Option 3
...	...

Operation	Cut	Unit	Total of average times [h/unit]
⊟Operation 1	⊟...	lot	40,5
Operation 1	⊟Cut 1	product	375,0
Operation 1	⊟Cut 2	cm	12,0
Operation 1	⊟Cut 3	product	2,1
Grand total			**107,4**

Fig. 4. Table of average production times on the example of operation 1

Next, the tables containing data for each technological operation (Fig. 5) are developed by taking data from pivot tables and completing the information with the number and workload of the employees within their executed activities. Dependence of time on parameters for each case are considered by means of formulae defining the relationship between duration (time) of a given cut and, for example, side length, number of holes, number of semi-products (“Quantity in FP”), etc. Introduced is also a division into value adding activities (VA) and non-value adding activities (NVA), as well as their types are determined.

Cut	Time [h/unit]	Unit	Quantity in FP	Time [h/FP]	No. of op.	Operator 1	Operator 2	...	Kind	Type	VA/ NVA
Cut 1	=...	=...		=...		x	x				
Cut 2	=...	=...		=...		x					
Cut 3	=...	=...		=...			x				
...	=...	=...		=...		x					

Fig. 5. Table of operations

On the grounds of the data concerning individual activities within technological operations, labour inputs of operators and cycle times (C/T) for each of them are calculated. Differences of production times between products with various parameters should be considered (Fig. 6).

Parameter 1: option 2 and Parameter 3: option 1	
Labour intensity of operator 1 [h]	=...
Labour intensity of operator 2 [h]	=...
...	=...
Labour intensity of operation [h]	=...
C/T of operation [h/FP]	=...
VA of operation [%]	=...

Parameter 2: option 1 and Parameter 3: option 3	
Labour intensity of operator 1 [h]	=...
Labour intensity of operator 2 [h]	=...
...	=...
Labour intensity of operation [h]	=...
C/T of operation [h/FP]	=...
VA of operation [%]	=...

Fig. 6. Tables of labour intensity of operations

After proper tables for all the operations within the technological process of a given product are prepared, the objective "calculator" is developed. On the grounds of relationships defined by suitable formulae, the calculator collects data on labour intensity after selecting proper configuration of parameters for the given product (Figs. 7 and 8).

Combination of product parameters	
Kind	
Width [h]	
Length [h]	
Height [h]	
Parameter 1	
Paramet	Option 1
	Option 2
Paramet	Option 3
...	...

Fig. 7. Panel of parameters

Operation	Operation 1	Operation 2	Operation 3	...	Total
No. of workstations					
No. of operators	=...	=...	=...	=...	=...
Labour intensity [h/FP]	=...	=...	=...	=...	=...
C/T [h/FP]	=...	=...	=...	=...	=...
Stock [h]	=...	=...	=...	=...	=...
L/T [h/FP]	=...	=...	=...	=...	=...
VA [%]	=...	=...	=...	=...	=...
NVA [%]	=...	=...	=...	=...	=...

Fig. 8. Table of the "labour intensity calculator" (L/T = lead time)

On the grounds of the described data and suitable relationships, the calculator makes it possible to evaluate labour inputs in the technological process of the product with selected parameters (Fig. 9).

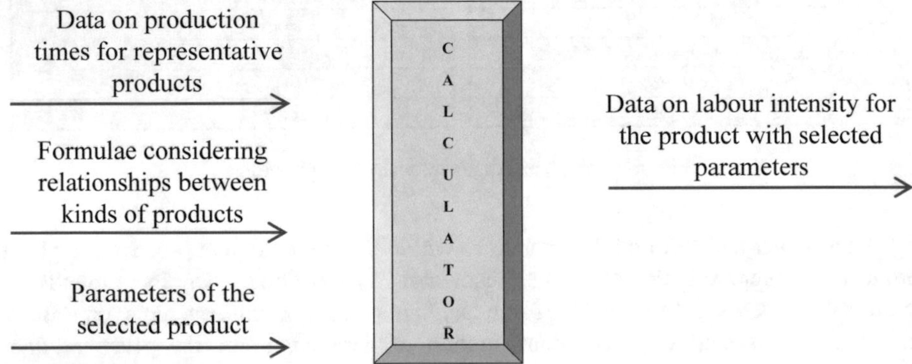

Fig. 9. Operation layout of the "labour intensity calculator"

3.2 Example of Calculator Application

Diversity of parameter configurations within the manufactured products requires from the companies a precise analysis of the processes from the viewpoint of the used labour inputs in order to adapt the management system to multivariant production, as well as in order to determine costs of the products, included in the customer price. In the frames of co-operation with the manufacturing company, the possibility of using the calculator for evaluation of labour input was investigated. Use of simulation was crucial for the company with regard to very fast development, implementation of new and innovative solutions, as well as for relatively young plants with not completely standardised management.

The program was developed for a few main groups of products with a number of available variants. The program made it possible to evaluate:

- labour input during manufacture of selected kinds of products,
- labour intensity for each executed technological operation,
- increase of labour inputs depending on individual dimensions of products,
- distribution of work among the operators (in the case of a few operators for the given operation),
- possibility to reduce labour intensity by increased number of workstations,
- cycle times for each executed technological operation,
- bottleneck of the given process,
- proportional value added for each operation,
- percentage of wastage in each operation.

Development of the program for calculation of labour intensity in the analysed case made it also possible to notice the problems occurring in the form of:

- uneven distribution of work among the employees (Fig. 10),
- relatively different cycle times of individual technological operations within the given manufacturing process (Fig. 11),
- high percentage of wastage in total working time (Fig. 12).

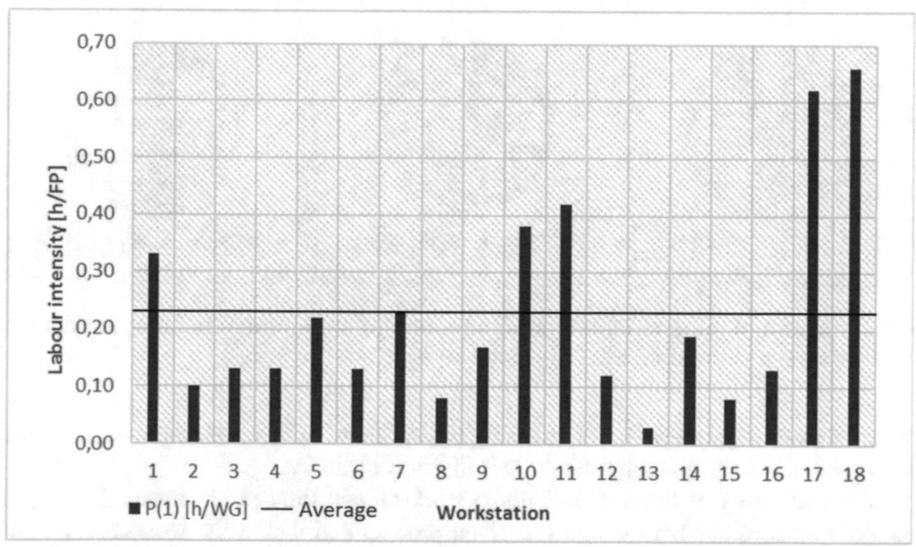

Fig. 10. Labour intensity of workstations within the production process on the example of a selected product variant

Analysis of labour intensities of workstations during execution of orders for various variants of products showed significant differences in workloads on subsequent stages of the manufacturing process. Calculation of labour intensity for various combinations

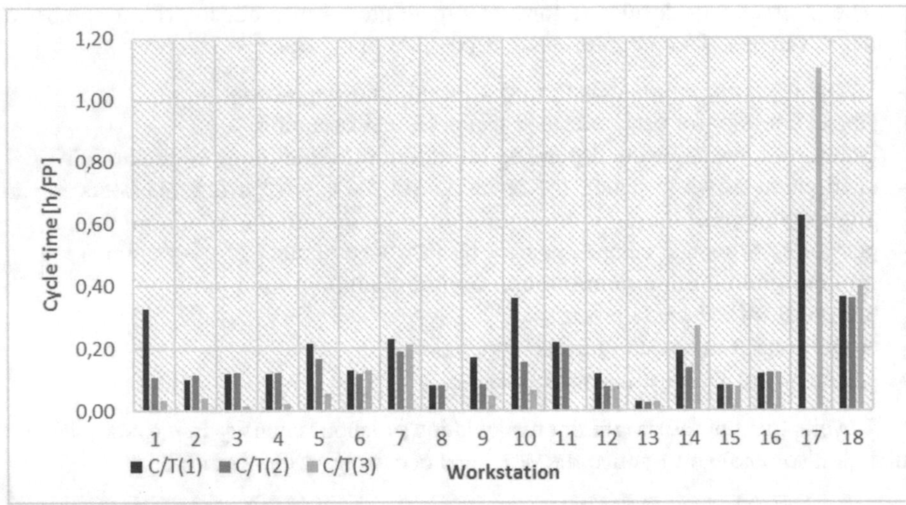

Fig. 11. Cycle time for workstations within the manufacturing process of 3 selected variants of the product

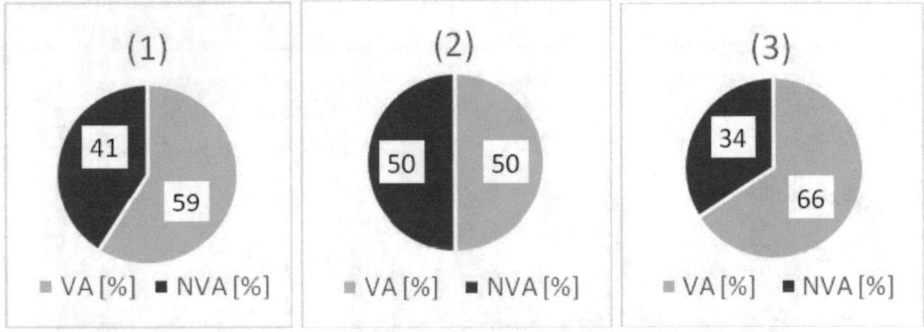

Fig. 12. Shares of value adding activities and non-value adding activities during execution of manufacturing processes of 3 selected product variants

of parameters makes a basis for selecting the optimum number of employees on each workstation and for planning the correct division of duties.

Analysis of cycle times of individual workstations showed significant differences between time values both within the given process and within the workstations. The results obtained using the simulation make a ground for planning division of duties among workstations, considering also variety of the executed cuts.

Analysis of time shares of the value adding and non-value adding activities showed not only high shares of wastages in manufacture, but also uneven distribution of these values among various workstations within these processes (Fig. 13).

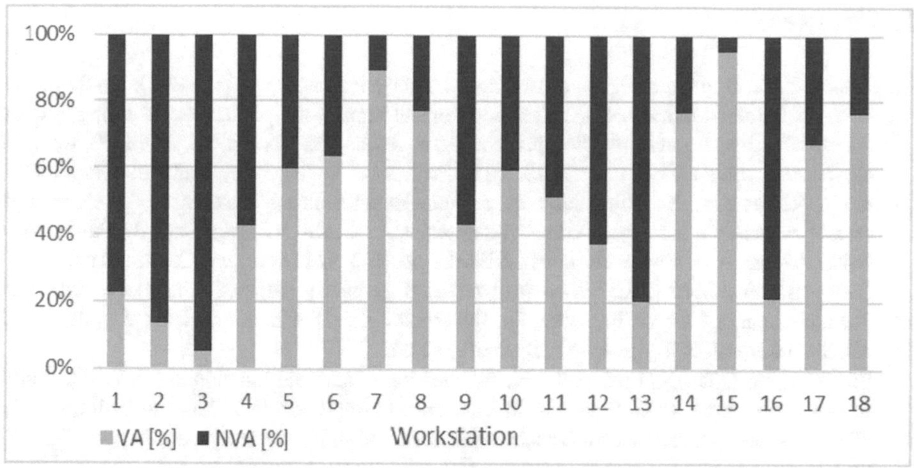

Fig. 13. Shares of value adding activities and non-value adding activities for workstations within the manufacturing process on the example of a selected product variant

Analysis of wastage shares at individual stages makes a starting point for the plan of improvements, indicating the points especially requiring corrections.

Use of the labour intensity calculator on an example of the real company showed its capabilities at calculation of production data together with identification and analysis of problems. Moreover, this tool can be also used for verification of effectiveness of potential improvements, utilising simulations based on rationalised data (e.g. increased number of employees, some activities shifted to other workstations, some unnecessary activities eliminated, etc.).

Results obtained for each kind of the offered products by use of the program for labour intensity calculation enabled the company to consider labour costs in the product prices, as well as to start working on improvements in the areas requiring corrections.

4 Summary

The paper is related to management of multivariant production that, favourably influencing competitive advantages of the company, poses also additional challenges in the fields of technology, organisation and management. As a tool for increasing the control of multivariant production, the "labour intensity calculator" is suggested. On the grounds of measurements of real data from representative processes and information included in the production documentation, using simulations of similar processes, the program estimates production data for products with preset configurations of parameters. The calculator can serve as a tool for analysis (of labour inputs in manufacture with preset parameters, cycle time, bottleneck, added value, wastages etc.) and control (of effectiveness of the analysed solutions in order to solve problems of uneven workload division among operators etc.) and thus as a basis for management of manufacturing processes.

References

1. Zipkin, P.: The limits of mass customization. MIT Sloan Manag. Rev. **42**(3), 81–87 (2001)
2. Kłos, S., Patalas-Maliszewskal, J.: The topological impact of discrete manufacturing systems on the effectiveness of production processes. In: Rocha, Á., Correia, A., Adeli, H., Reis, L., Costanzo, S. (eds.) WorldCIST 2017. AISC, vol. 571, pp. 441–452. Springer, Cham (2017)
3. Kluz, R., Antosz, K.: Simulation of flexible manufacturing systems as an element of education towards industry 40. In: Trojanowska, J., Ciszak, O., Machado, J., Pavlenko, I. (eds.) Advances in Manufacturing II. LNME, pp. 332–341. Springer, Cham (2019)
4. Czarnecka, A., Górski, A.: How can the fractal geometry help with analyze of the stock exchange market? In: Wilimowska, Z., Borzemski, L., Grzech, A., Świątek, J. (eds.) ISAT 2016. AISC, vol. 524, pp. 45–53. Springer, Cham (2017)
5. Sobaszek, Ł., Gola, A., Kozłowski, E.: Application of survival function in robust scheduling of production jobs. In: Ganzha, M., Maciaszek, M., Paprzycki, M. (eds) Proceedings of the 2017 Federated Conference on Computer Science and Information Systems **11**, pp. 575–578 (2017)
6. Rosienkiewicz, M., Gąbka, J., Helman, J., Kowalski, A., Susz, S.: Additive Manufacturing Technologies Cost Calculation as a Crucial Factor in Industry 40. In: Hamrol, A., Ciszak, O., Legutko, S., Jurczyk, M. (eds.) Advances in Manufacturing. LNME, pp. 171–183. Springer, Cham (2018)
7. Kempa, W.M., Paprocka, I., Kalinowski, K., Grabowik, C., Krenczyk, D.: Study on transient queueing delay in a single-channel queueing model with setup and closedown times. In: Dregvaite, G., Damasevicius, R. (eds.) ICIST 2016. CCIS, vol. 639, pp. 464–475. Springer, Cham (2016)
8. Grzybowska, K., Kovács, G.: The modelling and design process of coordination mechanisms in the supply chain. J. Appl. Logic **24**, 25–38 (2017)
9. Górnicka, D., Burduk, A.: Improvement of production processes with the use of simulation models. In: Wilimowska, Z., Borzemski, L., Świątek, J. (eds.) ISAT 2017. AISC, vol. 657, pp. 265–274. Springer, Cham (2018)
10. Kotowska, J., Burduk, A.: Optimization of production support processes with the use of simulation tools. In: Wilimowska, Z., Borzemski, L., Świątek, J. (eds.) ISAT 2017. AISC, vol. 657, pp. 275–284. Springer, Cham (2018)
11. Kochańska, J., Burduk, A.: Rationalization of retooling process with use of SMED and simulation tools. In: Wilimowska, Z., Borzemski, L., Świątek, J. (eds.) ISAT 2018. AISC, vol. 854, pp. 303–312. Springer, Cham (2019)
12. Antonelli, D., Stadnicka, D.: Combining factory simulation with value stream mapping: a critical discussion. Procedia CIRP **67**, 30–35 (2018)
13. Jarecki, W.: A concept of human capital. In: Kopycińska, D. (ed.) Human Capital in Economy. PTE, Szczecin (2003). (in Polish)
14. Stabryła, A.: Analysis and Design of Company Management Systems. Mfiles.pl, Kraków (2009). (in Polish)
15. Kutschenreiter-Praszkiewicz, I.: Selected problems of production planning. Ph.D. thesis. Lodz University of Technology (2000) (in Polish)
16. Saha, S.: Increasing organizational commitment strategically using Hard and Soft HRM. Int. J. Global Bus. Manag. Res. **2**(1), 62–63 (2013)

Production Resources Utilization Improvement with the Use of Simulation Modelling

Dagmara Górnicka, Joanna Kochańska, and Anna Burduk(✉)

Faculty of Mechanical Engineering, Wroclaw University of Science and Technology, Wybrzeze Wyspianskiego 27, 50-370 Wroclaw, Poland
{dagmara.gornicka,joanna.kochanska,
anna.burduk}@pwr.edu.pl

Abstract. In the paper, the problem of proper resources management is considered. The problem was analysed and the results of a research on production improvement are presented, concerning specifically utilization rates of machinery and operators, as well as increase of production capacity. These aspects affect the efficiency of the whole production process. Thus, the proper resources management is an additional challenge that needs to be done by every manufacturing company. Examinations were carried-out in a company from the automotive industry, where the utilization of resources of three production lines were considered. In this case, the operators and machines utilization rates were emphasized. The optimization criteria was to increase the utilization rate of resources and reduce the time needed to produce one piece of each analysed products. The methods used to perform the improvement were, among the others, kanban and supermarket systems. The results were verified using computer simulation models designed for both present state and the state after implementing the suggested solutions.

Keywords: Improvement of production process · Management of operators' work · Utilization of production machinery · Simulation model · Logistics

1 Introduction

The utilization of production resources allows to define how the company use production possibilities, which often turns out to be lower than conditions allow. It is usually caused by the organization of production. Thus, it is proper base to improve process without or with very low cost. The efficiency of the whole process is greatly affected by the utilization of human resources or machines [1, 8–10]. To analyse these parameters it can be used, among others, simulation modelling methods [4].

Simulation modelling is a method permitting experiments to be carried-out on models instead of on real objects [2, 3, 7]. Properties and construction of the object are modelled in a simplified way, according to the restrictions imposed by the program used for simulation. There are various solutions for production controlling and simulation modelling allows to analyse them without the need of interfere in production

J. Świątek et al. (Eds.): ISAT 2019, AISC 1051, pp. 41–50, 2020.
https://doi.org/10.1007/978-3-030-30604-5_4

process flow [5, 6]. Carrying-out a simulation permits the suggested improving solutions to be verified and analysed with no real intervention in the process course. This solution was used in the research works discussed in this paper and simulation models were developed using the program ProModel.

2 Company Details

Examinations were carried-out in a company manufacturing safety control systems in automotive industry. With regard to confidentiality policy, name and logo of the company were omitted in the paper.

2.1 Production Process

The process analysed in the paper is assembly of valves, executed on three assembly lines designated hereunder as line 1, line 2 and line 3. Each of these lines is equipped with two assembly workstations (housing assembly and screwing), adjustment station, two testers, final inspection station and failure analysis station. Within the described lines, two separate machines for assembly of the valve housing itself are located, which was named the pre-assembly process. Layout of machines within the considered lines is shown in Fig. 1.

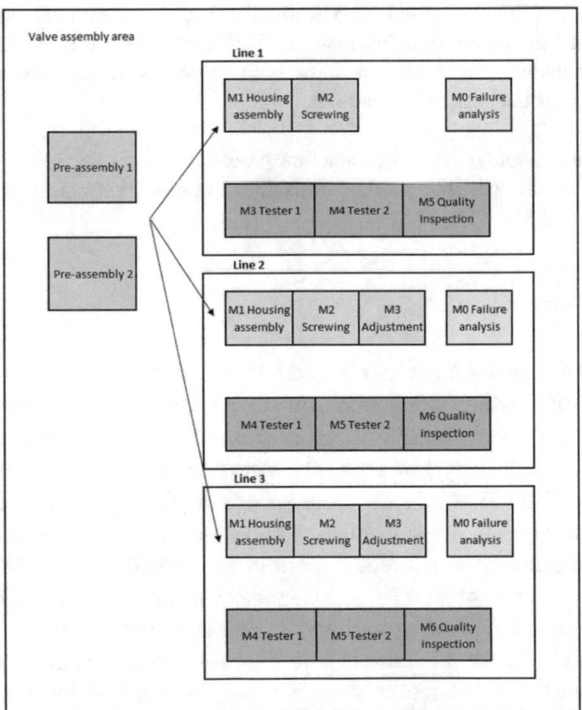

Fig. 1. Layout of valve assembly area

Production process is operated by eight operators, of that six ones work directly on assembly lines, one on the pre-assembly machines and one is a logistic worker. All the machines, including those of pre-assembly, are fed with components taken from the store just by the logistic worker.

In the valve assembly area, three types of valves are produced; two types on each line. These types are characterised by similar manufacturing processes, but with variable technological parameters. All kinds of valves are manufactured by three-stage pre-assembly, after that the housings are directed to individual production lines. There, actual assembly is carried-out, composed of assembly of housings and screwing, followed by adjustment and inspection of the products. These processes were modelled in the program ProModel that made it possible to consider not only the data related to the processes strictly productive, but also to auxiliary processes like e.g. logistics. Moreover, the program ProModel, after simulation, elaborates a series of diagrams permitting the selected criteria to be analysed.

3 Simulation Models

All the described production lines were defined in a simulation model, together with the pre-assembly machines. To build the model, the existing layout of the production area was used. The layout modelled with maintained distances between workstations is shown in Fig. 2.

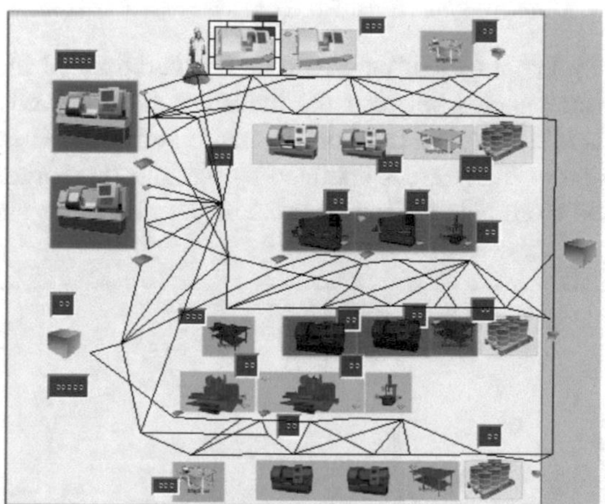

Fig. 2. Arrangement of production workstations in assembly lines [ProModel]

The model took into account real dimensions of the machines and distances between them, as well as technological and organisational data. Moreover, logistic processes making it possible to feed the assembly lines and numbers of operators together with their duties were considered.

3.1 Present-State Model

In order to identify the problems occurring in the valve assembly area, a present-state model was designed, on that the simulation was next carried-out. In consultation with the company, utilization rates of available operators (Fig. 3) and machinery (Fig. 4) were chosen for further analysis.

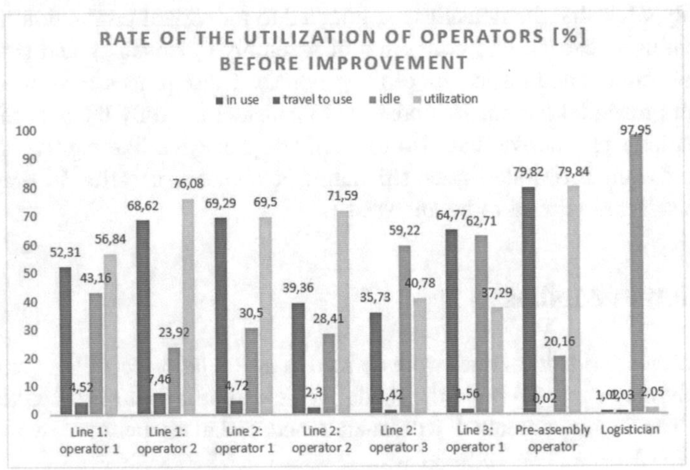

Fig. 3. Utilization rates of operators [ProModel] – present state

Figure 3 shows percentages of activities in the worktime of individual operators. Especially large discrepancies between the times can be observed in the case of the logistic worker. His idle time is much longer than other times. However, it should be stressed that only one of many areas operated by the logistic worker are considered in the model. Analysed were also utilization rates of the machines, shown in Fig. 4.

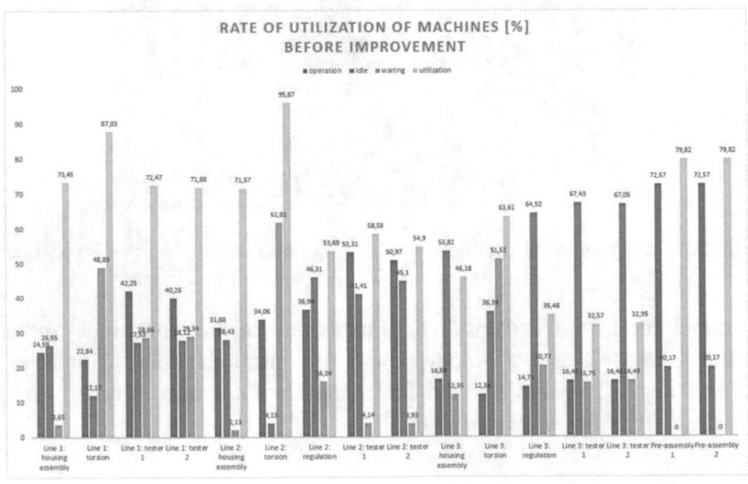

Fig. 4. Utilization rates of machines [ProModel] – present state

In analysis of utilization rates of the machines, workstations of final inspection and failure analysis were not considered. The reason is that these workstations present a small part of the entire process time, because operations are carried-out on them optionally only.

For further analysis, time required for manufacture of a single product in individual variants was measured, see Table 1.

Table 1. Production time of a single product – present state

Production line	Product	No. of operators	Production time of a single product [min]
Line 1	No. 10	2	6.7
	No. 13		5.0
Line 2	No. 12	3	8.7
	No. 15		4.7
Line 3	No. 11	1	10.1
	No. 14		8.0

According to the observations and analysis of simulation results, delays in production are mainly caused by shortage of components on the production lines. As simulation shows, awaiting times for components are very long for the operators, see Figs. 3 and 5. With this respect, rationalizations aimed at improvement of supply process of the production area were suggested. Next, a model of potential future state was developed, on that a new simulation was performed.

3.2 Model of Potential Future State

As mentioned before, shortage of components was identified as cause of delays, so implementation of the kanban system that facilitates communication between workstations, lines and the store was suggested. Within this improvement, a kanban board and supermarket of housings were introduced. According to an interview with the production manager, the housings are the most important elements of products with regard to shortage of components on assembly lines. Places of introducing the kanban board and the supermarket are shown in Fig. 5.

Introduction of the supermarket to the simulation model was executed by adjustment of the stock accumulated at the beginning of the simulation for individual types of housings. Simulation was carried-out on the so modified model and its results were presented like those for the present state, as diagrams of utilization rates of operators (Fig. 6) and machinery (Fig. 7).

It can be seen that almost all utilization rates of operators increased. An exception is the operator running the machine for pre-assembly of housings, whose utilization rate decreased. A separate issue is utilization rate of the logistic worker, who attends not only the considered lines but many more production areas, and his utilization is not considered in this analysis.

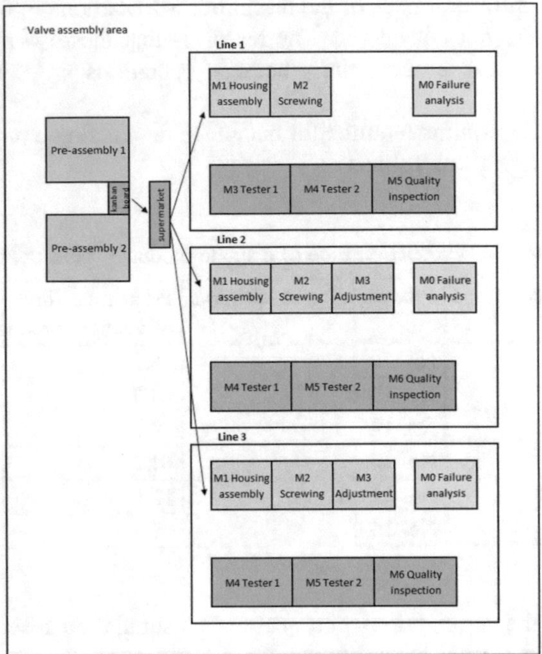

Fig. 5. Layout of valve assembly area with kanban board and supermarket

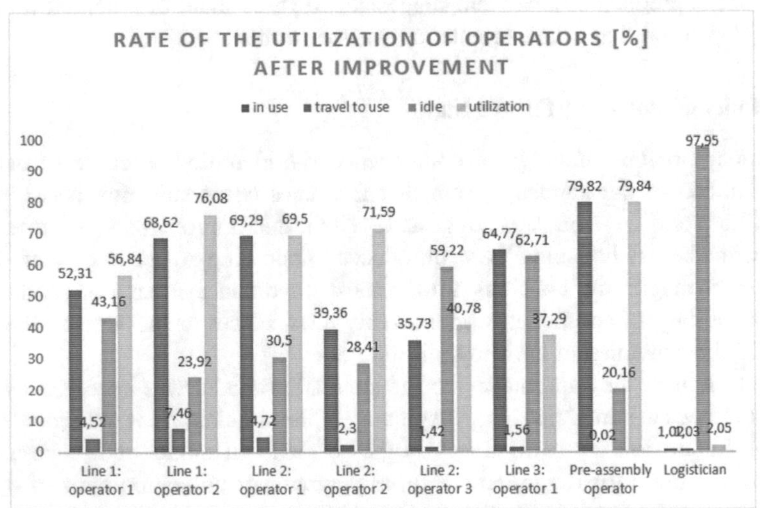

Fig. 6. Utilization rates of operators [ProModel] – potential future state

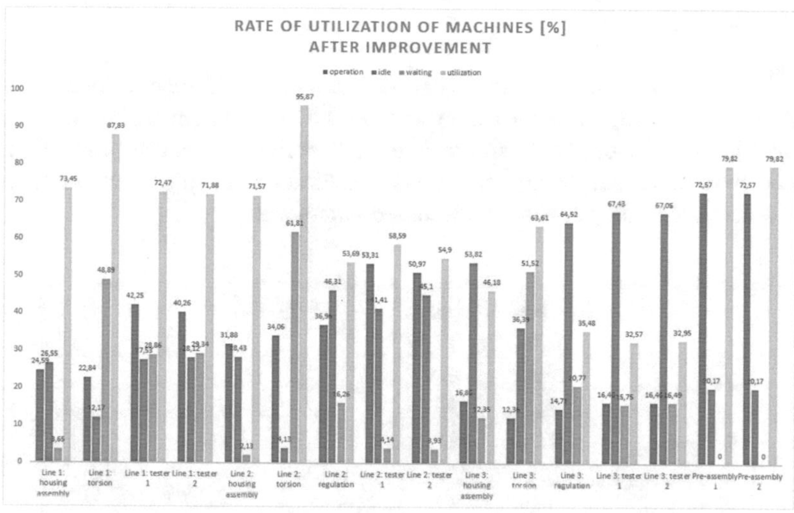

Fig. 7. Utilization rates of machines [ProModel] – potential future state

Utilization rates of machines and equipment were also changed, i.e. in most cases increased, which is discussed below in more details. The time necessary for manufacture of a single product was also calculated again and the results are given in Table 2.

Table 2. Production time of a single product – potential future state

Production line	Product	No. of operators	Production time of a single product [min]
Line 1	No. 10	2	5.1
	No. 13		4.8
Line 2	No. 12	3	5.2
	No. 15		4.2
Line 3	No. 11	1	9.5
	No. 14		7.9

In all of the six considered cases, time required for manufacture of a single finished product was reduced, which is highly desirable, since permits the company to increase production capacity. Results of the improvements are shown in details within the conclusions.

4 Conclusions

According to results of the simulations carried-out with use of the developed simulation models, introduction of a supermarket and the kanban system makes it possible to increase utilization of operators and machines, as well as to reduce production times. Analysed were exact data concerning profits resulting from improvement of the processes, according to the previously discussed parameters.

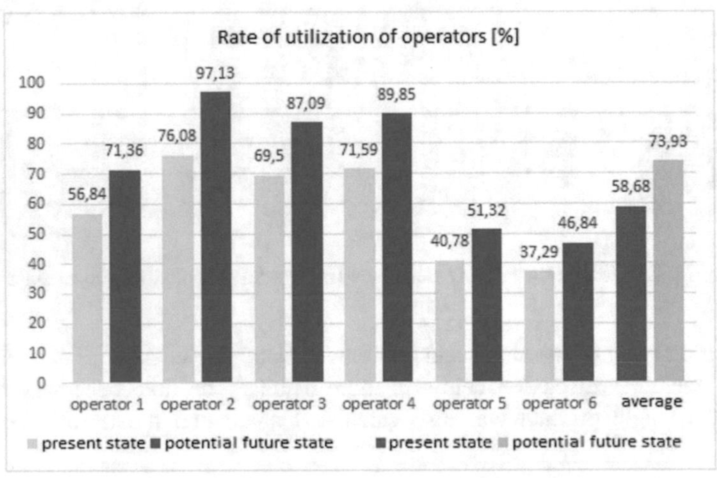

Fig. 8. Utilization rates of operators – comparative analysis

Utilization rates of operators are shown in Fig. 8. In the case of all the operators involved in assembly directly on the considered lines, utilization rates increased in average by over 15%, which means that the workers would spend larger parts of time on work than it was before implementation of the changes.

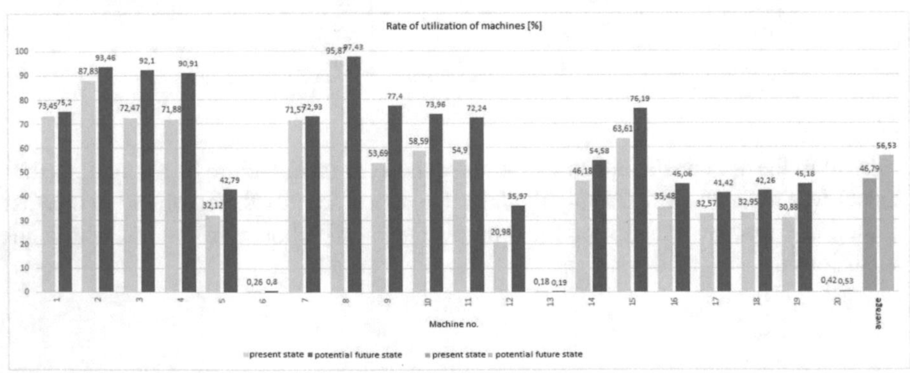

Fig. 9. Utilization rates of machines – comparative analysis

Utilization rates of machines, shown in Fig. 9, also indicate increase of production capacity. Thanks to implementation of the suggested improvements, these utilization should be higher, in average, by almost 10%. The machines No. 6, 13 and 20, distinguished in the diagram, show very small utilization rates because these workstations are designed for analysis of production errors, so they are used only when such errors emerge.

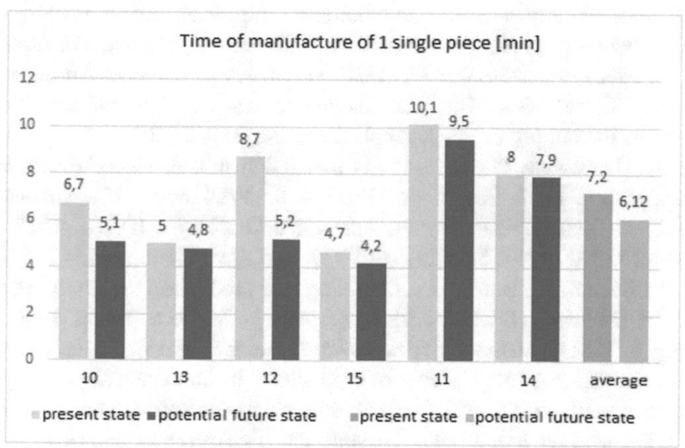

Fig. 10. Production time of a single product – comparative analysis

As a result of the suggested improvements, times required for manufacture of one piece of each product were also reduced, which is illustrated in Fig. 10. This means that the company manufactures the products in the time by 1 min and 8 s (in average) shorter, which presents reduction of production time by 15%.

Summarizing, the research was aimed at identifying reasons of delays occurring within production lines of valves and at suggesting some improving solutions. This aim was achieved; the reason of delays turned-out shortages of components in the production area, resulting, to a large extent, from communication difficulties. As a solution, implementation of the kanban system and a supermarket with the housing as its main component was suggested. According to the performed computer simulations, introduction of these changes should make it possible to reduce the delays and to increase utilization rates of production resources.

References

1. Ratnayake, R.M.C., Stadnicka, D., Antosz, K.: Deriving an empirical model for machinery prioritization: mechanical systems maintenance. In: 2013 IEEE International Conference on Industrial Engineering and Engineering Management, Bangkok (2013)

2. Kotowska, J., Burduk, A.: Optimization of production support processes with the use of simulation tools. In: Wilimowska, Z., Borzemski, L., Świątek, J. (eds) Information Systems Architecture and Technology: Proceedings of 38th International Conference on Information Systems Architecture and Technology – ISAT 2017. Advances in Intelligent Systems and Computing, vol. 657, pp. 275–284 (2018)
3. Drevetskyi, V., Kovela, I., Kutia, V.: A method of comparative evaluation of control systems by the set of performance measures. Informatyka, Automatyka, Pomiary w Gospodarce i Ochronie Środowiska, vol. 3 (2017)
4. Kłos, S., Patalas-Maliszewska, J.: The topological impact of discrete manufacturing systems on the effectiveness of production processes. In: Rocha, Á., Correia, A., Adeli, H., Reis, L., Costanzo, S. (eds.) WorldCIST 2017. AISC, vol. 571, pp. 441–452. Springer, Cham (2017)
5. Grzybowska, K., Kovács, G.: The modelling and design process of coordination mechanisms in the supply chain. J. Appl. Logic **24**, 25–38 (2017)
6. Krenczyk, D., Bocewicz, G.: Data-driven simulation model generation for ERP and DES systems integration. In: Jackowski, K., Burduk, R., Walkowiak, K., Wozniak, M., Yin, H. (eds) Intelligent Data Engineering and Automated Learning – IDEAL 2015. IDEAL 2015. Lecture Notes in Computer Science, vol 9375. Springer, Cham (2015)
7. Gwiazda, A., Sekala, A., Banas, W.: Modeling of a production system using the multiagent approach. In: Modtech International Conference - Modern Technologies in Industrial Engineering V, IOP Conference Series-Materials Science and Engineering, vol. 227 (2017)
8. Jasiulewicz-Kaczmarek, M.: The role of ergonomics in implementation of the social aspect of sustainability, illustrated with the example of maintenance. Occupational Safety and Hygiene, Occupational Safety and Hygiene, red. P. Arezes et al., CRC Press, Taylor & Francis Group, London (2013)
9. Rosienkiewicz, M., Gąbka, J., Helman, J., Kowalski, A., Susz, S.: Additive manufacturing technologies cost calculation as a crucial factor in industry 40. In: Hamrol, A., Ciszak, O., Legutko, S., Jurczyk, M. (eds.) Advances in Manufacturing. Lecture Notes in Mechanical Engineering. Springer, Cham (2018)
10. Kamińska, A.M., Parkitna, A., Górski, A.: Factors determining the development of small enterprises. In: Wilimowska, Z., Borzemski, L., Świątek, J. (eds) Information Systems Architecture and Technology: Proceedings of 38th International Conference on Information Systems Architecture and Technology – ISAT 2017. ISAT 2017. Advances in Intelligent Systems and Computing, vol 657. Springer, Cham (2018)

Numerical Studies on the Effectiveness of Optimization Algorithms to Identify the Parameters of Machine Tool Hybrid Body Structural Component Model

Paweł Dunaj(✉), Stefan Berczyński, and Sławomir Marczyński

West Pomeranian University of Technology Szczecin, Szczecin, Poland
{pawel.dunaj, stefan.berczynski, slawomir.marczynski}@zut.edu.pl

Abstract. Designing machine tools is an extremely complicated process, requiring thorough knowledge from the constructors in terms of statics, dynamics, contact mechanics, etc. Therefore, it is necessary to support this process by analyzing their models. However, only on the basis of experimentally identified models, reliable information about the object can be obtained. Presented paper contains numerical studies of the efficiency of optimization algorithms to identify parameters of FEM models. Numerical studies were carried out for the FEM model of the hybrid machine tool body structural component, consisting of steel and polymer concrete. In order to identify the parameters of the model, an interface was developed to enable data exchange between the FEM preprocessor of the Nastran FX and Matlab to cooperate. Numerical studies were then conducted using a developed identification algorithm based on accelerance functions, developed on the basis of reduced order FEM model and introduced correction parameters.

Keywords: Finite element method · Model order reduction · Hybrid machine tool body · Polymer concrete · Identification algorithm

1 Introduction

The machine tool bodies are usually made of a gray cast iron and they are characterized by a high vibration damping ability and a good machinability. Their disadvantages are relatively high mass and a significant limitations in terms of the possibility of modifying the existing structure.

The second most popular group are machine tool bodies made of steel. Compared to gray cast iron bodies, they have a much better mechanical properties, a lower mass (cast iron bodies are about twice as heavy while maintaining the same stiffness) also modifying their structure is easier and less expensive. However, machine tools steel bodies have a much lower damping ability.

Regardless of the well-established position of gray cast iron and steel in the construction of machine tool bodies, the restrictive requirements imposed on new machine tools, related to the increase in productivity, accuracy and simultaneous reduction of

J. Świątek et al. (Eds.): ISAT 2019, AISC 1051, pp. 51–61, 2020.
https://doi.org/10.1007/978-3-030-30604-5_5

energy consumption, forces the search for new solutions in the construction of machine tool bodies.

The solutions characterized by a good mechanical properties, a relatively low mass, and a high vibration damping abilities are hybrid machine tool bodies. This group is characterized by a large variety of construction solutions [3, 5, 9, 11, 12]. The main idea is the use of various materials for the synergistic use of their properties. The most commonly used hybrid solutions include a combination of materials such as steel, aluminum or gray cast iron with composite materials such as carbon fiber composites [14], sandwich composites or polymer concrete [13].

To effectively use synergistic properties of materials used in the construction of hybrid bodies, the design process should be supported by the analysis of their models, of which finite element models are the most often used. However, due to the innovativeness and uniqueness of hybrid solutions, the accuracy of currently used FEM models is insufficient. Therefore, the latest research focuses on the development of FEM modelling procedures to reliably evaluate the dynamic properties of hybrid machine tool bodies at the design stage.

Improving the accuracy of hybrid machine tool bodies FEM models can be achieved through experimental identification of basic structural components models parameters [1, 2, 4, 7, 8]. Determination of material constants on the basis of experimental tests is not always sufficient to obtain a reliable model. Therefore, experimental identification should be based e.g. on impact test analysis, static deflection analysis etc. This paper presents numerical considerations regarding the possibility of using optimization algorithms to identify the parameters of the steel-polymer concrete beam model being the basic structural component of the newly designed vertical lathe hybrid body.

2 Research Object

The vertical lathe hybrid body solution was based on a welded steel frame filled with polymer concrete [5, 6, 11]. The steel frame was made of square hollow cross-section steel profiles connected by welded joints. The steel frame provides the required stiffness and defines the geometrical structure of the machine tool. The use of polymer concrete was intended to increase the damping, while maintaining a low mass of the structure. Polymer concrete, consisted of an epoxy resin and mineral filling of various sizes. The composition of it was presented in Table 1. The concept of a hybrid machine tool body was shown in Fig. 1.

Table 1. Composition of polymer concrete (mass percentage)

	Epoxy resin	Ash	Small fraction (0.25–2 mm)	Medium fraction (2–10 mm)	Coarse fraction (8–16 mm)
Mass percentage	15%	1%	19%	15	50

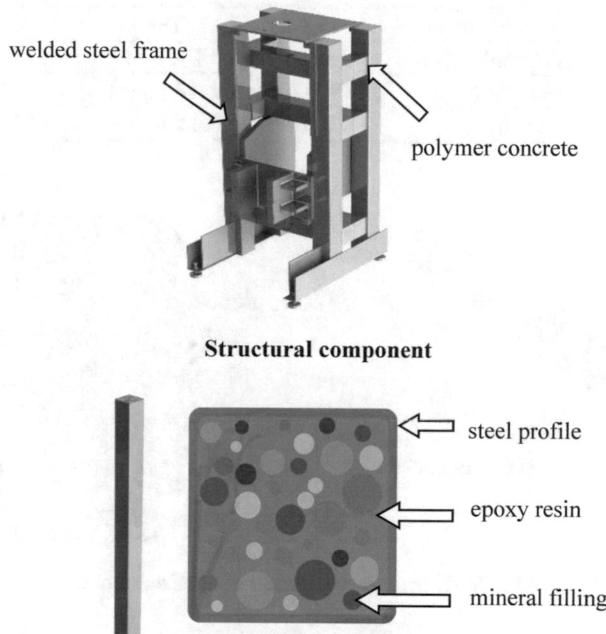

Fig. 1. Hybrid machine tool body composed of steel profiles filled with polymer concrete

In the presented paper an identification procedure was presented. It was based on structural component which was the beam composed of steel profile with a square hollow cross-section of 70 × 70 mm and a wall thickness of 3 mm and polymer concrete filling.

3 Finite Element Model of Structural Component

In the first stage a model of a single unconstrained beam was build using finite element method. Geometrical model was discretized using Nastran FX preprocessor. Both, steel coating and polymer concrete filling were discretized using six-sided isoparametric solid elements with 8 nodes (CHEXA) and five-sided isoparametric solid element with 6 nodes (CPENTA). To improve model efficiency structural mesh was used. Linear elastic models of materials were adopted, parameters which define each material were shown in Table 2. Contact between steel and polymer concrete was modelled as nodes coincidence. In summary, the developed model consisted of 19,600 elements and 21,985 degrees of freedom, the discretized model was presented in Fig. 2.

Table 2. Material properties used in model development

Parameter	Steel	Polymer concrete
Young's modulus E	205 GPa	17 GPa
Poisson's ratio ν	0.28	0.20
Density ρ	7800 kg/m^3	2118 kg/m^3
Loss factor η	0.0018	0.0048

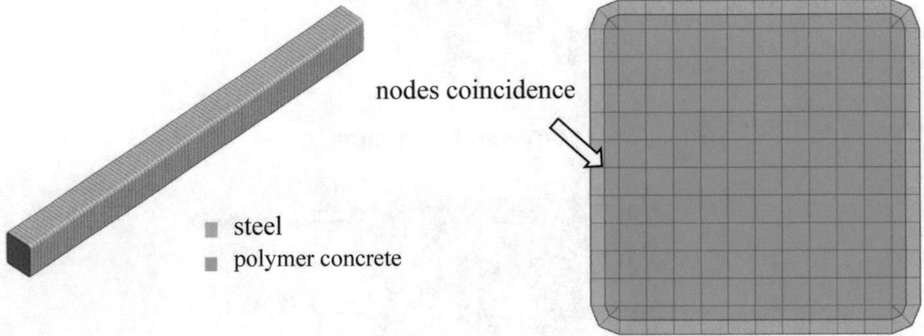

Fig. 2. Structural component discretized model

The damping of the structure was described by a complex stiffness; according to the adopted model, the damping matrix $\boldsymbol{C}$ can be expressed as follows:

$$\boldsymbol{C} = i\eta \boldsymbol{K} \tag{1}$$

where: i – imaginary unit; η – loss factor; $\boldsymbol{K}$ – stiffness matrix, $\boldsymbol{K} \in \mathbb{R}^{nxn}$.

Therefore, the component finite element model equation of motion takes the form:

$$\boldsymbol{M}\ddot{\boldsymbol{q}}(t) + (\boldsymbol{K} + \boldsymbol{C})\boldsymbol{q}(t) = \boldsymbol{f}(t) \tag{2}$$

or:

$$\boldsymbol{M}\ddot{\boldsymbol{q}}(t) + (1 + i\eta)\boldsymbol{K}\boldsymbol{q}(t) = \boldsymbol{f}(t) \tag{3}$$

where: $\boldsymbol{M}$ – mass matrix, $\boldsymbol{M} \in \mathbb{R}^{nxn}$; $\ddot{\boldsymbol{q}}, \boldsymbol{q}$ – acceleration and displacement vectors of the finite element assemblage, $\ddot{\boldsymbol{q}}, \boldsymbol{q} \in \mathbb{R}^{nx1}$; $\boldsymbol{f}$ – vector of externally applied loads, $\boldsymbol{f} \in \mathbb{R}^{nx1}$; t – time.

4 Model Parameters Identification Algorithm

The commercial Nastran FX software used in the modelling process did not allow automatic identification of model parameters. Therefore, an identification algorithm based on accelerance functions was developed. For this purpose, an external program that complement the Nastran FX software was developed. This program was developed using the Matlab environment and included the following modules: interface, model order reduction and identification.

4.1 Interface Module

The interface module was a connection between the Nastran FX preprocessor and the Matlab. Its operation consisted in importing matrices describing the component model: mass M and stiffness K. The import process was preceded by a number of preparatory activities, the first of which was the export of matrices from the Nastran FX to a text file in .bdf format. This was realized by exporting the model established in the preprocessor to the .nas file, which is its text form. Then the file .nas was modified by adding the EXTSEOUT (DMIGBDF) command in the case control section. Next using a NEi Nastran solver a .bdf file was created. On the basis of the received file (which is the record of the matrices describing the model), assigning subsequent elements of .bdf file to the elements of the matrices in Matlab, an import procedure was performed. Figure 3 shows the workflow of the interface module.

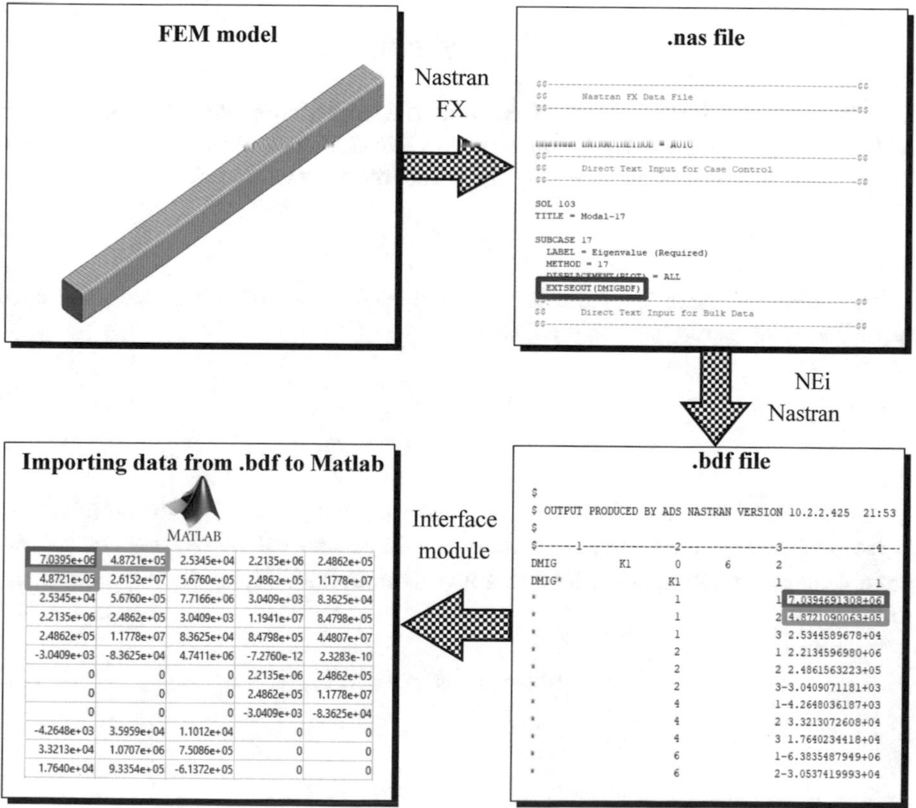

Fig. 3. Interface module workflow

4.2 Model Order Reduction Module

In the next – model order reduction module, on the basis of the imported mass and stiffness matrices, the reduction of model degrees of freedom (DOF) was carried out. This operation was aimed at shortening the calculation time associated with determining the acceleration function.

The purpose of model order reduction was to find the m DOFs model preserving the dynamic characteristics of the full – n DOFs model, in which $m << n$. It was realized by approximating the state vector by means of transformation:

$$\boldsymbol{q} = \boldsymbol{T}\boldsymbol{q}_R \tag{4}$$

where: $\boldsymbol{T}$ – transformation matrix $\boldsymbol{T} \in \mathbb{R}^{nxm}$ and $\boldsymbol{q}_R \in \mathbb{R}^{mx1}$. Hence, reduced model equation of motion can be expressed as:

$$\boldsymbol{M}_R\ddot{\boldsymbol{q}}_R + (1 + i\eta)\boldsymbol{K}_R\boldsymbol{q}_R = \boldsymbol{f}_R \tag{5}$$

where the reduced matrices $\boldsymbol{M}_R, \boldsymbol{K}_R$ are defined as:

$$\boldsymbol{M}_R = \boldsymbol{T}^T\boldsymbol{M}\boldsymbol{T} \tag{6}$$

$$\boldsymbol{K}_R = \boldsymbol{T}^T\boldsymbol{K}\boldsymbol{T} \tag{7}$$

Model order reduction was realized using a Kammer reduction which is based on normal mode shapes. Kammer reduction provides that all dynamic information contained in FEM model is preserved by the reduction process [6, 10].

$$\boldsymbol{q} = \boldsymbol{\phi}\xi \tag{8}$$

According to Kammer reduction method a state vector $\boldsymbol{q}$ can be partitioned according to master m and slave s DOFs:

$$\boldsymbol{q}_m = \boldsymbol{\phi}_m\xi \tag{9}$$

$$\boldsymbol{q}_s = \boldsymbol{\phi}_s\xi \tag{10}$$

To obtain a relationship between coordinates $\boldsymbol{q}_m$ and $\boldsymbol{q}_s$, Eq. (9) must be solved for ξ. However, due to the fact that the matrix $\boldsymbol{\phi}_m$ in general will be rectangular, it is not possible to solve this equation by direct inversion. Premultiplying Eq. (9) by $\boldsymbol{\phi}_m^T$, and solving it for ξ yields to:

$$\left[\boldsymbol{\phi}_m^T\boldsymbol{\phi}_m\right]^{-1}\boldsymbol{\phi}_m^T\boldsymbol{q}_m = \xi \tag{11}$$

where: $\left[\boldsymbol{\phi}_m^T\boldsymbol{\phi}_m\right]^{-1}\boldsymbol{\phi}_m^T$ represents the Moore-Penrose generalized inversof $\boldsymbol{\phi}_m$. Submitting Eq. (11) in to (10) gives:

$$\boldsymbol{q}_s = \boldsymbol{\phi}_s\left[\boldsymbol{\phi}_m^T\boldsymbol{\phi}_m\right]^{-1}\boldsymbol{\phi}_m^T\boldsymbol{q}_m \tag{12}$$

The transformation matrix is defined as follows:

$$\boldsymbol{T} = \begin{bmatrix} \boldsymbol{I} \\ \boldsymbol{\phi}_s\left[\boldsymbol{\phi}_m^T\boldsymbol{\phi}_m\right]^{-1}\boldsymbol{\phi}_m^T \end{bmatrix} \tag{13}$$

Model developed in Sect. 3 was subjected to a presented reduction method (reducing its size from 21,985 DOFs to 16 DOFs). Master degrees of freedom were shown in Fig. 4.

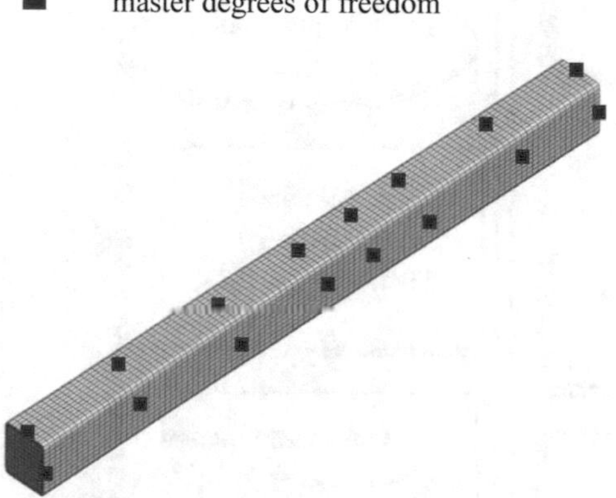

Fig. 4. Master degrees of freedom used in model order reduction

4.3 Identification Module

The last module was the identification module. The process of identifying model parameters has been reduced to the task of minimizing the objective function formulated in the following way:

$$\mathbf{Q} = \left(\boldsymbol{y}_{EXP} - \boldsymbol{y}_{FEM}\right)^T\left(\boldsymbol{y}_{EXP} - \boldsymbol{y}_{FEM}\right) \tag{14}$$

where: $\boldsymbol{y}_{EXP}$ – vector of amplitudes of the accelerance function determined on the basis of the experiment, $\boldsymbol{y}_{FEM}$ – vector of accelerance function amplitudes determined on the basis of the FEM model.

Decision variables, x_1, x_2, x_3 – correction coefficients of the parameters describing the model, respectively, Young's modulus E, loss coefficient η and density ρ were used.

Taking into account these coefficients, the equation describing the model takes the following form:

$$x_1 \boldsymbol{M}_R \ddot{\boldsymbol{q}}_R + x_2 \boldsymbol{C}_R \boldsymbol{q}_R + x_3 \boldsymbol{K}_R \boldsymbol{q}_R = 0 \tag{15}$$

where, the damping matrix $\boldsymbol{C}_R$ can be written as:

$$\boldsymbol{C}_R = i\eta \boldsymbol{K}_R \tag{16}$$

A schematic representation of the identification algorithm is included in Fig. 5.

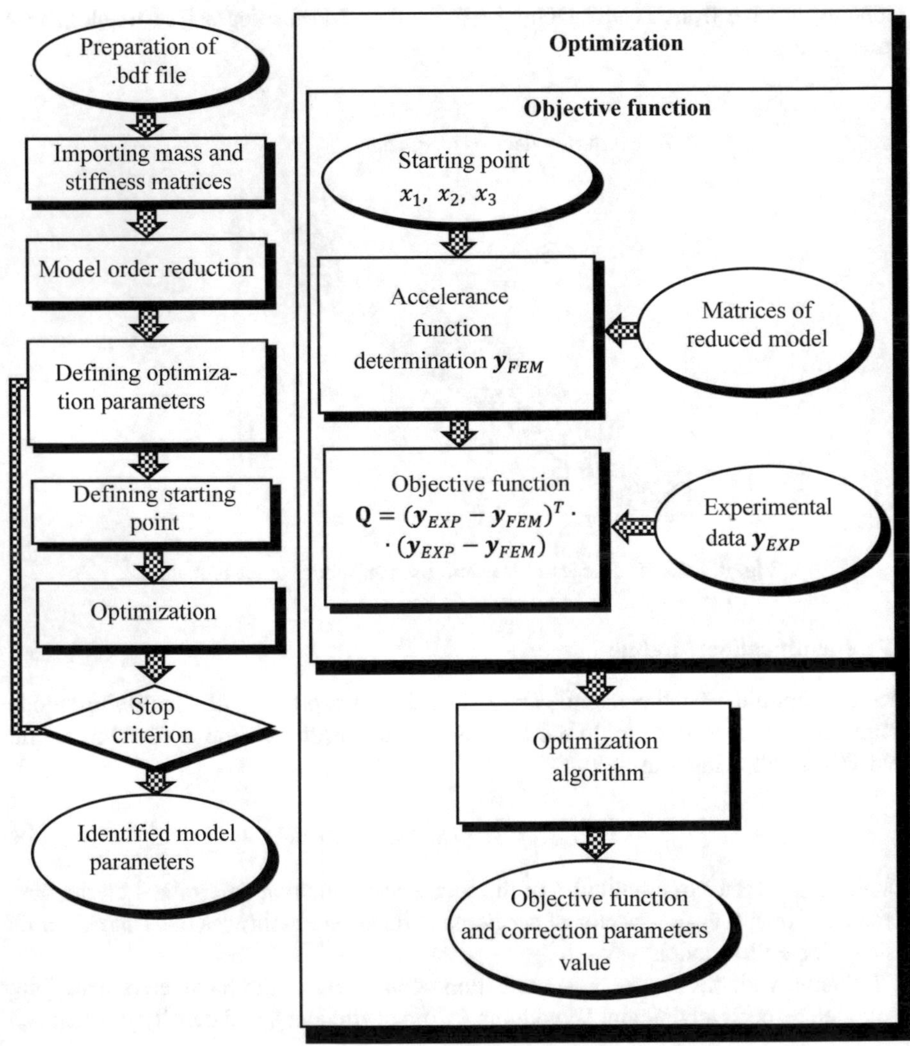

Fig. 5. Identification algorithm scheme

To examine the efficiency of the optimization algorithms used in parameter identification algorithm, vector $\boldsymbol{y}_{EXP}$ was determined by substituting following correction coefficients values:

$$\boldsymbol{x}_{\mathrm{EXP}} = [x_{1e}, x_{2e}, x_{3e}] = [1.5, 1.2, 1] \tag{17}$$

into model equation of motion (15) and on its basis determining accelerance function amplitudes.

5 Results

The optimization algorithms implemented in the Matlab environment were analyzed i.e.: Nelder-Mead, Sequential Quadratic Programming and Pattern Search. To obtain the best results, the optimization parameters were modified depending on the algorithm used. The results in the form of values of correction coefficient, values of objective function and calculation time are included in Table 3.

Table 3. Optimization results

Algorithm	Correction coefficient values	Objective function value	Calculation time
Nelder-Mead	[1.5, 1.2, 1]	$4.56 \cdot 10^{-12}$	100%
Sequential Quadratic Programming	[1.5004, 1.2005, 1.0098]	$1.71 \cdot 10^{-4}$	403%
Pattern Search	[1.2556, 1.0063, 0.6384]	6.7	325%

Figure 6 depicts selected accelerance function.

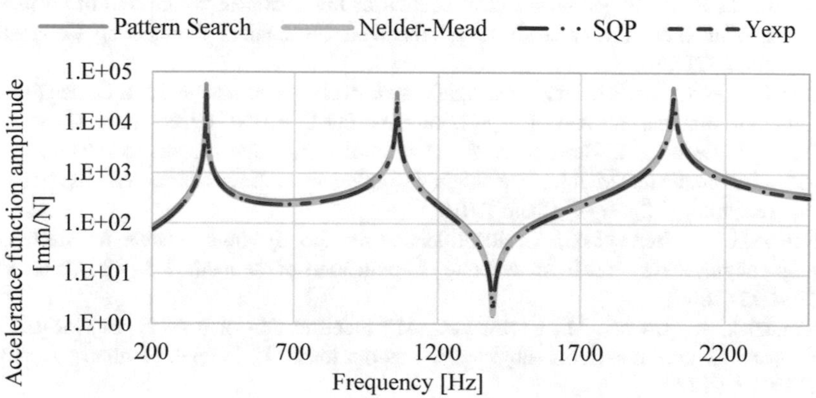

Fig. 6. Selected accelerance function

6 Findings

The use of the developed interface enables efficient connection of the Nastran FX preprocessor with Matlab, extending both its functionality and computing capabilities. More precisely, thanks to the interface used, it was possible to carry out the Kammer reduction (which by default is not available in the FEM software) and to carry out the identification process.

Research conducted on the possibility of using optimization algorithms implemented in Matlab to identify the parameters of the FEM model of steel-polymer concrete beams showed their high efficiency. Among the tested optimization algorithms, the Nelder-Mead algorithm proved to be the most accurate, both in terms of the value of correction coefficients and the value of the objective function, which is the actual indicator of fitting accuracy. In comparison to other algorithms, the time necessary to carry out calculations was much shorter.

In summary, the optimization algorithms examined are suitable for identifying the parameters of FEM models. Future work will focus on the study of their effectiveness using real experimental data.

References

1. Berczyński, S., Gutowski, P.: Identification of the dynamic models of machine tool supporting systems. Part I: an algorithm of the method. J. Vib. Control **12**(3), 257–277 (2006)
2. Bocian, M., et al.: Methods of identification of definite degenerated and nonlinear dynamic system using specially programmed nonharmonic enforce. J. Dyn. Syst. Meas. Control **139**(8), 081012 (2017)
3. Cho, S.-K., Kim, H.-J., Chang, S.-H.: The application of polymer composites to the table-top machine tool components for higher stiffness and reduced weight. Compos. Struct. **93**(2), 492–501 (2011)
4. Dhupia, J., et al.: Experimental identification of the nonlinear parameters of an industrial translational slide. In: Proceedings International Mechanical Engineering Congress and Exposition (2006)
5. Dunaj, P., et al.: Experimental investigations of steel welded machine tool bodies filled with composite material. In: Advances in Manufacturing II, pp. 61–69. Springer, Cham (2019)
6. Dunaj, P., Dolata, M., Berczyński, S.: Model order reduction adapted to steel beams filled with a composite material. In: International Conference on Information Systems Architecture and Technology. Springer, Cham (2018)
7. Gutowski, P., Berczyński, S.: Identification of the dynamic models of machine tool supporting systems. Part II: an example of application of the method. J. Vib. Control **12**(3), 279–295 (2006)
8. Jamroziak, K., Bocian, M., Kulisiewicz, M.: Identification of a subsystem located in the complex dynamical systems subjected to random loads. J. Comput. Nonlinear Dyn. **12**(1), 014501 (2017)
9. Jung, S.C., Lee, J.E., Chang, S.H.: Design of inspecting machine for next generation LCD glass panel with high modulus carbon/epoxy composites. Compos. Struct. **66**(1–4), 439–447 (2004)

10. Kammer, D.C.: Test-analysis model development using an exact modal reduction. Int. J. Anal. Exp. Modal Anal. **2**(4), 174–179 (1987)
11. Okulik, T., et al.: Determination of dynamic properties of a steel hollow section filled with composite mineral casting. In: Advances in Manufacturing II, pp. 561–571. Springer, Cham (2019)
12. Powałka, B., Okulik, T.: Dynamics of the guideway system founded on casting compound. Int. J. Adv. Manuf. Technol. **59**(1-4), 1–7 (2012)
13. Sonawane, H., Subramanian, T.: Improved dynamic characteristics for machine tools structure using filler materials. Proc. CIRP **58**, 399–404 (2017)
14. Suh, J.D., Lee, D.G., Kegg, R.: Composite machine tool structures for high speed milling machines. CIRP Ann. **51**(1), 285–288 (2002)

The Use of Knowledge-Based Engineering Systems and Artificial Intelligence in Product Development: A Snapshot

Stefan Plappert(✉), Paul Christoph Gembarski, and Roland Lachmayer

Institute of Product Development, Leibniz University of Hannover, Welfengarten 1a, 30167 Hannover, Germany
{plappert, gembarski, lachmayer}@ipeg.uni-hannover.de

Abstract. Beside the creative activities in product development, the design process involves multiple routine tasks that are subject to automation. Techniques like knowledge-based engineering, what is commonly understood as the merging of computer-aided design, object-oriented programming and artificial intelligence, have been discussed since years, but have not yet achieved a significant breakthrough. But in particular the actual debate on digitization and artificial intelligence draws much attention on fostering new automation potentials in design of products and services. This article aims at taking an actual snapshot in which fields of application knowledge-based engineering systems and artificial intelligence are used in product development. Therefore, the authors conducted a systematic literature review, limited to scientific literature of the last five years. The literature analysis and synthesis is condensed within a concept matrix that documents actual applications and shows further research potentials.

Keywords: Knowledge-based engineering · Artificial intelligence · Product development

1 Introduction

Product development and design problem solving are structured activities aimed at transforming technical and design requirements into a product specification, including geometric models, production data and assembly instructions [1]. It strongly depends on the experience and skills of the designer, i.e. finding suitable solution concepts, making an initial embodiment design and detailing parts and assemblies according to existing restrictions, e.g. from manufacturing or logistics [2]. Beside these creative tasks, the design process involves different routine tasks that are today subject of automation, like e.g. product configuration [3].

Techniques like knowledge-based engineering (KBE) have been discussed since years, but have not yet achieved a significant breakthrough beside single niche design activities like fixture design or applications in aerospace or automotive development [4].

J. Świątek et al. (Eds.): ISAT 2019, AISC 1051, pp. 62–73, 2020.
https://doi.org/10.1007/978-3-030-30604-5_6

Nonetheless, the vision of Chapman and Pinfold who understand KBE as "evolutionary step in computer-aided-engineering and (…) engineering method that represents a merging of object-oriented programming, artificial intelligence and CAD technologies, giving benefit to customized or variant design automation solutions" [5] is today more relevant than ever. A reason for this are the emerging methods and tools in the field of artificial intelligence.

This article aims to provide a snapshot how actual developments in knowledge-based engineering systems (KBES) and artificial intelligence disseminate in product development. Therefore, the authors conducted a systematic literature review, limited to scientific literature of the last five years. In the following Sect. 2, related work is presented in order to contextualize this study mainly in the field of knowledge-based engineering and product development. In Sect. 3, the methodology for the literature analysis is introduced. Afterwards in Sect. 4, the results of literature analysis and synthesis are presented and further discussed in Sect. 5. The final Sect. 6 concludes the article and presents a brief research agenda.

2 Related Work

KBE is founded on research on knowledge-based systems that have to be understood as computer aided problem solving tools. Problem solving behavior is generally modelled on that of a human expert, so the term *expert systems* developed as a synonym for knowledge-based systems of all kinds, particularly in the 1980s and 1990s. Examples include assistance or diagnostic systems in medicine, speech recognition tools and automatic classification systems [6].

In particular in engineering, expert systems were originally designed for product configuration [7] and design automation in special fields like in fixture design [8]. The rise of parametric CAD systems that allow defining variable geometric models led to new possibilities in knowledge integration within digital prototypes, like the implementation of design rules, dimensioning formulae or automated routines for geometry creation [2]. The modelling focus thus shifts from a single solution or product variant to a solution space or set of variants [9].

In order to support designers in the (commonly manual) formalization of knowledge and the creation of KBE applications, different process models and design methods, like e.g. KADS or MOKA have been proposed on the one hand [10, 11]. On the other hand, the use of artificial intelligence, like solving of constraint satisfaction problems, changed and extended the way solution spaces were modelled [12, 13].

Foundational research in knowledge-based systems and artificial intelligence dates back to the 1980s and 1990s where the available computing power was not sufficient to model and solve real world problems in the design engineering domain. Today, this lack of power seems overcome and emerging technologies like machine learning etc. extend the possibilities. Taking into account the burst of research in these fields of the last years, the authors want to investigate the dissemination of current research form KBE and artificial intelligence to product development, especially regarding the early phase of the product development process.

3 Methodology

In order to show which fields of application exist for knowledge-based engineering systems (KBES) and artificial intelligence in product development, the authors conducted a detailed literature search, following the methodological recommendations of vom Brocke et al. [14] as well as Webster and Watson [15]. The subsequent listing shows the different phases and their assignment to the sections: (I) definition of review scope, (II) conceptualization of topic, (III) literature search, (IV) literature analysis and synthesis and (V) research agenda.

Table 1. Search strategy and hits in Google Scholar

Search phrase	Product development	Production engineering	Manufacturing	Customization	Simulation	Optimization	roduct service system	Lifecycle	Modularization	Solution space	Product generation	Cyber physical system	Design	Modelling	Computer aided design	Σ	Pages
Knowledge-based	18	14	5	14	4	22	12	20	5	13	0	13	1	1	11	**153**	10
Artificial intelligence	15	12	9	4	4	4	6	2	8	0	4	2	4	1	2	**77**	10
Machine learning	3	3	2	0	2	4	2	1	2	0	2	2	4	1	1	**29**	3
Deep learning	0	0	1	0	0	0	0	0	0	0	0	1	0	1	0	**3**	3
Rule-based	3	3	1	0	0	3	0	0	0	1	0	0	4	0	0	**15**	1
Case-based	3	1	2	1	1	0	3	2	0	0	1	1	2	0	0	**17**	1
Constraint-based	4	0	0	0	0	0	1	2	1	2	0	1	0	0	0	**11**	1
Model-based	3	2	1	0	0	1	3	3	4	1	0	0	0	0	0	**18**	1
Agent-based	1	1	0	2	0	0	0	0	0	0	0	0	0	0	2	**6**	1
Cognitive	0	0	2	1	0	0	1	0	0	0	0	0	0	0	0	**4**	1
Ontology-based	0	1	0	1	0	0	2	0	0	1	0	0	0	0	0	**5**	1
Σ	**50**	**37**	**23**	**23**	**11**	**34**	**30**	**30**	**20**	**18**	**7**	**20**	**15**	**4**	**16**	**338**	

Additionally, the review was divided in two parts. First, suitable literature, with reference to artificial intelligence and knowledge-based engineering, was searched in the Scientific Society for Product Development (WiGeP).

In the analysis of the literature identified, the following search directions were distinguished: knowledge-based engineering (KBE), design optimization, simulation, product-service-systems (PSS), cyber physical production systems (CPPS), product generation and smart products. Based on these findings, the keywords were analyzed and combined to search phrases (Table 1). Second, a Google Scholar search was conducted using the search phrases to get a broad and interdisciplinary overview of the state of the art in computer science. The literature found was assessed according to its relevance for our research.

The search phrases consist of a method (vertical phrases) and an application (horizontal phrases). Based on the frequency of keywords in the WiGeP literature, the methods were divided into three levels to ensure a suitable search depth for Google Scholar. The first level comprises the generic phrases artificial intelligence and knowledge-based. The second level includes the common phrases machine learning and deep learning. The third level describes concrete methods such as case-based, agent-based or ontology-based. Due to a continuous, decreasing matching of the three method levels with our research purpose, we have adapted the search pages considered on Google Scholar, as described in Table 1. In order to get an actual snapshot, only literature of the last five years was included.

As shown in Table 1, product development plays an important role with 50 hits. Other important application areas are production engineering (n = 37) and optimization (n = 34). These results show a high importance for the practical application.

4 Literature Analysis and Synthesis

4.1 Classification of Results

We found 561 articles with the search method we used (Fig. 1). Of these, 223 were found among the members of the Scientific Society for Product Development (WiGeP). Another 338 articles were found by the method described in Sect. 3 on Google Scholar. As suggested by Webster and Watson [15], a complete keyword search and an evaluation of titles and abstracts was performed for each article (Evaluation I). Non-relevant articles were excluded for further consideration. The remaining 46 articles were verified as full text and selected after use in the early phase of the product development process (Evaluation II).

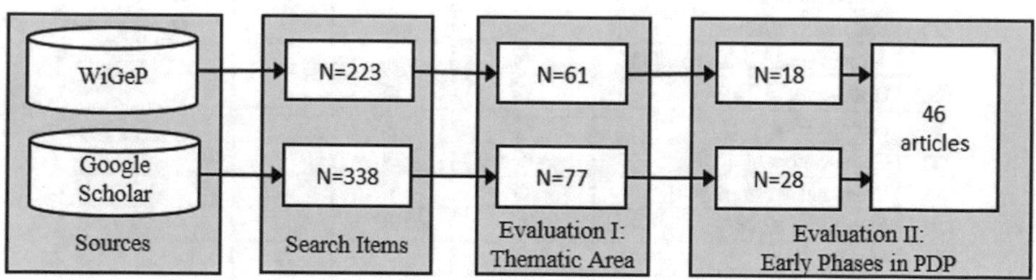

Fig. 1. Overview of search process

All relevant sources are listed in Table 2. Each article is provided with an *ID* (consecutive number), a *Reference* (article authors) and an assignment, whether the relevant articles present a *Methodology* (54%), show an *Application* (74%; further decomposed to conceptual modelling and presentation of relevant use cases) or present an *Algorithm* (22%). Multiple assignments were accepted.

Table 2. Overview of relevant articles

[ID]	Reference	Methodology	Application		Algorithm
			Conceptual	Use case	
01	Gembarski et al. [4]	-	-	●	-
02	Martins and Anderl [16]	-	-	●	●
03	Furian et al. [17]	-	-	●	-
04	Gembarski [9]	●	-	●	-
05	Konrad et al. [18]	●	-	-	-
06	Fender et al. [19]	●	-	-	●
07	Graff et al. [20]	●	-	-	-
08	Müller et al. [21]	●	-	●	-
09	Colombo et al. [22]	-	●	-	-
10	Chenchurin et al. [23]	-	●	-	-
11	Luft et al. [24]	●	●	-	-
12	Oellrich [25]	-	●	-	-
13	Hjertberg et al. [26]	●	-	-	-
14	Relich et al. [27]	-	●	-	-
15	Gembarski et al. [28]	-	●	-	-
16	Zhang et al. [29]	●	-	-	-
17	Zeng et al. [30]	-	-	●	-
18	Levandowski et al. [31]	●	-	●	-
19	Borjesson and Hölttä-Otto [32]	-	●	-	●
20	Garg [33]	-	●	-	●
21	Baykasoğlu and Ozsoydan [34]	-	●	-	●
22	Temple et al. [35]	-	●	-	●
23	Fuge et al. [36]	-	-	-	●
24	Abdeen et al. [37]	-	-	-	●
25	Debreceni et al. [38]	-	-	-	●
26	Zhu et al. [39]	-	-	-	●
27	Althuizen and Wierenga [40]	●	-	-	-
28	Hashemi et al. [41]	●	●	-	-
29	Moreno et al. [42]	●	-	-	-
30	Gembarski et al. [43]	●	●	-	-
31	Brem and Wolfram [44]	●	-	-	-
32	Biskjaer et al. [45]	-	●	-	-
33	Münzer [46]	●	-	●	-
34	Wang and Yu [47]	●	-	●	-
35	Yu et al. [48]	●	-	-	-
36	Pan et al. [49]	●	-	●	-
37	Trehan et al. [50]	-	●	-	-
38	Hagenreiner and Köhler [51]	●	-	●	-
39	Relich [52]	●	-	-	-
40	Hu et al. [53]	●	-	●	-
41	Chen et al. [54]	-	●	-	-
42	Fougères and Ostrosi [55]	●	●	-	-
43	Siqueira et al. [56]	●	●	-	-
44	Gembarski et al. [57]	●	●	-	-
45	Brockmöller et al. [58]	●	●	●	-
46	Bibani et al. [59]	-	●	●	-

4.2 Synthesis of Applications

Next, we classified the articles using a concept matrix (Table 3) divided into three dimensions: *Category/Applications* (determined from the articles), *Product Model* and

Table 3. Resulting concept matrix.

Category/Applications		Product model							Appearance in reference [ID]
		①	②	③	④	⑤	⑥	⑦	
Methods	Constraint-Satisfaction-Problem (CSP)	●	-	-	●	-	●	-	14, 22, 33, 39
	Theory of Inventive Problem Solving (TRIZ)	-	-	-	-	●	●	-	10
	Hazard Analysis	●	●	-	●	●	-	-	8
	Design Structure Matrix (DSM)	-	-	-	●	-	-	-	18, 19
	Modular Function Deployment (MFD)	●	●	-	-	-	-	-	19
	Case-based Reasoning (CBR)	-	-	-	●	●	-	●	26, 27, 28, 40, 43, 46
	Solution Space Exploration	-	-	-	●	●	●	-	4, 6, 7, 16, 24, 25, 32, 33, 41
	Design Methods	●	●	●	-	-	-	-	12, 23, 29, 30, 31, 45
Tools	*Software*								
	Computer-Aided Design (CAD)	-	-	-	●	●	●	●	2, 4, 9, 10, 11, 15, 17, 30, 36, 38, 42, 43, 44, 45, 46
	Finite Element Method (FEM)	-	-	-	-	●	●	-	36, 43
	Algorithm								
	Feature Recognition	-	-	-	●	●	-	-	2, 34, 42
	Artificial Bee Colony (ABC) Algorithm	-	-	-	-	●	-	-	20
	Firefly Algorithm	-	-	-	-	●	-	-	21
	Genetic Algorithm	-	-	-	-	●	-	-	24
	Fuzzy Neural Networks (FNN)	●	-	-	-	-	-	-	14, 39
	Artificial Neural Networks (ANN)	●	-	-	-	-	-	-	39
	Cluster Algorithm	-	-	-	-	●	-	-	26
	Data Mining	-	●	-	-	-	-	-	5
Processes	Knowledge-Based Engineering (KBE)	-	-	-	●	●	●	●	1, 4, 9, 11, 15, 30, 37, 38, 44, 45, 46
	Variant Management	-	●	-	●	●	●	-	4, 5
	Product Portfolio Management	●	●	-	●	-	-	-	14, 19
	Product Lines	-	-	-	●	●	●	-	17, 22
	Configuration Management	-	●	-	-	●	-	-	5, 16
	Knowledge Management	-	-	-	-	●	●	●	3, 9, 11, 13, 45
Design support	Decision Making	-	-	-	-	●	●	●	3, 45
	Problem Solving	-	-	-	-	●	●	●	3
	Design Automation	-	-	-	●	●	●	-	1, 34, 37
	Parametrization	-	-	-	●	●	●	-	2, 15, 33, 35, 40, 43
	Modularization	-	-	-	●	-	-	-	18
	Design Optimization	-	-	-	-	●	●	-	20, 21, 24, 38
	Assistance System	-	-	-	●	●	●	●	3, 10
	Agent-Based Modeling	-	-	●	●	●	-	-	41, 42
	Coverage (Count)	7	7	2	16	23	15	7	

① List of Requirements; ② Function, Function Structure; ③ Principle of Action, Structure of Action; ④ Building Structure; ⑤ Preliminary Design; ⑥ Overall Design; ⑦ Part and Assembly Drawings, other Documentation.

Appearance in Reference (source of the applications). Each of the 32 applications can be described by characteristics and were assigned to one of the following groups: *Methods*, *Tools*, *Processes* and *Design Support*.

The product models were chosen according to VDI Guideline 2221:1993 [60], which structures the product development process into four phases and seven product models: *Task Clarification* (List of Requirements), *Concept* (Function, Function Structure, Principle of Action, Structure of Action), *Embodiment Design* (Building Structure, Preliminary Design, Overall Design) and *Detailed Design* (Part and Assembly Drawings, other Documentation).

The concept matrix shows that 70% of applications for KBES and artificial intelligence in product development are used in the embodiment design phase. The concept phase accounts for 12% of the applications, whereby the Principles of Action and Structures of Action are strongly underrepresented with 3%. The detailed design phase seems underrepresented as well with only 9% of appearance.

5 Discussion

Interestingly, the methods block is the only one with a well-balanced distribution where all phases and product models are addressed. Regarding tools, a lot of CAD-centric articles were found but only a few which document holistic engineering environments that encompass synthesis as well as analysis tools, like FEM simulations. With respect to the application of distinct algorithms, some are used for task clarification, where the goal is to predict a new product success (e.g. refer to [27, 52]). Others support the preliminary design, e.g. as optimization algorithms.

In design support, the designer is usually supported during the embodiment and detailed design phase. An exception to this is agent-based modeling, which supports conceptual design synthesis [54].

So, the concept matrix indicates that all phases of the design process are represented in literature oriented to KBES and artificial intelligence. Nonetheless, it appears that the early phase, in particular the finding of concepts, is strongly underrepresented. The tasks that are carried out in this phase belong to the "real" engineering tasks of design problem solving which are characterized by open, ill-structured problems that are hard to code in rule sets or in models. Interestingly, no actual literature was found on decision support.

Another point that was surprising is the lack of design automation systems for detailed design. The considered literature indeed discusses design automation, but only on level of the embodiment design. The designer is ought to use the results and further detail the design relying on his experience. Design automation systems that deliver a complete set of drawings and manufacturable artifacts could not be identified.

Regarding algorithms, most of the literature describes newly developed algorithms and their benchmarks. Those are commonly simple standardized experiments like dimensioning and optimizing machine elements, e.g. tension/compression springs, pressure vessels or welded beams [33, 34]. An application to more complex real world problems or multi criteria optimization is usually not covered.

This literature research is not free of limitations. For the literature research we have oriented ourselves on the methodical approach of vom Brocke et al. [14], because in our opinion it supports a traceable and expandable literature review. Even if the literature search follows a systematic methodology, the decisions for the selection of the search phrases and literature sources, as well as their classification, are limited and according to subjective aspects. For a further identification of literature, especially with regard to the late phases of the product development process, a search with further keywords and sources can be helpful. Additionally, the search was a breadth search mainly using Google scholar. A depth search in relevant journals would be complementary. Relaxing the time constraint would also lead to different results.

6 Conclusion and Future Research

This contribution provides a snapshot for the applications of KBES and artificial intelligence in product development that were documented in the last five years in scientific literature. The selected literature focuses on the early phase of the product development process.

With the help of a concept matrix we have assigned 46 relevant articles to the categories methods, tools, processes and design support. By classifying the articles to the product models, a transition to the product development process according to the VDI guideline 2221:1993 is achieved. Our contribution can be used by researchers who are interested in the application of KBES and artificial intelligence in product development, for example to classify their research within the product development process or to derive new research questions with regard to the computer support of developers in the search for solution principles and their structures.

The future research can be oriented at two points from the literature search. Firstly, the concept matrix shows that there is a need for research in order to examine the principles and structures of action using methods of KBES and artificial intelligence. If we consider the entire concept phase of the product development process, functional orientation becomes more and more important. An interesting approach is the transition from function to design.

The second approach aims at the continuous process support of the product development process. Often the support by KBES and artificial intelligence in the literature refers to the embodiment design phase. The detailed design phase is often excluded. In practice, the detailed design phase plays a decisive role in establishing a continuous value-added process.

References

1. Ullman, D.G.: The Mechanical Design Process, 4th edn. Mcgraw-Hill, New York (2009)
2. Vajna, S.: CAx für Ingenieure: eine praxisbezogene Einführung, 2nd edn. Springer, Heidelberg (2009)

3. Verhagen, W.J.C., Bermell-Garcia, P., van Dijk, R.E.C., Curran, R.: A critical review of Knowledge-Based Engineering: an identification or research challenges. Adv. Eng. Inform. **26**(1), 5–15 (2012)
4. Gembarski, P.C., Li, H., Lachmayer, R.: Template-based modelling of structural components. Int. J. Mech. Eng. Robot. Res. **6**(5), 336–342 (2017)
5. Chapman, C.B., Pinfold, M.: The application of a knowledge based engineering approach to the rapid design and analysis of an automotive structure. Adv. Eng. Softw. **32**(12), 903–912 (2001)
6. Milton, N.R.: Knowledge Technologies, 3rd edn. Polimetrica sas, Monza (2008)
7. Sabin, D., Weigel, R.: Product configuration frameworks - a survey. IEEE Intell. Syst. Appl. **13**(4), 42–49 (1998)
8. Boyle, Y., Brown, D.C.: A review and analysis of current computer-aided fixture design approaches. Robot. Comput. Integr. Manuf. **27**(1), 1–12 (2011)
9. Gembarski, P.C.: Komplexitätsmanagement mittels wissensbasiertem CAD – Ein Ansatz zum unternehmenstypologischen Management konstruktiver Lösungsräume. TEWISS, Garbsen (2018)
10. Schreiber, G., Wielinga, B., de Hoog, R., Akkermans, H., Van de Velde, W.: CommonKADS: a comprehensive methodology for KBS development. IEEE Expert **9**(6), 28–37 (1994)
11. Stokes, M.: Managing Engineering Knowledge: MOKA: Methodology for Knowledge Based Engineering Applications. Wiley-Blackwell, London (2001)
12. Barták, R., Salido, M.A., Rossi, F.: Constraint satisfaction techniques in planning and scheduling. J. Intell. Manuf. **21**(1), 5–15 (2010)
13. Felfernig, A., Hotz, L., Bagley, C., Tiihonen, J.: Knowledge-Based Configuration: From Research to Business Cases. Newnes. Morgan Kaufmann, Amsterdam (2014)
14. vom Brocke, J., Simons, A., Niehaves, B., Riemer, K., Plattfaut, R., Cleven, A.: Reconstructing the giant: on the importance of rigour in documenting the literature search process. In: Proceedings of the European Conference on Information Systems (ECIS), Verona, Italy, pp. 2206–2217 (2009)
15. Webster, J., Watson, R.T.: Analyzing the past to prepare the future: writing a literature review. MIS Q. xiii–xxiii (2002)
16. Martins, T.W., Anderl, R.: Feature recognition and parameterization methods for algorithm-based product development process. In: 37th Computers and Information in Engineering Conference, pp. 1–11. The American Society of Mechanical Engineers, Cleveland (2017)
17. Furian, R., Von Lacroix, F., Correia, A., Faltus, S., Flores, M., Grote, K.-H.: Evaluation of a new concept of a knowledge based environment. In: The 3rd International Conference on Design Engineering and Science, Pilsen, Czech Republic, pp. 186–191 (2014)
18. Konrad, C., Löwer, M., Schmidt, W.: Varianzsteuerung integraler Produkte durch den Prozessbegleitenden Einsatz von Data-Mining Werkzeugen. In: Brökel, K., et al. (eds.) Gemeinsames Kolloquium Konstruktionstechnik, DuEPublico, vol. 15, pp. 213–222 (2017)
19. Fender, J., Duddeck, F., Zimmermann, M.: Direct computation of solution spaces. Struct. Multidiscip. Optim. **55**(5), 1787–1796 (2017)
20. Graff, L., Harbrecht, H., Zimmermann, M.: On the computation of solution spaces in high dimensions. Struct. Multidiscip. Optim. **54**(4), 811–829 (2016)
21. Müller, M., Roth, M., Lindemann, U.: The hazard analysis profile: linking safety analysis and SysML. In: Annual IEEE Systems Conference, Orlando, USA, pp. 1–7 (2016)
22. Colombo, G., Pugliese, D., Klein, P., Lützemnberger, J.: A study for neutral format to exchange and reuse engineering knowledge in KBE applications. In: International Conference on Engineering, Technology and Innovation, Bergamo, Italien, pp. 1–10 (2014)

23. Chechurin, L.S., Wits, W.W., Bakker, H.M., Vaneker, T.H.J.: Introducing trimming and function ranking to solidworks based on function analysis. In: Cavallucci, D., et al. (eds.) Procedia Engineering, vol. 131, pp. 184–193. Elsevier
24. Luft, T., Roth, D., Binz, H., Wartzack, S.: A new "knowledge-based engineering" guideline. In: 21st International Conference on Engineering Design, Vancouver, Canada, pp. 207–216 (2017)
25. Oellrich, M.: Webbasierte Konstruktionsmethoden-Unterstützung in der frühen Phase der Produktentwicklung (Dissertation), Helmut-Schmidt-Universität/Universität der Bundeswehr Hamburg, Hamburg (2015)
26. Hjertberg, T., Stolt, R., Poorkiany, M., Johansson, J., Elgh, F.: Implementation and management of design systems for highly customized products – state of practice and future research. In: Curran, R., et al. (eds.) Transdisciplinary Lifecycle Analysis of Systems, pp. 165–174. IOS Press, Amsterdam (2015)
27. Relich, M., Świć, A., Gola, A.: A knowledge-based approach to product concept screening. In: Omatu, S., et al. (eds.) Distributed Computing and Artificial Intelligence, 12th International Conference. Advances in Intelligent Systems and Computing, vol. 373. Springer, Cham (2015)
28. Gembarski, P.C., Li, H., Lachmayer, R.: KBE-modeling techniques in standard CAD-systems: case study – autodesk inventor professional. In: Proceedings of the 8th World Conference on Mass Customization, Personalization, and Co-Creation, MCPC 2015, pp. 215–233. Springer, Cham (2015)
29. Zhang, L.L., Chen, X., Falkner, A., Chu, C.: Open configuration: a new approach to product customization. In: Felfernig, A., Forza, C., Haag, A. (eds.) 16th International Configuration Workshop, pp. 75–79. Novi Sad, Serbia (2014)
30. Zeng, F., Li, B., Zheng, P., Xie, S. (S.Q.): A modularized generic product model in support of product family modeling in one-of-a-kind production. In: 2014 IEEE International Conference on Mechatronics and Automation, pp. 786–791. IEEE, Tianjin (2014)
31. Levandowski, C., Müller, J.R., Isaksson, O.: Modularization in concept development using functional modeling. In: Borsato, M., et al. (eds.) Transdisciplinary Engineering: Crossing Boundaries, pp. 117–126. IOS Press, Amsterdam (2016)
32. Borjesson, F., Hölttä-Otto, K.: A module generation algorithm for product architecture based on component interactions and strategic drivers. Res. Eng. Design **25**(1), 31–51 (2014)
33. Garg, H.: Solving structural engineering design optimization problems using an artificial bee colony algorithm. J. Ind. Manag. Optim. **10**(3), 777–794 (2014)
34. Baykasoğlu, A., Ozsoydan, F.B.: Adaptive firefly algorithm with chaos for mechanical design optimization problems. Appl. Soft Comput. **36**(11), 152–164 (2015)
35. Temple, P., Galindo, J., Jézéquel, J.-M., Acher, M.: Using machine learning to infer constraints for product lines. In: SPLC 2016 Proceedings of the 20th International Systems and Software Product Line Conference, pp. 209–218. ACM, New York (2016)
36. Fuge, M., Peters, B., Agogino, A.: Machine learning algorithms for recommending design methods. J. Mech. Des. **136**(10), 101103 (2014)
37. Abdeen, H., Varró, D., Sahraoui, H., Nagy, A.S., Hegedüs, Á., Horváth, Á.: Multi-objective optimization in rule-based design space exploration. In: ASE 2014 Proceedings of the 29th ACM/IEEE International Conference on Automated Software Engineering, pp. 289–300. ACM, New York (2014)
38. Debreceni, C., Ráth, I., Varró, D., De Carlos, X., Mendialdua, X., Trujillo, S.: Automated model merge by design space exploration. In: Stevens, P., Wąsowski, A. (eds.) Fundamental Approaches to Software Engineering. Lecture Notes in Computer Science, vol. 9633. Springer, Heidelberg (2016)

39. Zhu, G.N., Hu, J., Qi, J., Ma, J., Peng, Y.-H.: An integrated feature selection and cluster analysis techniques for case-based reasoning. In: Engineering Applications of Artificial Intelligence, vol. 39, pp. 14–22. Elsevier (2015)
40. Althuizen, N., Wierenga, B.: Supporting creative problem solving with a casebased reasoning system. J. Manag. Inf. Syst. **31**(1), 309–340 (2014)
41. Hashemi, H., Shaharoun, A.M., Sudin, I.: A case-based reasoning approach for design of machining fixture. Int. J. Adv. Manuf. Technol. **74**(1–4), 113–124 (2014)
42. Moreno, D.P., Yang, M.C., Hernández, A.A., Linsey, J.S., Wood, K.L.: A step beyond to overcome design fixation: a design-by-analogy approach. In: Gero, J.S., Hanna, S. (eds.) Design Computing and Cognition 2014, pp. 607–624. Springer, Cham (2014)
43. Gembarski, P.C., Bibani, M., Lachmayer, R.: Design catalogues: knowledge repositories for knowledge-based engineering applications. In: Marjanovic, D., Storga, M., Pavkovic, N., Bojcetic, N., Skec, S. (eds.) DS 84: Proceedings of the DESIGN 2016 14th International Design Conference, pp. 2007–2015. The Design Society, Dubrovnik (2016)
44. Brem, A., Wolfram, P.: Research and development from the bottom up - introduction of terminologies for new product development in emerging markets. J. Innov. Entrepreneurship. Syst. View Time Space **3**(9), 1–22 (2014)
45. Biskjaer, M.M., Dalsgaard, P., Halskov, K.: A constraint-based understanding of design spaces. In: DIS 2014 Proceedings of the 2014 Conference on Designing Interactive Systems, pp. 453–462. ACM, New York (2014)
46. Münzer, C.: Constraint-based methods for automated computational design synthesis of solution spaces (Dissertation). ETH Zürich, Zürich, Switzerland (2015)
47. Wang, Q., Yu, X.: Ontology based automatic feature recognition framework. Comput. Ind. **65**(7), 1041–1052 (2014)
48. Yu, R., Gu, N., Ostwald, M., Gero, J.S.: Empirical support for problem–solution coevolution in a parametric design environment. Artif. Intell. Eng. Des. Anal. Manuf. **29**(1), 33–44 (2015)
49. Pan, Z., Wang, X., Teng, R., Cao, X.: Computer-aided design-while-engineering technology in top-down modeling of mechanical product. Comput. Ind. **75**, 151–161 (2016)
50. Trehan, V., Chapman, C., Raju, P.: Informal and formal modelling of engineering processes for design automation using knowledge based engineering. J. Zhejiang Univ. Sci. A **16**(9), 706–723 (2015)
51. Hagenreiner, T., Köhler, P.: Concept development of design driven parts regarding multidisciplinary design optimization. Comput. Aided Des. Appl. **12**(2), 208–217 (2015)
52. Relich, M.: A computational intelligence approach to predicting new product success. In: Proceedings of the 11th International Conference on Strategic Management and its Support by Information Systems, pp. 142–150 (2015)
53. Hu, J., Qi, J., Peng, Y.: New CBR adaptation method combining with problem–solution relational analysis for mechanical design. Comput. Ind. **66**, 41–51 (2015)
54. Chen, Y., Liu, Z.-L., Xie, Y.-B.: A multi-agent-based approach for conceptual design synthesis of multi-disciplinary systems. Int. J. Prod. Res. **52**(6), 1681–1694 (2014)
55. Fougères, A.-J., Ostrosi, E.: Intelligent agents for feature modelling in computer aided design. J. Comput. Des. Eng. **5**(1), 19–40 (2018)
56. Siqueira, R., Bibani, M., Duran, D., Mozgova, I., Lachmayer, R., Behrens, B.-A.: An adapted case-based reasoning system for design and manufacturing of tailored forming multi-material components. Int. J. Interact. Des. Manuf. (IJIDeM), 1–10 (2019)
57. Gembarski, P.C., Sauthoff, B., Brockmöller, T., Lachmayer, R.: Operationalization of manufacturing restrictions for CAD and KBE-systems. In: Marjanovic, D., et al. (eds.) DS 84: Proceedings of the DESIGN 2016 14th International Design Conference, pp. 621–630. The Design Society, Dubrovnik (2016)

58. Brockmöller, T., Gembarski, P.C., Mozgova, I., Lachmayer, R.: Design catalogue in a CAE environment for the illustration of tailored forming. In: Engineering for a Changing World, vol. 59. ilmedia, Ilmenau (2017)
59. Bibani, M., Gembarski, P.C., Lachmayer, R.: Ein wissensbasiertes System zur Konstruktion von Staubabscheidern. In: Krause, D. et al. (eds.) Proceedings of the 28th Symposium Design for X, pp. 165–176. The Design Society, Bamberg (2017)
60. VDI: VDI Guideline 2221 - Systematic approach to the development and design of technical systems and products, Beuth, Berlin (1993)

State Machine of a Redundant Computing Unit Operating as a Cyber-Physical System Control Node with Hot-Standby Redundancy

Jacek Stój(✉)

Silesian University of Technology, Gliwice, Poland
jacek.stoj@polsl.pl

Abstract. Cyber-physical systems CPS are computer systems interconnected with physical world. In case of industrial applications they are called also Industrial Control Systems ICS. Most characteristic feature of ICS is operation in real-time. It means that they must satisfy strictly defined temporal constraints. Another important aspect is reliability. It is expected that ICS system operates 24/7. Some systems have to be also fault-tolerant, which means that a failure of one or more components should not disable proper operation of the system. To satisfy the latter demand, redundancy is used. One of examples of redundancy application is redundancy of computing units (like for example *Programmable Logic Controllers* PLCs in industrial control systems). It may be realized using dedicated components but also in more cost-effective way, using standard elements. In the second case, proper implementation of software redundancy routines is crucial. The paper shows a state machine used for implementation of computing units redundancy operating in hot-standby mode.

Keywords: Cyber-physical systems · Industrial Control Systems · Networked Control Systems · Communication network · Real-time · Temporal determinism · Redundancy · Hot-standby · Real-Time Ethernet protocol · EtherCAT · Programmable Logic Controller · Control switchover

1 Introduction

Cyber-physical systems CPS are computer systems tightly interconnected with physical world. They encompass hardware and software components that are intertwined with the real word. Computations executed by software, with the usage of hardware, is performed for sensing the real word state, processing it, and reacting to it by performing control through some physical components, i.e. actuators [1]. The term *cyber-physical systems* in industrial environment applies to Industrial Control Systems ICS.

There are two most important features of ICS: operation in real-time and reliability. The former feature is satisfied by using for realization of ICS only real-time components, starting with computing units operating in real-time (e.g. *Programmable Logic Controller* PLC) and network protocols that are temporally determined, while creating distributed control systems called Networked Control Systems NCS [2]. Only usage of real-time elements may provide satisfaction of time constraints defined for a given

J. Świątek et al. (Eds.): ISAT 2019, AISC 1051, pp. 74–85, 2020.
https://doi.org/10.1007/978-3-030-30604-5_7

system, which makes it possible to respond in a timely manner to requests coming from outside – the real world.

Industrial Control Systems are expected to be highly reliable as most often they have to operate 24/7. A failure of ICS may cause production downtime which is costly. Moreover, misoperation may lead to a damage of equipment, environment or injuries and death of people when Safety-Critical Systems are concerned [3].

Redundancy is used to make ICS invulnerable to failures of its components. Most crucial elements of a system are multiplicated (duplicated or triplicated in most cases) in order to let the system continue its operation in spite of a failure of one or more of redundant elements.

There are many possible architectures with redundancy because redundancy may be applied to every element. One of them may be computing unit which is responsible for realization of control algorithms [4, 5]. There are some ready-to-use solutions for implementation of this kind of redundancy. It is provided by vendors like Siemens as Fail-safe automation system, i.e. F-System [6] or GE Automation named PACSystems High Availability [7]. They are robust solutions. The main drawback is however their price. Hardware components needed for the above solutions are expensive and may be out of the budged of many applications.

Computing unit redundancy may be also realized by standard components, that are not dedicated for application in high reliable systems, in a more cost-effective way. In that case however, the redundancy mechanisms has to be implemented in the user program because they are non-existent in the operating system or in the firmware of common programmable devices [8, 9]. The paper shows an example of application of standard devices for realization of computing units redundancy and describes a state machine which is implemented in the redundant units.

The paper is organized as follows. Section 2 includes brief description of redundancy of computing units in Industrial Control Systems. Next, in Sect. 3 the cost-effective redundancy idea is presented together with modes of operation of redundant computing units, the hardware used for experimental research and the created system interconnections. Section 4 includes description of implemented state machine. It is followed by a summary of experimental research performed on the created testbed in Sect. 5. Last section summarizes the presented research work and mentions future works.

2 Computing Unit Redundancy

Industrial Computer Systems consist mainly of one or more input modules, output modules and computing units. The input modules acquire the state of computer system inputs which are the signals from sensors giving information about the real world – in this case the controlled industrial process. They may be analogue inputs providing values of current measurements like temperature or pressure. They may also be binary inputs for example with states of valves (open/closed), switches (turned on/turned off) or information about some limits exceeded (level high/not high).

Current state of the system inputs is processed by a computing unit CU. It makes decisions about requested reaction to the current inputs state. For example, when the level is too high, pumps are turned on in order to pump out water from a tank. So the CU decides on the requested state of the system outputs to a given state of the system inputs. It is passed over to the output modules which energize actuators performing action on the controlled industrial process, e.g. by actual turning the pump *on* or *off*. It is done by output signals like binary requests (turn on/off a pump, open/close a valve etc.) or analogue requests (e.g. position to which a regulation valve should be set).

Input and output modules may be installed as extension modules of a computing unit or as remote input/output modules connected to the computing unit by a communication network, as depicted in Fig. 1 – parts (a) and (b) of the figure accordingly. Systems without redundancy are presented on the left side of the figure and systems with computing units redundancy on the right side. The remote input/output modules together with the communication interface module are referred as a Remote I/O station RIO.

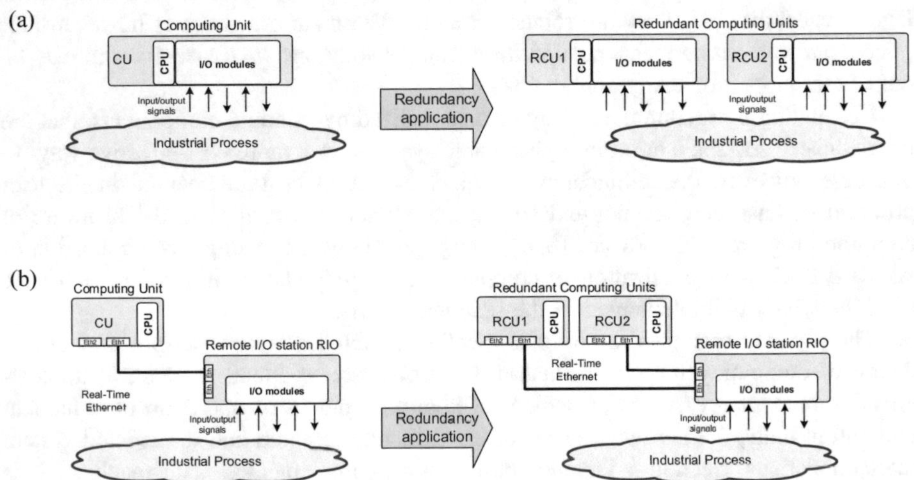

Fig. 1. Industrial Control System: (a) centralized (no communication networks), (b) distributed (I/O modules connected as a remote I/O station)

In case of distributed control systems, the remote input/output modules may be grouped in one or more remote input/output stations RIO. In the Fig. 1 however, only one RIO is presented in order to keep the figure clear. Additional RIO stations may be connected one to another creating a line topology (providing that every RIO station has got build-in Ethernet switch). Network switches may also be used for creating e.g. star topology.

The communication network connecting the RIO station to the CU must be a real-time network in order to realize the data exchange in such a way, that all the real-time constrains defined for the system are satisfied. There may be of course more than one RIO stations. The number of RIO stations does not influence operation of the RCU from the functional point of view.

3 Cost-Effective Computing Units Redundancy

Implementation of redundancy of computing units increases their reliability and the system as a whole. It refers to both centralized and distributed systems (see: Fig. 1), though in centralized system it significantly increased the cost of control cabinet as every I/O module together with I/O signals have to be multiplied (usually duplicated). Therefore, more often redundancy in realized together with input/output modules organized as remote RIO stations (see: Fig. 1b). In that case RIO stations uses non-redundant input/output modules. That solution is less expensive but vulnerable to I/O modules failure. However, the required architecture should always be a result of not cost calculation, but of risk analysis (which is out of scope of this paper).

As mentioned in the introduction, for systems with computing units redundancy, some dedicated solutions may be used like Fail-safe automation system by Siemens [6] or PACSystems High Availability Systems by GE Automation [7]. These solutions are expensive and therefore one may consider using standard devices for that purpose (with no redundant functions included by vendor). This kind of redundancy application is possible as described in [9] and presented in Fig. 2.

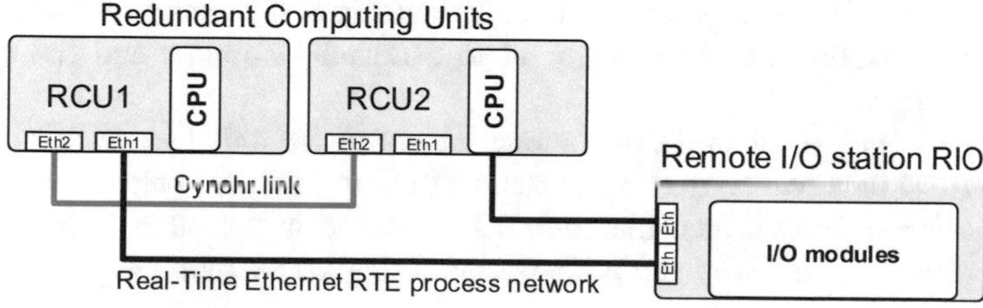

Fig. 2. Example of a cost-effective system with computing unit redundancy

Figure 2 presents a system with computing unit redundancy. The redundant computing units are later called RCU. There are two of them, as double redundancy is considered. Both RCUs have got the same hardware configuration and user program. They may be standard programmable devices such as Programmable Logic Controllers PLC.

The RCUs receive the state of the controlled system from RIO station which serves as an interface between the ICS and the industrial process. The RIO is connected to both RCUs by Ethernet based communication network with Real-Time Ethernet protocol by Beckhoff.

3.1 Hardware

The redundant computing units RCUs were CX5140 embedded devices by Beckhoff with no local I/O modules used for process control and no additional communication interfaces apart from two built-in Ethernet ports. The Remote Input/Output RIO station consisted of BK9100 Ethernet TCP/IP Bus Coupler by Beckhoff with the following I/O modules: 24VDC digital inputs KL1408 (2 pieces, 8 points each), 24VDC digital outputs KL2408 (2 pieces, 8 points each), 0-20 mA analogue inputs KL3448 (8 channels), 4-20 mA analogue outputs KL4424 (4 channels). That gives 36 bytes of user data, including status information about the state of analogue inputs.

The RIO station was connected with RCUs by Ethernet network and Real-Time Ethernet protocol by Beckhoff [9]. The protocol allows communication of the station with many computing units independently. Additionally, the RCUs were connected with each other via another Ethernet interface and EtherCAT communication protocol [11, 12] which provides efficient data exchange between RCUs.

3.2 Modes of Operation

In the considered type of redundancy, the redundant computing units may operate in two modes: *duplex* and *hot standby*[1]. In both modes the computing units execute basically the same operations: they receive the state of system inputs, process it and respond to it. The response is sent to the system outputs. The difference between the two above modes is in the way the outputs are controlled. In case of *duplex* operation, the outputs states obtained from both RCUs are compared. When they are convergent (e.g. both RCUs request to open a given valve etc.) then the outputs are energized according to the requested outputs state. If there is a divergence between outputs received from RCUs, the outputs are set to a default/safe state and the controlled process is stopped.

In *hot standby* control mode, the outputs are controlled only by one RCU at a time. When the RCU fails (which may be a cause of the RCU failure itself or a failure in the communication network), then the control is switched to the other RCU which was until that moment in *standby* and it starts controlling the systems outputs.

3.3 Hot-Standby Operation

In this paper, system operating in *hot standby* mode is considered. In that case both RCUs receive data about the current state of RIO inputs, but only one unit may control outputs of the RIO station. Usage of Real-Time Ethernet by Beckhoff communication protocol gives the possibility to do so. It may be freely chosen during run-time whether a given RCU should send the output data and control the system outputs.

Both RCUs diagnose all the time the activity of the other RCU and the RIO station. They decide independently from each other, but taking into consideration the other device state, whether they should operate in *standby* role or should be *active* and take control over the systems outputs (controlled process). The decision making is described in the following sections.

[1] Other modes are possible (*warm standby* and *cold standby*) but they are not considered here as not interesting from the point of view of computer system operation.

3.4 System Interconnections

For proper operation of every industrial computer system, an effective diagnostic is necessary. In case of the presented system, the most important diagnostic problem was detection of operation of redundant RCU, i.e. whether it is *healthy* or *faulty*. For that purpose a communication link was provided between the RCUs in order to exchange diagnostic data. However, that would not be enough. Having the communication link broken, the RCU which is in *standby* mode, does not know the cause of the fault, i.e. whether it is caused by a fault in the communication link itself or fault of the other, *active* RCU. Therefore, apart from diagnostic information exchange, the activity of the RCUs was being detected by monitoring of one of the RIO outputs with a *heartbeat* signal (see: Fig. 3).

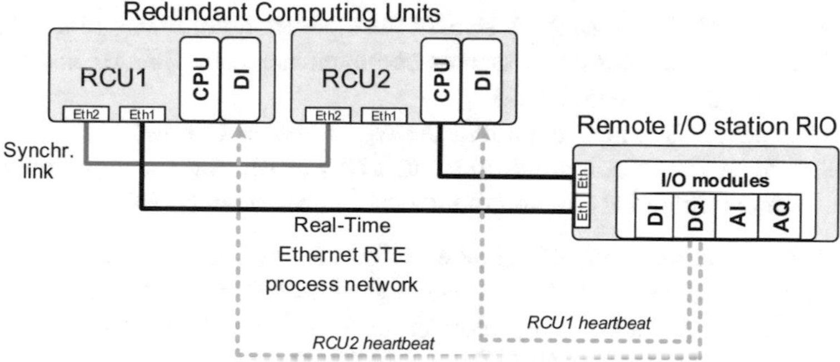

Fig. 3. *Heartbeat* signal connection for CU failure detection

For the needs of failure detection of the *active* RCU, one of the binary outputs of the RIO (separate for both RCU) is being constantly toggled. The output is connected to a local input module of the other RCU. No change of the input signal means that the *active* RCU does not control the outputs of the RIO. That is the trigger for taking control of the system by the RCU being in *standby*.

The communication link between RCUs was also used for exchanging some process related data, e.g.:

- state of the RCU (communication status with RIO station among other things),
- state of inputs received by the other RCU,
- state of the controlled devices – equipment of the controlled process,
- operator requests, system parameters and set points.

For that reason, on the Fig. 3 it is called *synchronization link* – it was used to synchronize internal states of the redundant RCUs. It is similar to synchronization links available in commercial solutions, like Redundancy Memory Exchange modules offered by GE Automation.

4 Redundant Computing Unit State Machine

The redundant computing units RCUs operating in the considered type of redundancy may be in two basic states: *active* (when the given RCU controls the system outputs) and *ready* (then the RCU is ready to take over the control over system outputs). In case the other RCU is *active*, the *ready* RCU is considered to be in *standby* operation mode). Additionally, the above states may happen in one RCU when the other is fault-free or faulted. All the possible states distinguished in the presented solution are listed below:

- `ACTIVE_2CPU`: the given RCU is *active*, the remote RCU is *ready*.
- `READY_2CPU`: the RCU is *ready*, the remote RCU is either *ready* or *active* (it is allowed that both RCUs are *ready* and none *active*),
- `ACTIVE`, the RCU is *active* and controls the system outputs, the state of other RCU is unknown, usually it is neither *ready* nor *active*,
- `READY`: given RCU is *ready*, the state of the other (remote) RCU unknown, usually is neither *ready* nor *active* (it cannot communicate with RIO or is unreachable through the synchronization link),
- `NO_RIO`: the RCU cannot communicate with remote I/O station,
- `NO_SYNCH`: refers to communication fault between RCUs causing units synchronization impossible (there is no communication between the RCUs).

In one RCU, the *readiness* of the other RCU is determined by communicating with it via the synchronization link. The *activity* (states `ACTIVE_2CPU` and `ACTIVE`) of the other RCU is also detected by monitoring a *heartbeat* signal provided via local input module of the given RCU as depicted in Fig. 3.

The transitions between the RCUs states are depicted in Fig. 4. Transition to the states `NO_RIO` and `NO_SYNCH` are however not presented because they occur from every other state if only there is no communication with the RIO station or with the other RCU (accordingly). Table 1 presents the possible coincidences of different RCU states in both units.

After restart the RCU is in the `NO_RIO` state until communication with the remote input/output RIO station is established. From the `NO_RIO` state it goes to the `READY` state unless there is no communication with the other RCU in which case it goes to the `NO_SYNCH` state instead. Whenever both RCUs are in the `READY` state, they goes to the `READY_2CPU` state. Similarly, when one RCU is `READY` and the other is `ACTIVE`.

When the RCUs are in `NO_SYNCH` state, it would be expected that one of the RCU will go the `ACTIVE` state and start controlling the industrial process. It is crucial that when it happens, the other RCU remains in `READY` state. Therefore it has to be predefined which RCU may perform control tasks in case of lack of synchronization. That RCU is flagged as PRIMARY RCU.

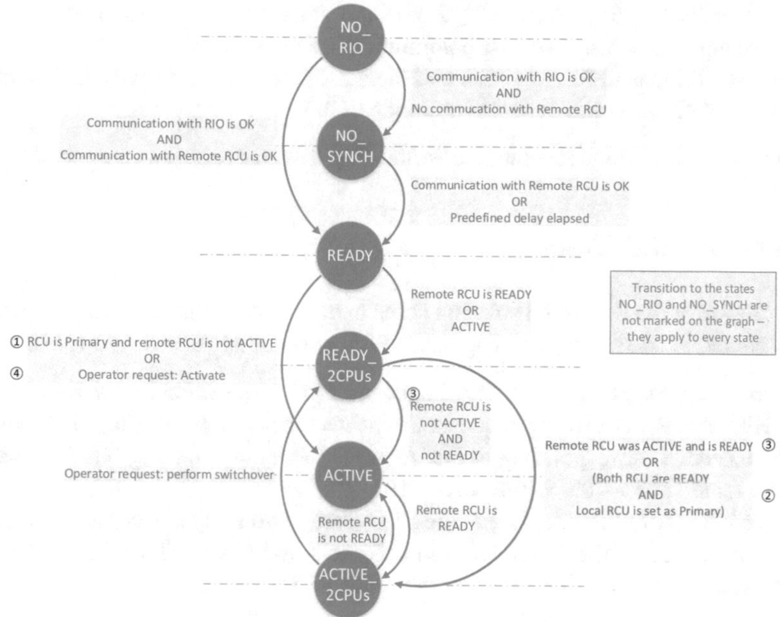

Fig. 4. The Redundant Computing Units RCUs state machine

Table 1. Coincidence of RCU state machines in computing units

The state of one RCU	Possible states of the other RCU					
	ACTIVE_2CPU	READY_2CPU	ACTIVE	READY	NO_RIO	NO_SYNCH
ACTIVE_2CPU		Yes				
READY_2CPU	Yes	Yes				
ACTIVE				Yes[a]	Yes	Yes
READY			Yes[b]	Yes	Yes	Yes
NO_RIO			Yes	Yes	Yes	Yes
NO_SYNCH			Yes	Yes	Yes	Yes

[a]See Footnote 1.
[b]When one RCU is *active* and the other is *ready* and there is communication between the RCUs, then the RCUs goes to READY_2CPU and ACTIVE_2CPU states accordingly.

A RCU may go from READY or READY_2CPU state to ACTIVE or ACTIVE_2CPU state when:

1. No RCU is *active* and the given RCU is *ready* and set as PRIMARY[2],
2. Both RCU are *ready*, and the given RCU is set as PRIMARY,

[2] One RCU has greater priority in going to the ACTIVE state to avoid situation when both RCU sets to be ACTIVET. That RCU is considered as PRIMARY RCU.

3. The other RCU was in `ACTIVE_2CPU` and changed its state to another state
4. (for example as a result of lost communication with RIO),
5. The given RCU is in `READY` state and there is operator request to make it *active* (the RCU doesn't know the state of the other RCU).

The above conditions are marked in the Fig. 4 with according numbers.

5 Experimental Tests

In order to verify the described solution some experimental research was performed. To the system depicted in Fig. 3 a diagnostic PLC was added. It had two roles:

1. *Response time measure*ment: the diagnostic PLC was cyclically energizing one of the RIO inputs performing requests and it was measuring the time the considered system needed for responding to that requests (which was setting one of the outputs to the same state as the inputs was set).
2. *Simulate RTUs faults*: the diagnostic PLC, using two relays, was disconnecting the *heartbeat* signal of the active RCU in order to simulate its failure. Then, the system was supposed to switch control to the other RCU.

During the experimental tests the system response time was measured by the diagnostic PLC in two scenarios: without and with RCU fault simulation. Results are presented in Fig. 5.

While analyzing temporal characteristics we should first of all consider the worst case. In the testbed he longest response time occurs when the RCU needs two automata cycles in order to generate a response (when a request occurs just after inputs acquisition performed at the beginning of automata cycle, then the RCU will get the request during the next automata cycle and only then a response may be generated). Moreover, the response has to be acquired by the diagnostic PLC – again 2 automata cycles are needed. Additionally, there is a delay in detecting changes of inputs by the binary input modules. It is defined as "input filter time". Both for inputs of the RCU (request signal acquisition) and the diagnostic PLC (response signal acquisition) the filtering time was 3 ms. The cycle time of the RIO station was below 0.5 ms in our configuration has little significance. Similarly, time needed for outputs update. Taking into consideration that the RCU controller automata cycle was 20 ms, and diagnostic PLC automata cycle was 1 ms, and after summing the above values we get accordingly: 2*20 ms + 2*1 ms + 2*3 ms = 48 ms. In Fig. 5 it is visible that the worst case response time is never exceeded – the greatest measurements are grouped around 44–45 ms.

In Fig. 5a there is also another group of measurements around 33–34 ms. It is when the input signals acquisition (request and response signal) and the automata cycle of the diagnostic PLC overlap the RCU automata cycle. Then the response time is by around 8 ms shorter than the values of the other measurement group.

In graph presented in Fig. 5b there is yet another group measurements around 37–38 ms. That kind of measurement happens when during realization of the switchover routines only part of the above 8 ms overlaps with the RCU automata cycle.

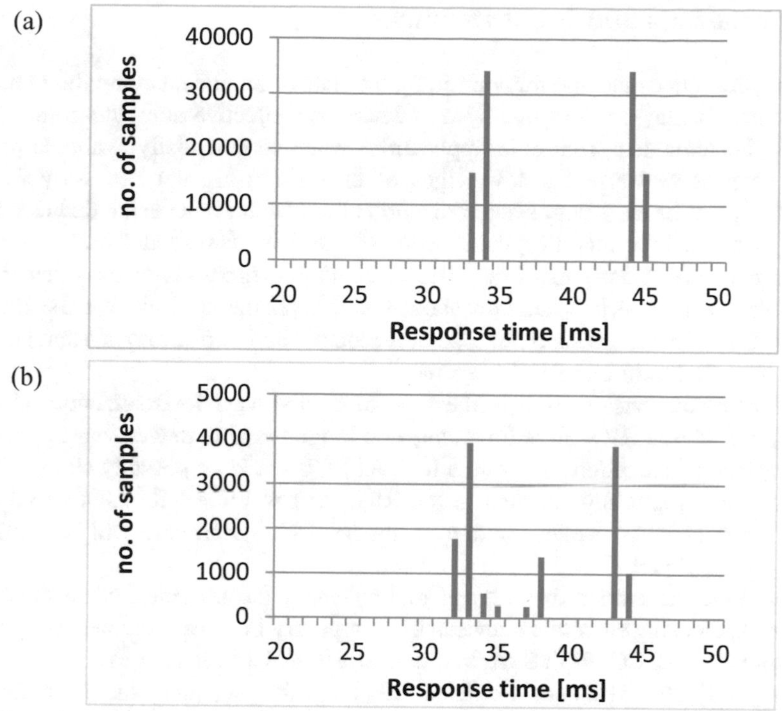

Fig. 5. Experimental research results: (a) system without redundancy, (b) system with redundancy during realization of control *switchover* routines

However the above description may be vague, more detailed discussion about the results is out of scope of this paper. The goal of the experimental test was not the measurement and analysis of the response time, but it was checking whether realization of switchover routines changes the response time in comparison to system when only one RCU performs process control tasks.

The experiments showed that the switchover routines were performed according to the presented state machine. For every request there was an appropriate response in the expected time in spite of RCU fault simulation. That means that a failure of the *active* RCU was always detected by the other RCU which then became *active* and provided expected response.

The experiments showed also that in case of the considered system there are no delays associated to the redundancy implementation. However, proper operation of that system was possible only after setting appropriate timeout delays concerning the RCU fault detection using the *hearbeat*. Moreover, other experiments (not described in this paper) showed that the RCU being in *standby* mode has to have communication with RIO station disabled. Otherwise, communication done by two RCUs interfere with each other causing occasional errors in communication (the error occurred once per 1–2 h).

6 Conclusions and Final Remarks

In the paper, a state machine for controlling the state of redundant computing unit RCU is presented. It may be used in a system where cost-effective computing units redundancy is implemented. That is in applications when commercially available products are too expensive while the advantages of their using are not necessary from the functional point of view (e.g. synchronization of redundant units using dedicated fiber-optic synchronization modules, certification of the redundant modules).

The most important part of the state machine is diagnostics of the other RCU in order to make appropriate decision about taking over the control over the industrial process. Other issues are servicing operator requests and performing system startup in such a way, that only one RCU is *active*.

The point of consideration is the type of device used for realization of remote input/output station RIO. Here a regular, non-programmable device was applied. The state machine of the redundant computing unit RCU could be possibly simpler if there could be some logic implemented in the RIO. That way the RIO itself could decide where it should be controlled by one or another RCU. That case will be considered during latter research.

Moreover, the author intention is to implement the presented state machine in RCUs realized using embedded devices like Raspberry Pi or Beaglebone. They may be programmed using CODESYS environment which is dedicated for PLC programming according to IEC61131-3 standard. Especially interesting would be the latter embedded platform from the point of view of real-time operation. Beaglebone has got a PRU unit – Programmable Real-Time Unit [13]. It could be used for implementing the synchronization routines and fast data exchange between the redundant embedded devices.

Acknowledgment. The research work financed by BK-213/RAU2/2018.

References

1. Colombo, A.W., Karnouskos, S., Kaynak, O., Shi, Y., Yin, S.: Industrial cyberphysical systems: a backbone of the fourth industrial revolution. IEEE Ind. Electron. Mag. **11**(1), 6–16 (2017). https://doi.org/10.1109/MIE.2017.2648857
2. Zhang, D., Nguang, S.K., Yu, L.: Distributed control of large-scale networked control systems with communication constraints and topology switching. IEEE Trans. Syst., Man, Cybern.: Syst. **47**(7), 1746–1757 (2017). https://doi.org/10.1109/TSMC.2017.2681702
3. Martins, L.E.G., Gorschek, T.: Requirements engineering for safety-critical systems: overview and challenges. IEEE Softw. **34**(4), 49–57 (2017). https://doi.org/10.1109/ms.2017.94
4. Kwiecień, A., Stój, J.: The cost of redundancy in distributed real-time systems in steady state. In: Kwiecień, A., Gaj, P., Stera, P. (eds.) CN 2010. CCIS, vol. 79, pp. 106–120. Springer, Heidelberg (2010). https://doi.org/10.1007/978-3-642-13861-4_11
5. Stój, J., Kwiecień, A.: Temporal costs of computing unit redundancy in steady and transient state. In: Kosiuczenko, P., Madeyski, L. (eds.) Towards a Synergistic Combination of Research and Practice in Software Engineering. SCI, vol. 733, pp. 1–14. Springer, Cham (2018). https://doi.org/10.1007/978-3-319-65208-5_1

6. Safety Engineering in SIMATIC S7, Siemens Simatic System Manual, April 2006
7. PACSystems™ Hot Standby CPU Redundancy User's Manual, GE Fanuc Intelligent Platforms, Programmable Control Products, GFK-2308C, March 2009
8. Zuloaga, A., Astarloa, A., Jiménez, J., Lázaro, J., Araujo, J.A.: Cost-effective redundancy for Ethernet train communications using HSR. In: 2014 IEEE 23rd International Symposium on Industrial Electronics (ISIE), pp. 1117–1122, 1–4 June 2014
9. Stój, J.: Cost effective computing unit redundancy in networked control systems using real-time ethernet protocol. In: Borzemski, L., Świątek, J., Wilimowska, Z. (eds.) ISAT 2018. AISC, vol. 852, pp. 43–53. Springer, Cham (2019). https://doi.org/10.1007/978-3-319-99981-4_5
10. Real-time Ethernet with TwinCAT network variables, Application Note DK9322-0110-0024, Beckhoff, January 2010
11. Maruyama, T., Yamada, T.: Spatial-temporal communication redundancy for high performance EtherCAT master. In: 22nd IEEE International Conference on Emerging Technologies and Factory Automation, ETFA, Limassol, pp. 1–6 (2017)
12. Beckhoff: EtherCAT System Documentation. Version 5.3, 29 June 2018
13. Smołka, I.: Embedded systems based on raspberry Pi programmed with CODESYS. Studia Informatica Journal, 39(1), Gliwice 2018. https://doi.org/10.21936/si2018_v39.n1.839

Systems Analysis and Modeling

Estimation of the Characteristics of the Stochastic Interconnection of Meteorological Elements of an Aerodrome for Solving Problems of Forecasting Flight Conditions

N. F. Khalimon[1], D. P. Kucherov[1](✉), and I. V. Ogirko[2]

[1] National Aviation University, Kiev, Ukraine
{natalyhl208, d_kucherov}@ukr.net
[2] Ukrainian Academy of Printing, Lviv, Ukraine
ogirko@gmail.com

Abstract. The mixed central moments of the lowest layer of clouds and the horizontal runway visibility range in the region of the airfield located in the zone of temperate continental climate have been investigated. The data of long-term observations and the climatic characteristic of the aerodrome based on them were used as the initial statistical material. The regression equations of horizontal runway visibility range have been constructed on the height of the lowest layer of clouds. The developed methodology for studying the characteristics of a stochastic relationship can be used to predict the factors influencing the meteorological conditions on flight safety and automating the construction of climatic characteristics of aerodromes.

Keywords: Aerodrome meteorological elements · Stochastic interconnection · Flight conditions forecast

1 Introduction

Measurements of meteorological parameters at aerodromes are one of the most important elements of the meteorological support system for taking off and landing of aircraft and, ultimately, ensuring flight safety [1]. In accordance with this, there are increased requirements for the volume, efficiency, and reliability of measuring meteorological information and for used technical facilities [2–4].

The general trend in the development and implementation of aerodrome measuring and information systems at aerodromes is the expansion of the capabilities of systems, the range of tasks to be solved, i.e. increasing the level of automation of weather support at the aerodrome. Automation allows to simplify the solution of the tasks of the climatic description of an aerodrome and to increase the reliability and objectivity of the data, and also makes it possible to visually present them in the form of graphs, diagrams, etc. First of all, it should be noted that the possibilities of meteorological measurements are enhanced.

J. Świątek et al. (Eds.): ISAT 2019, AISC 1051, pp. 89–100, 2020.
https://doi.org/10.1007/978-3-030-30604-5_8

Important additional tasks are the automation of operational observations and the formation of climatic characteristics of the aerodrome. To obtain the climatic characteristics of an aerodrome, it is required to form long-term archives of weather-dangerous meteorological conditions and phenomena, the formation of monthly and annual tables of the frequency of these parameters and phenomena, their daily and annual variations, duration, etc.

Thus, the problem of investigation is the insufficient accuracy of measuring the height of the lowest layer of clouds (LLC) and the horizontal runway visibility range (RVR), which cause aviation accidents in the airfield. It is necessary to conduct regular research of the variability of the height of the LLC and horizontal RVR, weather conditions of varying degrees of complexity, hazardous weather, wind, etc. [5]. Taking into account the relatively slow change of meteorological elements, it was proposed to recalculate the climatic characteristics of airfields in five years [6].

To solve the problems of automating measurements, forecasting changes in some meteorological elements, based on the results of the analysis of others, it is necessary to have software systems for processing statistical data and analyze the characteristics of the stochastic interconnection of the most significant meteorological elements.

The aim of the investigation is the development of a method for estimating and forecasting the meteorological situation, namely runway visibility range in the area of the aerodrome to ensure the safe landing and take-off of aircraft. The objective of the paper is to improve the accuracy of measuring the horizontal RVR of airfield facilities. This work is devoted to the study of these issues.

2 Statistical Characteristics of the Most Important Meteorological Elements of the Aerodrome

One of the most important meteorological elements of the aerodrome is the height of LLC and RVR [5, 7]. According to the results of the analysis and forecast of their average values, frequency of occurrence over the time of year and day, you can more accurately plan the aerodrome handling capacity, the degree of keeping the schedule and, ultimately, provide the necessary level of safety.

The height of the LLC is constantly changing, and this layer itself is a wavy surface with an amplitude of oscillations up to 50–100 m/h. The course of changes in the RVR value is very close to the course of changes in LLC (see Figs. 1 and 2).

Judging by the graphs, the processes of changing LLC and RVR, on the one hand, are essentially non-stationary, and, on the other, the trends of their changes are very similar. Therefore, it is of interest to study the characteristics of their stochastic relationship. This interest is not only theoretical but also practical. The measurement procedure for RVR, in contrast to measurements of the height of an LLC, is very time-consuming, and the measurement results have a very low accuracy [7]. It is difficult to set the theoretical dependence of the RVR on weather conditions since various atmospheric phenomena affect on RVR in different ways, their contribution to the overall result is inconsistent and poorly formalized [8].

The multiple correlation coefficients and multiple regression are used as the main characteristics of the stochastic relationship [9, 10]. Based on the physical nature of the problem, it is logical to choose the RVR as the dependent variable of the regression equation, and the LLC height as the independent variable (argument). First of all, it is necessary to calculate the sample correlation coefficients. If they are much less than one, then the forecast of the RVR depending on the results of measuring the height of LLC with using the regression equation will be unreliable.

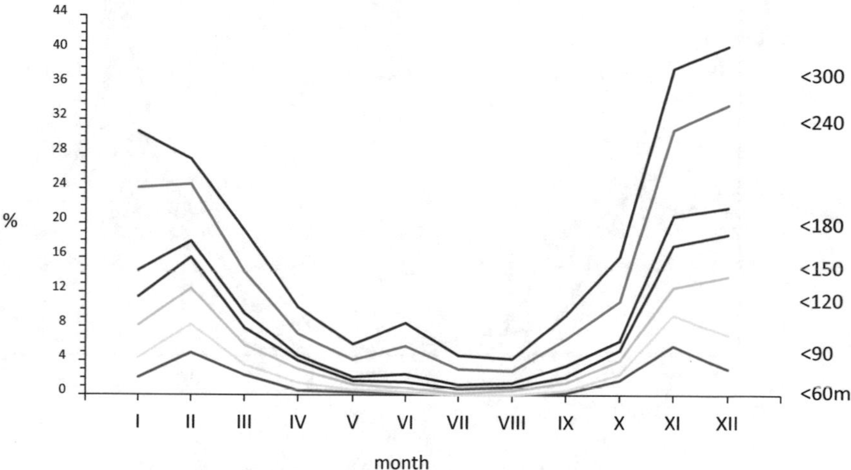

Fig. 1. Changes of the height of LLC by months within the specified limits (a percentage of the total number of measurements)

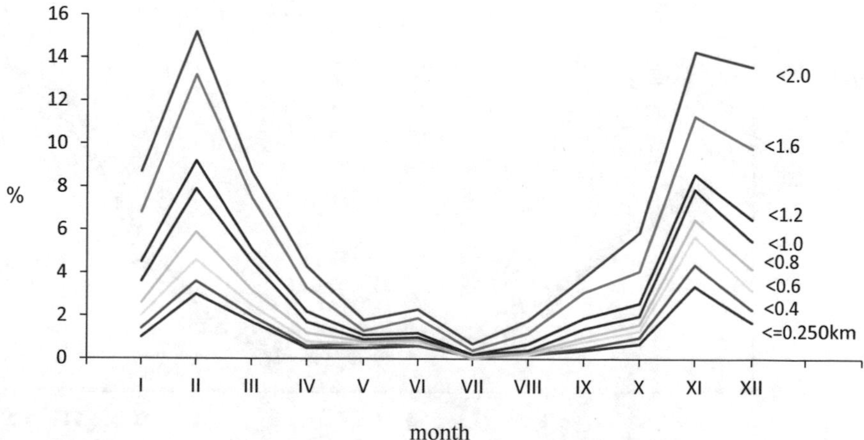

Fig. 2. RVR changes by months within the specified limits (a percentage of the total number of measurements)

In addition, to automate measurements and calculations, it is necessary to choose an approximation method of the repeatability curves for the height of LLC and RVR. The most flexible and accurate method is the least-squares polynomial approximation.

In Figs. 3, 4, 5 and 6 graphs of experimentally obtained dependences and their polynomial approximation of the 6th and 10th degree are shown.

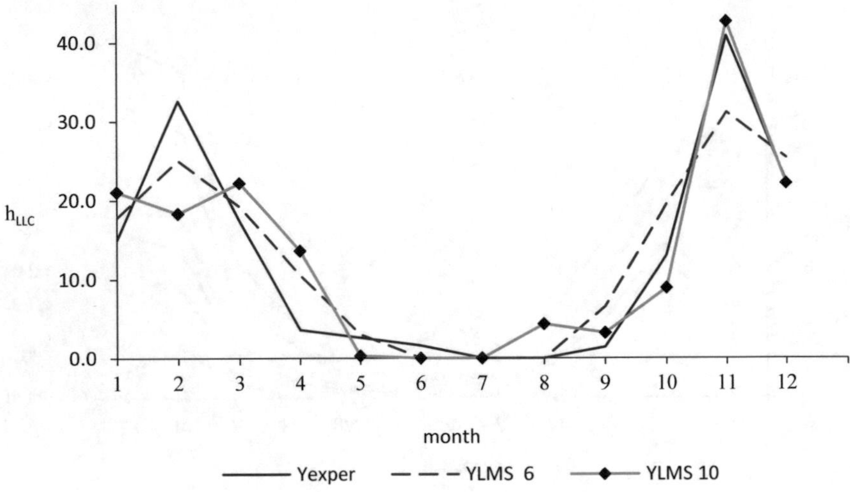

Fig. 3. Polynomial approximation of 6 and 10° of the experimental curve of the height repeatability LLC < 60 m

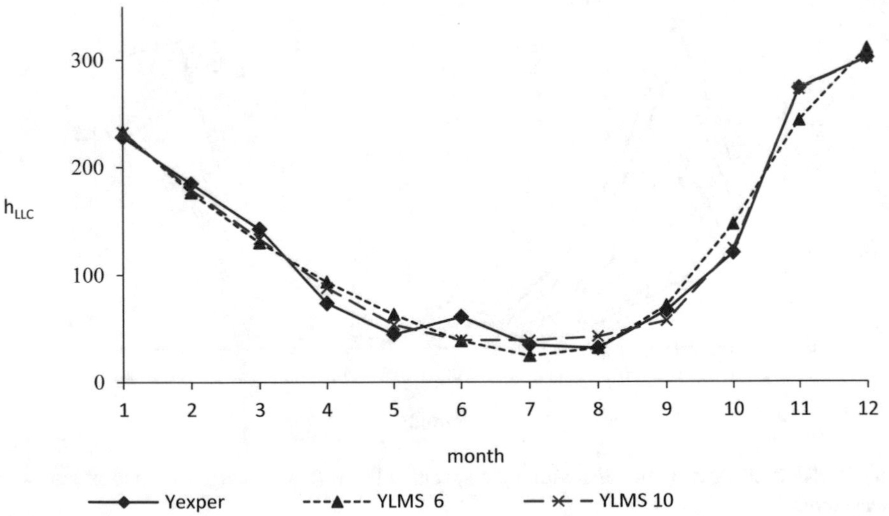

Fig. 4. Polynomial approximation of 6 and 10° of the experimental curve of the height repeatability LLC < 300 m

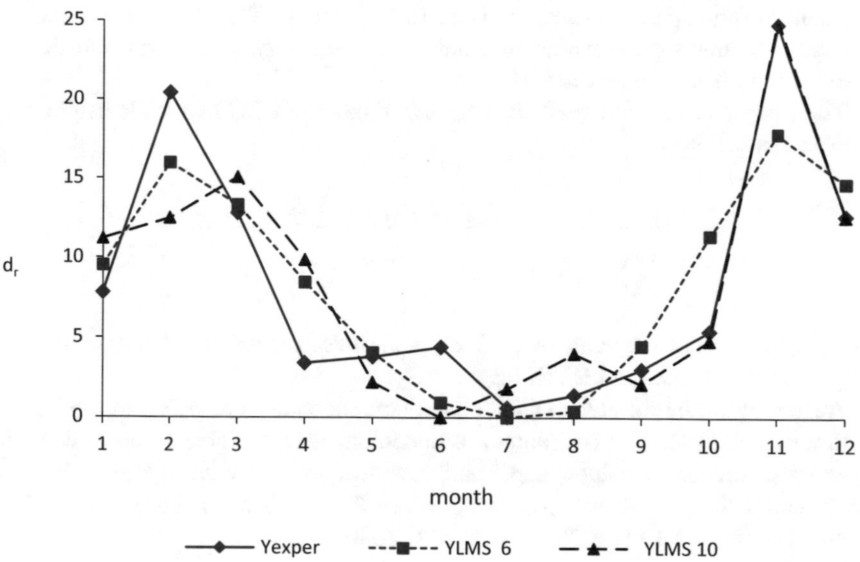

Fig. 5. Polynomial approximation of 6 and 10° of the experimental curve of the height repeatability RVR < 250 m

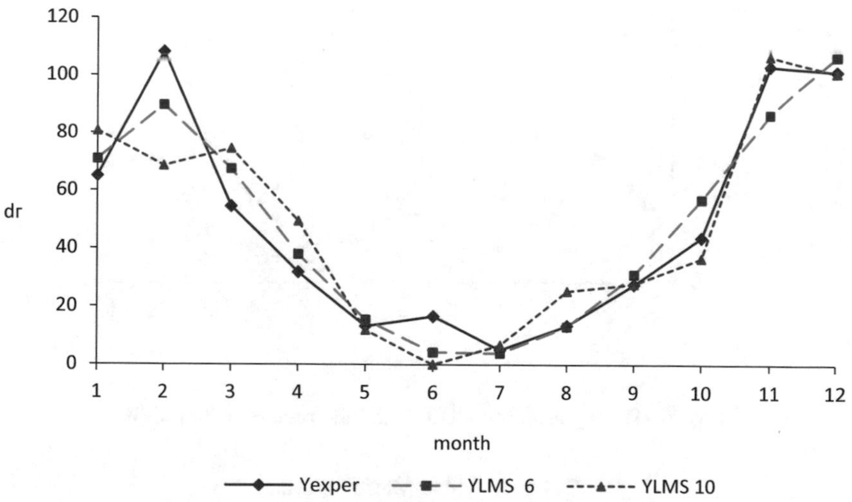

Fig. 6. Polynomial approximation of 6 and 10° of the experimental curve of the height repeatability RVR < 2 km

In the process of performing the calculations, it has been established that polynomial approximation of even order gives higher accuracy than a polynomial approximation of odd order, which is close to the corresponding even order. In general, the order of a polynomial at which an acceptable accuracy is achieved strongly depends on the fluency of the approximable curve (see Figs. 3 and 4). Therefore, in addition to the calculation of the approximating functions, it is necessary to perform statistical processing of the measurement results.

The sample correlation coefficient r_{hd} for N pairs of LLC and RVR height can be estimated as follows:

$$r_{hd} = \frac{\sum_{i=1}^{N} h_{li} d_{vi} - N\bar{h}_l \bar{d}_v}{\left[\left(\sum_{i=1}^{N} (h_{li})^2 - N(\bar{h}_l)^2\right)\left(\sum_{i=1}^{N} (d_{vi})^2 - N(\bar{d}_v)^2\right)\right]^{1/2}} \quad (1)$$

where h_{li} is the selective value of LLC; d_{vi} is the selective value of RVR; $\bar{h}_l$, $\bar{d}_v$ are the sample values of LLC and RVR, respectively.

According to the formula (1), the correlation coefficients of pair samples LLC and RVR were calculated from minimum to maximum values. Table 1 shows data about the correspondence of values, and Table 2 shows some results of calculations. The Fig. 7 shows the graphs of a pair samples are the numbers of repeatability of measurement results for LLC < 300 and RVR < 2000.

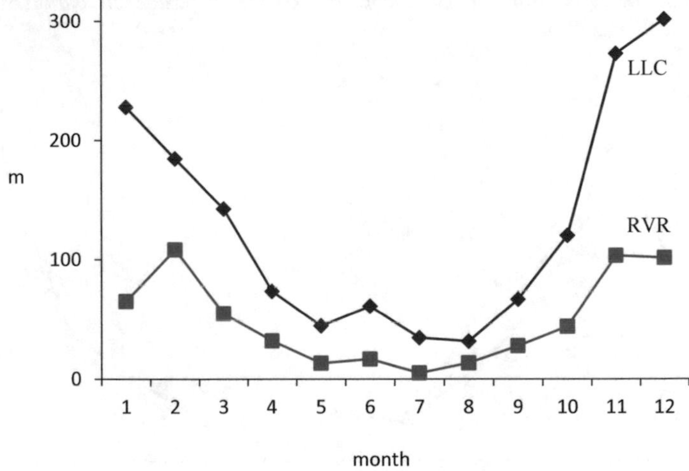

Fig. 7. The repeatability of LLC and the corresponding RVR

Table 1. Value Matching Data

LLC, m	<60	<90	<120	<150	<180	<240	<300
RVR, m	<250	<400	<600	<800	<1000	<1200	<2000

Table 2. Some results of calculations

Sample	LLC <60 & RVR <250	LLC <150 & RVR <800	LLC <300 & RVR <2000
correlation coefficient r_{hv}	≈0.86	≈0.93	≈0.93

In these measurements, ejections are observed, which are caused by random factors that are not related to the regularities of meta-elements changing. For smoothing out the ejections, we order the sequences by presenting them in the form of a non-decreasing series (Fig. 8).

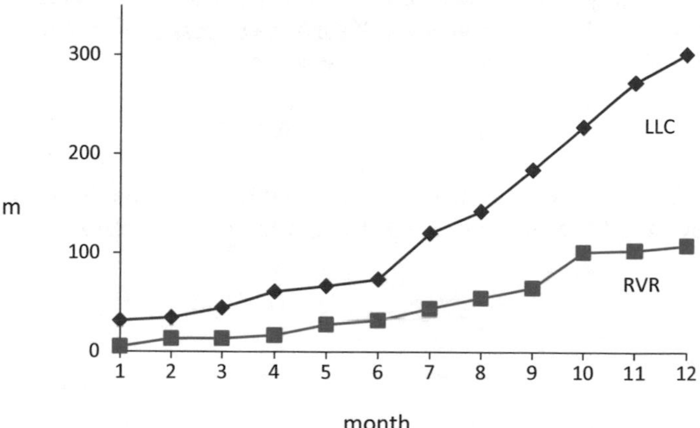

Fig. 8. Ordering in non-decreasing sequences

With such a smoothing of the data, the correlation coefficients of pairs of sample values become even closer to one, as can be seen from Table 3.

Table 3. Correlation coefficients of pairs

Selected data	LLC <60 & RVR <250	LLC <150 & RVR <800	LLC <300 & RVR <2000
correlation coefficient r_{hv}	$\approx$0.988	$\approx$0.99	$\approx$0.975

When calculated by the direct method and with preliminary smoothing of the data, a strong correlation of the researched samples is found as a result; therefore, one should expect a high accuracy of the prediction using the regression analysis method.

3 Creating Regression Models and Calculating Regression Coefficients

In the work, the linear regression of a random variable RVR (function) for a random value of the height of the LLC (argument) was investigated:

$$d_v = A + Bh_l \tag{2}$$

where A is the free member, B is the coefficient (the tangent of the slope of the regression line).

If the value of the correlation coefficient r_{hd} is one, there is essentially a deterministic functional relationship between h_l and d_v. In this case, the prediction error $\tilde{d}_v$ of the measured value h_l will be zero.

In practice, always $r_{hd} < 1$, and there will be some scatter of points on the graph. In Fig. 9 depicts a set of pairs of values (X, Y) and the regression line Y on X. Deviations of the measured values from the predicted ones are

$$d_{vi} - \tilde{d}_{vi} = d_{vi} - (A + bh_l) \tag{3}$$

To calculate the coefficients and with the minimum error, the least squares method is also usually used and the sum of squared deviations is minimized

$$\delta_{\sum}^2 = \sum_{i=1}^{N} (d_{vi} - A - bh_l)^2 \tag{4}$$

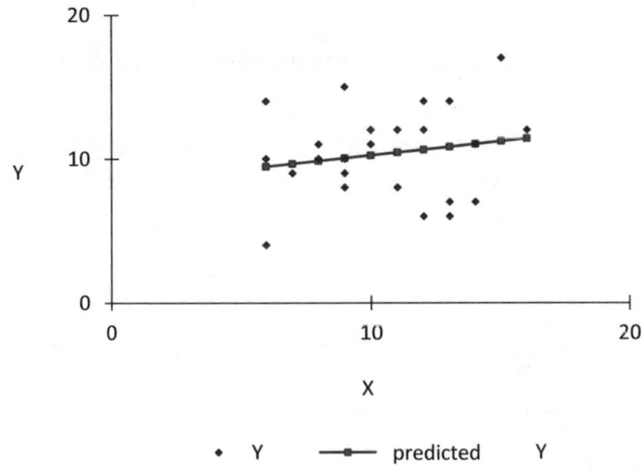

Fig. 9. Y regression on X (X, Y) is the aggregate of random numbers distributed according to Poisson's law, zero cross-correlation

The minimum is reached at such values of the coefficients A and B, which satisfy the condition

$$\frac{\partial \delta_{\Sigma}^2}{\partial A} = \frac{\partial \delta_{\Sigma}^2}{\partial B} = 0 \tag{5}$$

For samples of limited volume N, condition (5) allows obtaining only estimates of the coefficients A and B, which we denote a and b, respectively. Substituting expression (4) into (5) and solving Eq. (5) by the method of indefinite coefficients, equating the coefficients with the same degrees, we obtain the corresponding equations for the estimates:

$$\left.\begin{aligned} a &= \bar{d}_v - b\bar{h}_l, \\ b &= \frac{\sum_{i=1}^{N} h_{li} d_{vi} - N\bar{h}_l \bar{d}_v}{\sum_{i=1}^{N} h_l^2 - N\left(\bar{h}_l\right)^2} \end{aligned}\right\} \tag{6}$$

Using the equations for the estimates (6), we can finally obtain an expression for the predicted values d_v by the measured h_l:

$$\hat{d}_v = a + bh_l = (\bar{d}_v - b\bar{h}_l) + bh_l = \bar{d}_v + b(h_l - \bar{h}_l) \tag{7}$$

which essentially is a regression equation on.

Figures 10, 11, 12 and 13 shows the graphs of the regression lines d_v on h_l for two pairs of special values of the height of the LLC and RVR, and Table 4 shows the calculated values of the estimates of the coefficients A and B. Table 5 shows the values of mean-square errors of the coefficient estimation.

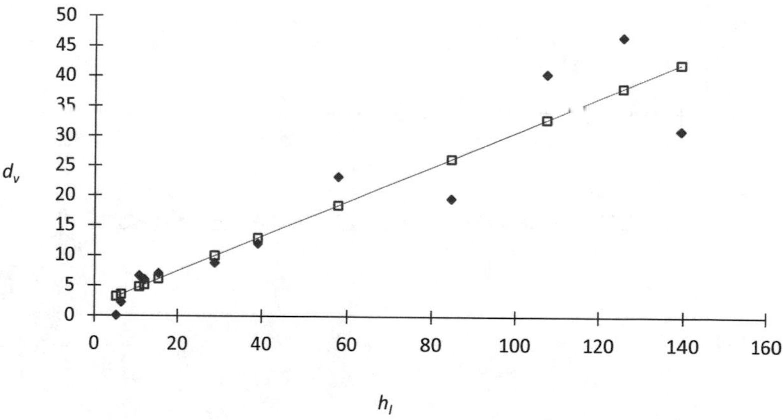

Fig. 10. Regression line d_v on h_l: LLC <150 and RVR <800.

The mean-square errors of the coefficient B estimate is relatively small, so one can expect that the prediction accuracy of the slope of the regression line is very high. As for the standard deviation mean-square errors of the coefficient A, it is much more than for B. However, the coefficient of variation d_v is much less than one; therefore, estimation errors A have practically no effect on the resulting forecast accuracy.

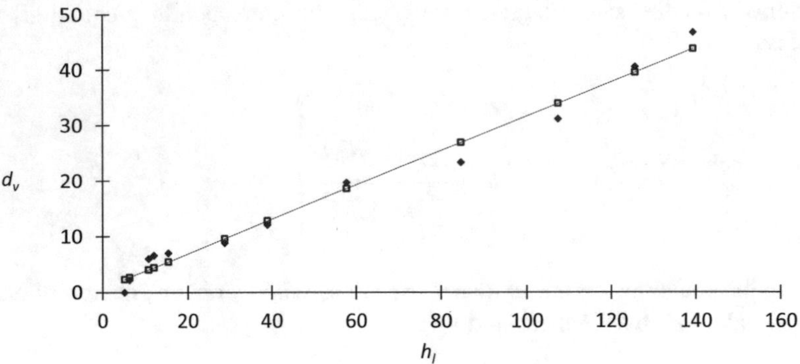

Fig. 11. Regression line d_v on h_l: LLC <150 and RVR <800, non-decreasing series

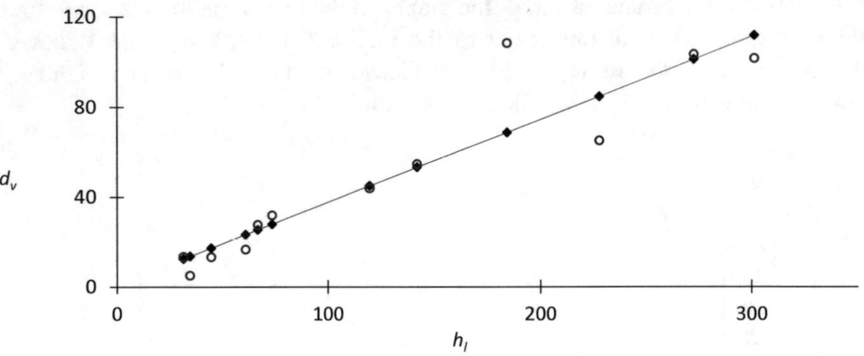

Fig. 12. Regression line d_v on h_l: LLC <300 and RVR <2000

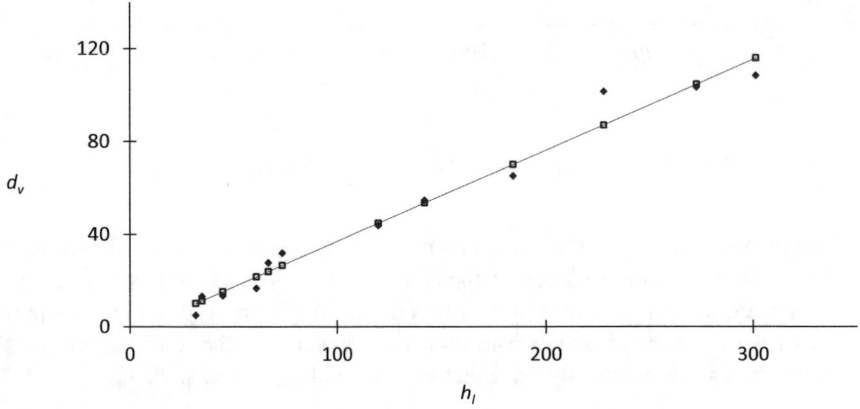

Fig. 13. Regression line d_v on h_l: LLC <300 and RVR <2000, non-decreasing series

Table 4. The values of the coefficient estimates *A* and *B*

Equation coefficients	LLC <150 & RVR <800	LLC <150 & RVR <800 non-decreasing	LLC <300 & RVR <2000	LLC <300 & RVR <2000 non-decreasing
A	1.67	0.72	0.93	−2.17
b	0.29	0.31	0.37	0.39

Table 5. Ratio estimates

Mean-square errors	LLC <150 & RVR <800	LLC <150 & RVR <800 non-decreasing	LLC <300 & RVR <2000	LLC <300 & RVR <2000 non-decreasing
A	2.54	0.96	7.46	3.14
B	0.035	0.014	0.047	0.02

The hypothesis of normal deviations was also tested and confirmed. Confidence intervals for the 5% level of significance were calculated. The values of these intervals are significantly less than the absolute values of the coefficients. Consequently, the probability of a gross error such as jumping from one curve to another when predicting is the second order of smallness, and such an event itself can be attributed to rare ones.

4 Conclusions

The climatic characteristic of an aerodrome is one of the most important components of ensuring flight safety, the rhythm of work and maintaining a schedule (or a planned flight table). The variability of the main meteorological elements is small. Therefore, recalculation of climatic characteristics can be carried out at sufficiently long intervals and, more importantly, the results of long-term measurements can be used to obtain highly accurate estimates of climatic characteristics over a long period.

According to the results of the statistical analysis of climate data in the area of the aerodrome, it has been established that there is a strong correlation between the height of the LLC and the RVR. Therefore, it is possible to predict the change in one parameter from the measured data for another parameter. This technique can also be used “in the opposite direction” - to predict the height of an LLC based on the results of an RVR measurement, if, for example, the RVR measurement has high accuracy and can be obtained with less labor.

In the paper a technique has been proposed for finding horizontal RVR based on the refinement of the coefficients of the regression equation based on measurements of the height of the LLC, which makes it possible to eliminate gross prediction errors.

The developed method of estimation and forecasting can be used to create firmware and software for control and testing instrumentation of flight safety meteorological equipment [11, 12].

References

1. Bolelov, E.A.: Meteorologicheskoe obespechenie poletov grazhdanskoj aviacii: problemy i puti ih reshenija [Meteorological service for civil aviation: problems and ways of their solution]. Scientific Bulletin of Moscow State Technical University of Civil Aviation, 21 (05), 117–129 (2018)https://doi.org/10.26467/2079-0619-2018-21-5-117-129
2. Seleznev, V.P.: Meteorologicheskoe obespechenie poletov [Meteorological support of flights]. LIBROKOM, Moscow, 190 (2018). (in Russian)
3. Kucherov, D.P., Berezkin, A.L.: Identification approach for determining radio signal frequency. In: XI International Conference on Antenna Theory and Techniques (ICATT), Kyiv, Ukraine, pp 379–382 (2017). https://doi.org/10.1109/icatt.2017.7972668
4. Mahringer, G.: Terminal aerodrome forecast verification in Austro Control using time windows and ranges of forecast conditions. Meteorol. Appl. **15**, 113–123 (2008). https://doi.org/10.1002/met.62
5. Shakina, N.P., Nudel'man, L.A.: Trebovanija k sostavleniju klimaticheskogo opisanija ajerodroma [Requirements for the climatic description of the airfield]. Meteoagentstvo Rosgidrometa, Moscow, 48 (2007). (in Russian)
6. Chernaya, O.O.: Aviameteoobespechenie i pravo [Supply of meteorological data and law]. Dashkov and K°, Moscow, 336 (2016). (in Russian)
7. Zemljanko, N.I., Khalimon, N.F., Zhadan, L.P.i dr.: Klimaticheskaja harakteristika aeroporta Kiev-ZHuljany [Climate characteristics of the airport Kiev-Zhuliany]. Kiev International University of Civil Aviation, Kiev, 172 (2000). (in Russian)
8. Belov, V.V., Veretennikov, V.V., Zuev, V.E.: Teorija sistem v optike dispersnyh sred [Systems with applications in scattering media]. Spectr, Institute of Atmospheric Optics of Siberian Branch of the Russian Academy of Science (IAO SB RAS), Tomsk, 402 (1997). (in Russian)
9. Bendat, J.S., Piersol, A.G.: Izmerenie i analiz sluchajnyh processov [Analysis and Measurement Procedures] Mir, Moscow, 540 (1989). (in Russian)
10. Mirskij, G.J.A.: Harakteristiki stohasticheskoj vzaimosvjazi i ih izmerenija [Characteristics of the stochastic relationship and their measurement]. Energoizdat, Moscow, 320 (1982). (in Russian)
11. Khalimon, N.F.: Informacionnye tehnologii ocenivanija harakteristik stohasticheskoj vzaimosvjazi meteorologicheskih javlenij [Information technologies for evaluating the characteristics of the stochastic relationship of meteorological phenomena]. Scientific Bulletin of Ukrainian scientific-research institute of telecommunications, 4(12), Kiev, pp 85–90 (2009). (in Russian)
12. Hansen, B.A.: Fuzzy Logic–Based Analog Forecasting System for Ceiling and Visibility Weather and Forecasting. vol. 22, pp. 1319–1330. AMS Publication, Boston (2007). https://doi.org/10.1175/2007waf2006017.1

From Interactive Simulations Towards Decision Patterns in Conflict Situations

Andrzej Najgebauer(✉)

Faculty of Cybernetics, Military University of Technology,
Gen. S. Kaliskiego 2, 00-908 Warsaw, Poland
andrzej.najgebauer@wat.edu.pl

Abstract. The researches cover the process of construction and testing of “intelligent” software using the interactive simulation system, towards knowledge extraction. This type of simulation can be considered as a type of common intelligence of experts taking part in many experiments basing on scenarios of conflict real or hypothetical. The limitations of such intelligence are presented. The experiences of the interactive simulation systems construction and experimenting in many war games have resulted in elaboration the technological line of conflict analysis with different computer tools utilisation. The main thread of the paper is focused on interactive simulation for land operations and searching the best course of action by the sides of conflict on the bases of recognised state of battlefield and the intuition or knowledge of the sides’ commanders. The recognition of this intuition and knowledge of commanders and coding it in the form of decision patterns is the main goal of the research. In the process of the experimental environment construction the requirements for information system were described. On the basis of own methodology for interactive simulation system designing and development, prototype simulation system - MSCombat and professional systems - Złocień and SWDT were built and tested. The directions of transformation of the interactive simulation environments into the artificial intelligence system are proposed.

Keywords: Interactive simulation · Decision patterns · Conflict analysis and modelling

1 Introduction

The main topic of the paper focuses on interactive simulation of land operations and the search for the best way of action of conflicting parties based on the recognized state of the battlefield and the intuition or knowledge of the commanders of the parties. Recognition of this intuition and knowledge of the commanders and encoding it in the form of decision patterns is the main goal of the research. However as the first phase of the research the process of modelling, designing and development the interactive simulation environment is presented as well as a big challenge.

Roger Penrose in his excellent book (Penrose R. 1989. *The Emperor’s New Mind. Concerning Computers, Minds, and the Laws of Physics.* Oxford University Press 1989) mentioned that simulation can be considered as a type of intelligence, where

J. Świątek et al. (Eds.): ISAT 2019, AISC 1051, pp. 101–114, 2020.
https://doi.org/10.1007/978-3-030-30604-5_9

software there is a mind and hardware there is a body. Additionally another author (Ilachinski J., 1996 - *Land Warfare and Complexity, Part I: Mathematical Background and Technical Sourcebook*, Centre for Naval Analyses, Alexandria, Virginia, USA, 1996.) wrote, that military conflict can be concerned as a complex adaptive system. Integrating these two observations, one can attempt to build a simulation interactive environment, reproducing the adaptive system with decisions introduced by the players, adequate to the situation encountered on the simulated battlefield. One of the important conclusions in relation to armed conflict outside the non-linear order, hierarchical structure of command and control of the fight is adjustment, collectivism and self-organization in a certain time perspective. The main conclusion from these considerations is that there is an intuition of the commander in the battlefield as the ability to perceive certain patterns - this is the basis for thinking about separating the principles of decision making in armed conflicts.

In order to test these hypotheses, it is necessary to conduct complicated research. Research includes the process of building, testing and experimenting with a simulation system, towards the extraction of knowledge.

A certain scheme of IT support for the analysis of conflicts in the context of decision making by the parties to these conflicts is presented. The key element of IT support is an interactive simulation environment for mapping military operations and their decision-making processes. In order to build a simulation environment reflecting conflict situations of military character it is necessary to develop a formal model, preceded by identification of the modelling goal and specification of requirements for an interactive system. This may be facilitated by an attempt to answer a number of questions that arise at the beginning. How to identify and reflect the real conflict situation? What steps should be taken to solve the conflict? What decisions can be made by the conflicting parties? How to identify the best decisions in an uncertain situation based on defined criteria? What type of conflict model should we use in a given situation? Are there any behavioural patterns in the decision-making situation? How to show the dynamics and changes in a conflict situation? After many years of experience in modelling, implementing and applying such systems, some answers to the questions have been obtained and, as a result of the gained experience, an outline of the methodology for constructing, testing and using an interactive simulation environment has been formulated, both in decision support and in the training of decision-makers.

The following phases of the above mentioned methodology of creating interactive simulation environments for mapping military conflict situations have been identified [1]:

- identification of conflict and its general description
- conflict model as a non-cooperative n-player game (as a starting point)
- model of the combat process - multidimensional stochastic process
- decision model - a multi-stage problem of stochastic optimisation with risk criteria
- computer environment for simulating combat processes
- experimentation, monitoring and visualisation
- post-simulation analysis

Simulations in military domain can be divided into three types- live, virtual and constructive. (1) Live – real people operating real equipment, platforms or systems, usually on tactical or weapons ranges. (2) Virtual – real people operating simulated systems. The latter may include any mix of equipment, system or platform simulators, rigs or testbeds, computer presentations/visualisations or prototypes of the same, factories and production lines, and/or constructive models of forces or force elements. (3) Constructive – simulated people operating simulated equipment, or systems. Computer models represent the actions of people and/or equipment using mathematical equations and/or logical algorithms.

The presented model is built for constructive simulation. Interactive distributed simulation is a technology developed by military experts to facilitate the integration and use of many simulations and simulators in a distributed, shared computing environment reflects the decision situations and activities as a result of the decisions made during combat process. The paper is focused on land operations, where the commanders and their staffs should be trained, so it is proposed the construction specific computer environment on the basis of operational game model [1]:

$$\boldsymbol{\Gamma}_S = <\boldsymbol{N}, M_{SW}, SC_W> \tag{1}$$

where: $\boldsymbol{N}$ - a set of players (parties), the component of the set is determined by their role (head, trainee, observer, analyst, expert, opposing forces commander), personal data, team structure, if group participant. M_{SW} – campaign process model, which can be described by $S(t)$ [1] and set of complementary parameters, SC_W - scenario of combat, which is defined as initial state of campaign process $S(t_0)$, termination time T^Y ($Y = A, B, C, \ldots$-participant of game notation) and mission from higher level T^Y,

The phases of game construction are presented in [1].

2 Problem Definition

The representation of the armed situation usually takes place with a certain accuracy dictated by the command level. There is a rule of control of subordinate troops and observation of the terrain by the commander two levels below. The basic issue here is the decision making processes in armed situations, and the subject of assessment is the decision-maker who makes the decisions. The decision-making process in an armed conflict is very complex due to uncertainty, incompleteness of information and the pace of the fight. All these aspects can be carefully reproduced in an environment of interactive distributed simulation. The simplification is used for the purpose of synthetic presentation of transformation problem and the way of modelling in the paper. In the process of modelling the fight, only those elements that are important from the point of view of decision making should be distinguished. We will focus on the combat units of land-based troops at division or brigade level, as an example of a combat process where the decision-making process is very complex. The conflict simulation can be shown in the diagram [Fig. 1]. Interaction in a computer environment is one of the methods of updating decisions when the decision-maker tries to react to changing combat conditions. In [1], the general decision making procedure is proposed as a multi-stage stochastic programming task. The results of the decision can be observed. Strategies of decision makers (players) are formulated on the basis of a combat process model, which is defined as a multidimensional stochastic process. The stochastic model expresses uncertainty in a conflict situation. In an interactive simulation environment it is possible to observe the battlefield and make decisions on the basis of these measurements. After the implementation of the MSCombat prototype solution, the team made a professional Złocień system, in which more sophisticated models were implemented.

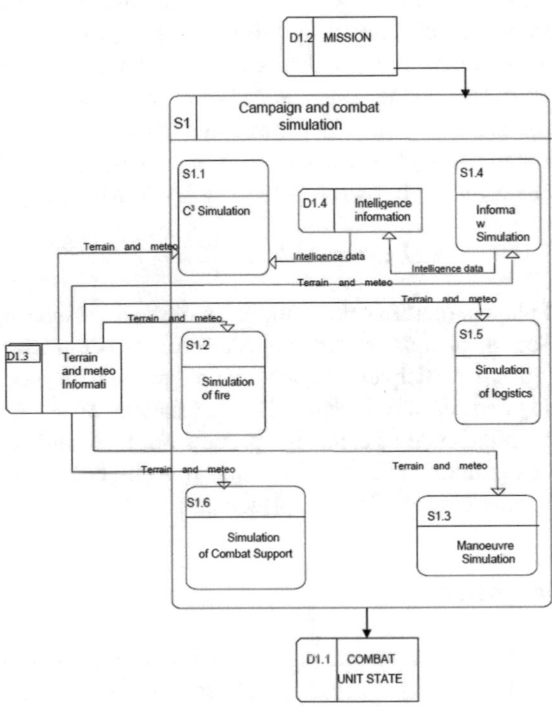

Fig. 1. Data flow diagram of campaign simulation process (2nd level) own source, [9]

Model of Conflict

The distinguished components of the multidimensional stochastic process describe in a certain simplification the course of changes in the values of the characteristics of the combat process, including both the fighting parties, conventionally called A and B, and the so-called battlefield, constituting the environment for the fighting enemies. The description refers to characteristics such as the number of ready to fight combat units of both fighting sides at any time, the combat value of these units at any time, the state of the command and communication system of both sides, the degree of efficiency of the system of securing operations, the degree of detection of the opponent through the system of reconnaissance of both sides, the location of combat units of both sides, a set of passable routes on the battlefield, the weather at any time. Probably the mentioned set of characteristics does not constitute a complete list of quantities that could be distinguished in the description of the battle process of land-based troops of tactical level. However, it seems sufficient to show the influence of superior combat tasks and decisions taken at the level of the tactical union on the combat process.

Let $\{\boldsymbol{S}(\boldsymbol{t}), \boldsymbol{t} \in [\boldsymbol{0}, \boldsymbol{T}]\}$ be the multidimensional stochastic process [1], where:

$$\{S(t) = (S_{1Y}(t), S_{2Y}(t), S_{3Y}(t), S_{4Y}(t), S_{5Y}(t), S_{6Y}(t), S_{1PW}(t), S_{2PW}(t)), t \in [0, T], Y = A, B\} \tag{2}$$

$S_{1Y}(t)$ - number of combat units of side Y $(A,\ B)$(they are ready to fight), $S_{1Y}(t) \in \{0, 1, 2, \ldots, M^Y\}$, $S_{2Y}(t)$ - combat potential of side Y

$S_{2Y}(t) = (S^1_{2Y}(t), S^2_{2Y}(t)) \quad S_{2Y}(t) \in \{\{0, \bar{S}^1_{2Y}\} \times \{0, \bar{S}^2_{2Y}\}\}$,

$S^1_{2Y}(t)$- combat potential of side Y in an attack, $S^2_{2Y}(t)$- combat potential of side Y in a defence, $S_{3Y}(t)$ - state of readiness, $S_{3Y}(t) \in \{0,\ 1\}$, $S_{4Y}(t)$ - efficacy degree of logistics system, $S_{4Y}(t) \in \{1,\ 2,\ 3,\ 4\}$, the levels of combat potential renewal in a period of time, $S_{5Y}(t)$ - surveillance level, $S_{5Y}(t) \in \{1, 2, \ldots, 7\}$, the levels of enemy recognition, $S_{6Y}(t)$ - combat units placement,

$S_{6Y}(t) = \left(S^1_{6Y}(t), S^2_{6Y}(t), \cdots, S^i_{6Y}(t), \cdots, S^{H^Y}_{6Y}(t)\right), S^i_{6Y}(t) \in H^Y = \{1, 2, 3, 4, 5, 6\}$,

$S_{1PW}(t)$- the set of battlefield routes, $S_{1PW}(t) \subset L = \{l_1, l_2, \cdots, l_k, \cdots, l_K\}$, $S_{2PW}(t)$- weather at the time t, $S_{2PW}(t) \in \{1, 2, 3, 4, 5, 6, 7\}$- set of numbers of different types of weather.

Each component of a multidimensional process depends on many random factors, and the transition from state to state is caused by the implementation of combat tasks of the superior level, decisions of the parties to the conflict and the implementation of these decisions (manoeuvring and fighting). In addition, transitions to certain states are triggered by the change of weather conditions, the passage of time, destruction or masking of elements of the battlefield. The phase space has a complex character. We had to apply some simplifications in the description, leading to the discretization of this space.

The terrain model is discrete and can be expressed as follows:

$$\mathbf{Z} = <\mathbf{G}, \{\varphi_1\}_{1=1,2,3,}\{\varnothing\}> \tag{3}$$

where $\mathbf{G} = <W, U>$ - Berge graph, W – set of vertices, U – set of arcs, $U \subset W \times W$

$\varphi_1 : W \rightarrow \boldsymbol{C}$ – function, which describes a location of the zone, represented by vertex $w \in W$, $\boldsymbol{C} \subset R^3$, $\boldsymbol{C}$ describes the whole battlefield area of the conflict, $\varphi_2 : W \rightarrow 2^{\boldsymbol{L}}$ function, which describes an assignment of roads to node w (these roads are accessible for mobility in a zone w), $\boldsymbol{L}$- set of all road numbers in a battlefield, $\varphi_3 : W \times J^{A(B)} \rightarrow 2^{\boldsymbol{TU}}$ – assignment function of formation type subset to node w, $\boldsymbol{TU}$ – set of all possible types of combat units formation, $\varphi_4 : W \times J^{A(B)} \rightarrow R^n$ – the determination of highest velocities of combat units on roads of battlefield, $n = |\boldsymbol{L}|$. Arcs of the graph represent possibility of close transition between zones. The process is discrete in states and continuous in time (DC). Combat actions, decisions and their realizations, natural phenomena and so on causes the transitions between states. The phase space X of the process $S(t)$ can be represented as follows:

$$\begin{aligned}\mathbf{X} = [&\{0,1,\ldots,M\} \times \{0,1,\ldots,\overline{S^1_{2Y}}\} \times \{0,1,\ldots,\overline{S^2_{2Y}}\} \times \\ &\times \{0,1\} \times \{1,2,3,4\} \times \{1,2,3,4,5,6,7\} \times (\mathbf{H}^M)^Y]^2 \times \\ &\times 2^{\mathbf{L}} \times \{1,2,3,4,5,6,7\}\end{aligned} \quad (4)$$

Estimation of the number of possible states of the combat process *S(t):*

$$|\boldsymbol{X}| = 21952 \cdot (\overline{S^1_{2A}}+1)(\overline{S^2_{2A}}+1) \cdot M \cdot N \cdot 36^{M+N}(\overline{S^1_{2B}}+1)(\overline{S^2_{2B}}+1) \cdot 2^{|\boldsymbol{L}|} \quad (5)$$

Any element of the combat system that represents one side or battlefield can be represented as an object. Transitions in the battle process are linked to events on the battlefield and their time is randomly variable. One important source of randomness on the battlefield is the random outcome of a local battle. We have proposed many stochastic models of local conflict in the local sense as a closed combat process between two basic combat units (the process of destroying the resources of combat units). Depending on the type of forces we use different models - for land forces we use modified stochastic Lanchester models for mobile forces (tanks or armoured transporters) and for artillery units we use Salvo1, Salvo2 models. Similarly for maritime units. Air forces are dominated by duel models (for air combat) or zone models of overcoming air defence.

So the next model in the evolutionary process of professional simulation system designing there is [9]:

$$SPW(t) = \langle S_A(t), S_B(t), R(t), S_O(t) \rangle \quad (6)$$

where:

$S_A(t), S_B(t)$ status vectors of the conflict sides;

$R(t) = (R_1(t), R_2(t), \ldots, R_{Ln}(t))$- vector of states of other conflict participants (e.g. state of civilian population, state of guerrillas, etc.);

$S_O(t) = (Z_1(t), Z_2(t))$- the state of the conflict environment describing: field conditions, meteorological conditions, engineering conditions, contamination, electromagnetic conditions, etc. This model is based on square nets (200 m x 200 m granularity).

3 The Interactive Simulation Environments

In our team we designed and developed several interactive simulation environments and as the first prototype MSCombat [6] based on the stochastic model *S(t)* (2) and 2 advanced (implemented) simulation systems - Złocień [9, 12] and SWDT [11], which are object-oriented and based on the model of the battlefield (6). The environments were built to test possible solutions in land operations simulation. The main direction in the technological process is the use of High-Level Architecture principles, because one of the most important requirements for Polish environments is the possibility of cooperation with simulation systems of NATO countries. HLA has become a standard for communication and synchronization of simulation environments.

The HLA consists of a number of interrelated components. The three defining components of HLA are: (1) HLA principles, which define interoperability and the capabilities that a simulation must have in order to achieve it within the HLA; (2) Template of an object model, which is a semiformal methodology for determining classes, attributes and interactions of simulation and object federation; and (3) Interface specification, a precise specification of functional activities that a combat simulator may cause or may cause during an exercise with the use of HLA.

SWDT [11] is a discrete time – driven simulator and models a two-sided land conflict of military units on the company/battalion level. The models concern the couple processes of firing interaction and movement executed by a single military unit. These two complementary models use a terrain model described by a network of square areas which aggregates movement characteristics with 200 m × 200 m granularity. The course of each process depends on many factors among them: terrain and weather conditions, states and parameters of weapons the units are equipped, type of executed units activities (attack, defence), distance between opponent units.

4 The Experimentation and Training

The professional system [9, 12] consists of a number of software applications, which should be mentioned: (1) Combat Simulator - the main part of the software responsible for simulating the situation on the battlefield. (2) Procedures for playing exercises - a package of procedures for playing exercises is an extension of the simulator that allows you to play the course of previously performed exercises. (3) Calibration parameters editor - software used to modify the values of calibration parameters of models of actions and behaviours of active simulation objects (military units, elements of the battlefield). (4) Operational database - is embedded in Oracle database server (OLTP class). The database model has been defined using ORM technique for the object-oriented model of the information structure of the Złocień system. (5) Map database - is embedded in Oracle9i database server (OLTP class). It contains tables describing such elements as point objects (e.g. bridges), linear objects (e.g. road sections, river sections), surface objects (forested areas, reservoirs), which were loaded from data available in VPF format, DTED height grid, as well as data describing squares converted on this basis (200 × 200 size, about 12.5 million squares in total) on which the combat simulator is based (Figs. 2, 3 and 4).

Both environments Złocień and SWDT enable management very wide experimentation process and collecting a lot of data connected with decision analysis.

The After Action Review tool enables checking selected game characteristic values off-line, after simulation. AAR tool supports the characteristics' analysis where the set is divided into a few categories as follows: state/resource levels of units (weapons, personnel, materials, potential), losses distribution of units regarding weapons, personnel, materials, potential (time distribution, structural distribution, time-structural distribution), schedule for units recognized by other units, task realization

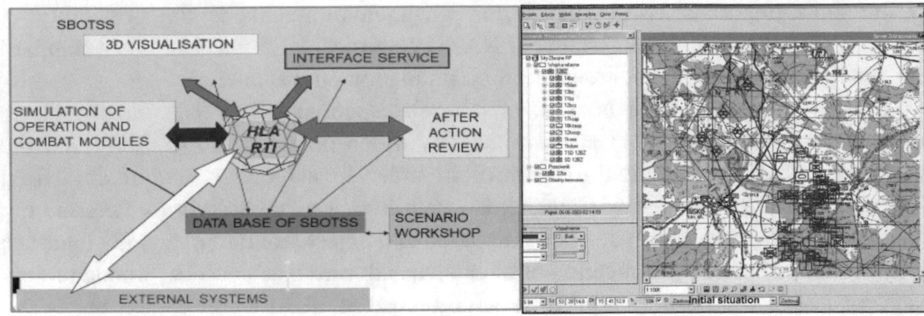

Fig. 2. Złocień – main components and exercise manager screen own source, [9]

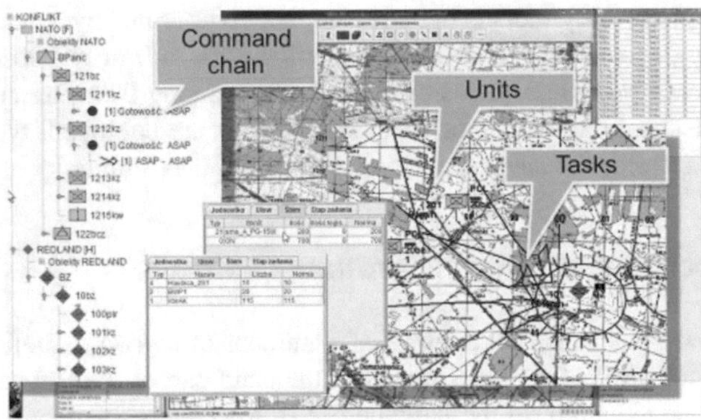

Fig. 3. SWDT- scenario editor [11]

degree, real action speed, relation between real action speed and normative one. AAR tool allows presentation of simulation results using tables and graphs and can help in testing and learning new expert tools. The analysis scheme is as follows:

- the construction of combat scenario,
- the experimenting with players of both sides monitoring of chosen characteristics (decisions, results of combat, movement, state of units),
- the statistical analysis due to pattern of action and reaction recognition for different parameters of battlefield,
- the comparison of battle results for different scenarios and identification of similarity of decision situations.

After Action Reviews are conducted during training exercises as logical predetermined events (based on the exercise scenario) are completed and at the conclusion of the exercise to assist the commander in answering many questions and gathering the interested characteristics of conflict (Table 1).

This data are the basis of detailed analysis of results the battle due to classification of decisions made by parties. The next point is identification the decision patterns.

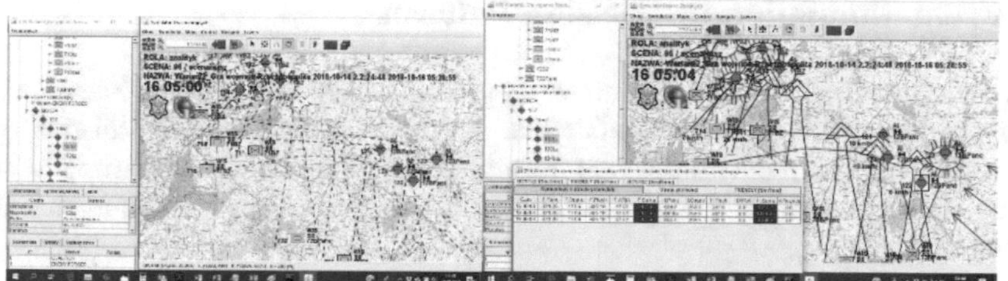

Fig. 4. SWDT – Simulation of combat – COA determination and realization

Table 1. State of combat units for the selected time.

Time	Side	Unit	Weapon/mun	Code	State
0	0	711bz	SPW	BWP-1	31
0	0	711bz	SPW	KBKAK	310
0	0	711bz	SPW	RPG-7	60
0	0	711bz	SPW	HS-2S1	8
0	0	711bz	AMO	NB-PG-15W-73mm	1224
0	0	711bz	AMO	NB-7.62x54mm	102000
0	0	711bz	AMO	PK-9M14-125mm	204
0	0	711bz	AMO	NB-OG-15W-73mm	816
0	0	711bz	AMO	NB-7.62x39mm	37200
0	0	711bz	AMO	NB-PG-7M-40mm	300
0	0	711bz	AMO	NB-OF-24-122mm	350
0	0	711bz	AMO	NB-BK-6M-122mm	50
0	0	711bz	MPS	ON	30960
0	0	712bz	SPW	BWP-1	31
0	0	712bz	SPW	KBKAK	500
0	0	712bz	SPW	RPG-7	60
0	0	712bz	SPW	HS-2S1	8
0	0	712bz	AMO	NB-PG-15W-73mm	1224
0	0	712bz	AMO	NB-7.62x54mm	102000
0	0	712bz	AMO	PK-9M14-125mm	204
0	0	712bz	AMO	NB-OG-15W-73mm	816
0	0	712bz	AMO	NB-7.62x39mm	60000
0	0	712bz	AMO	NB-PG-7M-40mm	300
0	0	712bz	AMO	NB-OF-24-122mm	420
0	0	712bz	AMO	NB-BK-6M-122mm	60
0	0	712bz	MPS	ON	32460
780	1	203bz	SPW	BWP-2	31
780	1	203bz	SPW	KBKAK	310
780	1	203bz	AMO	NB-HE-30x165mm	20500
780	1	203bz	AMO	NB-7.62x54mm	82000
780	1	203bz	AMO	PK-9M14-125mm	164
780	1	203bz	AMO	NB-7.62x39mm	37200
780	1	203bz	MPS	ON	20452
780	1	204bcz	SPW	T-80U	31
780	1	204bcz	AMO	NB-3WBM6-125mm	1640
780	1	204bcz	AMO	NB-7.62x54mm	82000

5 From Simulation to Decision Patterns

The transformation of interactive simulation environment into expert system (AI system) there is a challenge. I wouldn't like to present complete transformation of the interactive simulation environment at the article but, however it is interesting to obtain these decision patterns in a conflict situation. The pattern can be defined as an ordered pair of vectors:

$$\xi = <\alpha, \beta>, \tag{7}$$

where:

$\boldsymbol{\alpha}$ - vector of input magnitudes of the conflict situation, which fully identify the current state of the controlled system of A(B)-decision maker,

$\boldsymbol{\beta}$ - vector of possible actions and its evaluations (risk, cost, effectiveness and so on)

One of possible pattern acquisition there is an assortment of tendency model: $X(t) = f(t) + x(t)$ - model of chosen characteristics

- side *A(B)* losses in a moment t, probability of wining in a moment t, e.t.c.
- *f(t)*- trend function (linear, lognormal, logistic,…)
- *x(t)* - random trend variation. The scheme of model trend assortment is as follows:
  - registration of characteristic realisation,
  - estimation of trend model parameters,
  - statistical verification of models,
  - choice of the best model verified

Quite interesting problem of knowledge generalization there is looking for fast response of local combat [9]. The local combat is defined as a clash of two formations, which consists in direct fire of two sides under optical visibility.

Coming to decision patterns our team has proposed an approach to acquisition of the patterns [13, 15]. If We have the set of decision situations patterns in the figure:

$$DSS = \{SD : SD = (SD_r)_{r=1,..,8}\} \tag{8}$$

The vector *SD* represents decision situation which is described by the following eight elements: SD_1 - commanding level of opposite forces, SD_2 - type of task of opposite forces (e.g. attack, defence), SD_3 - commanding level of own forces, SD_4 - type of task of own forces (e.g. attack, defence), SD_5 - net of squares as a model of activities (interest) area $SD_5 = \left[SD^5_{ij}\right]_{\substack{i=1,..,SD_7 \\ j=1,..,SD_8}}$, $SD^5_{ij} = (SD^{5,k}_{ij})_{k=1,..,7}$

$PDSS = \{PS : PS \in DSS\}$, so for current decision situation we have to find the most similar situation from the set of patterns. Using the similarity measure function (4) we can evaluate distances between two different decision situations especially the current and the pattern. There are several methods of finding the most matched pattern situation to the current, which can be used. We propose two main approaches deal with following measures: distance vectors measure, weighted graphs similarity measure.

We determine the subset of decision situation patterns $PDSS_{CS}$ which are generally similar to the current situation considering such elements like: task type, command level of own and opposite units and own units potential:

$$PDSS_{CS} = \{PS = (PS_i)_{i=1,..,6} \in PDSS : PS_i = CS_i,\\ i = 1,..,4, dist_{potwl}(CS, PS) \le \Delta Pot\} \tag{9}$$

where

$$dist_{potwl}(CS, PS) = \max\{|CS_k^6 - PS_k^6|, k = 1,..4\}$$

ΔPot - the maximum difference of own forces potential. Formulating and solving the multi-criteria problem we can find the best pattern for the current situation.

Another approach there is construction and application the combat results generator. The simulation process is time-consuming and the time of running depends on accuracy and the "granularity" of combat model. The proposal consists in building an expert tool that allows us to generate the result of the fight very quickly. The generated combat result for the set initial conditions (as a scenario) should provide information about the time of the fight and the assessment of the condition of the participant in the conflict.

The element of the table is the probability distribution vector (empirical) and/or the regression function of the output characteristics. Input quantities are processes that have multiple states. The individual element of the table can be identified by an index *(i,j,...k)* that describes the number of states of the input quantities. This generator can be used as a tool to quickly obtain local battle results by random selection according to a specific output size distribution; a knowledge base element for decision support and simulation systems; a method to compare different battle models; a method to assess the pay-off a theoretical game model (in armed conflict). This kind of knowledge can also be used in the CAST logic method to assess the success or loss of decision makers [16]. The tool there is procedure of service of multidimensional table $GW = [gw_{ij...k}]_{I \times J \times ... \times K}$ The element of the table $gw_{ij...k}$ there is vector of probability distributions (empirical) and/ or regression function of output magnitudes (Fig. 5).

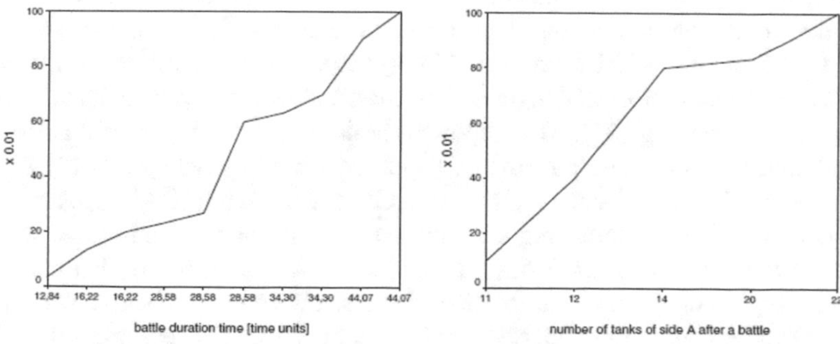

Fig. 5. The probability distribution of battle duration and number of tanks side A [8]

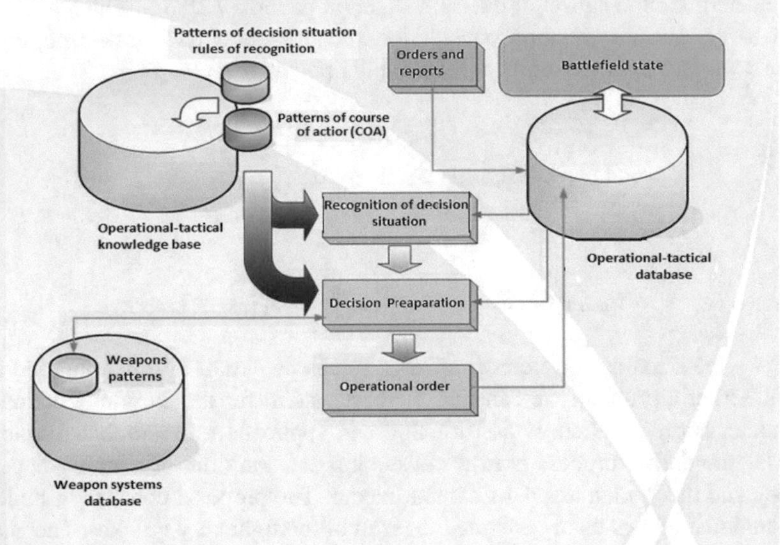

Fig. 6. The idea of knowledge-based DSS (own sources- system "Guru" [10])

6 Conclusions

Modelling and simulation with computer simulation environment presented have many applications within the defence and other than war processes. The principle domain of interactive simulation environment is (see Clarke 1995, Hawkins 2000 as well):

- as a training device for use at the individual, command team, and strategic levels,
- as a tool for the conduct of campaign and mission rehearsal, and the analysis of operations,
- as a device to assist the development of defence policy and doctrine, and the formulation of the equipment programme,
- as a tool for testing the military equipment.

The environment proposed is built as an opened system and can be developed and improved into expert system. The characteristics of battle process are being monitored during the simulation process and their statistical analysis allows combat actions predicting for different conflict situations (Fig. 6) [10]. Some interesting results are presented as the concept of CAST logic and stochastic PERT analysis to support military joint operation planning [21]. The important issue is a quantitative evaluation of the prepared plan (COA). There are papers, which deal with modeling and evaluation of COA based on Timed Influence Nets used in CAusal STrength (CAST) logic [17–19]. They are appropriate for modeling situations such as military situations, in which the estimate of the conditional probability is hard and subjective. So in plan recognition, the CAST logic could break the bottleneck of knowledge acquisition and give the uncertainty modelling a good interface. Falzon in the paper [20] describes a modelling framework based on the causal relationships among the critical capabilities and requirements for an operation. The framework is subsequently used as a basis for the construction, population and analysis of Bayesian networks to support a rigorous and systematic approach to COG analysis. Authors of the paper [22] show that problems of military operation planning can be successfully approached by real-time heuristic search algorithms, operating on a formulation of the problem as a Markov decision process. The system was built as a prototype knowledge-based decision support system and was tested in many computer assisted exercises organised in Polish Armed Forces (2007–2016). Author of the paper took part as an expert in the Strategic Defence Review and used described tools for the analysis of Armed Forces capabilities and also in the process of predicting of Armed Forces structure development.

References

1. Najgebauer, A.: Decision support systems for conflict situations. Models, Methods and the Interactive Simulation Environments (in Polish). Ed. Bulletin WAT. Warsaw 1999, Poland, p. 294 (1999). ISBN 83-908620-6-9
2. Clarke, T.L.: Distributed Interactive Simulation Systems for Simulation and Training in the Aerospace Environment. SPIE Optical Engineering Press, Orlando (1995)
3. Dockery, J., Woodcock, A.E.R.: The Military Landscape, Mathematical Models of Combat. Woodhead Publishing Ltd., Cambridge (1993)
4. Hawkins, G.P.: The UK MoD Requirement for Modelling and Simulation. Simulation for Military Planning, British Crown Copyright 2000. London (2000)
5. Ilachinski, J.: Land Warfare and Complexity, Part I: Mathematical Background and Technical Sourcebook. Centre for Naval Analyses, Alexandria (1996)
6. Najgebauer, A., Pierzchała, D., Rulka, J.: The simulation researches of decision processes in a conflict situation with opposite objectives. In: SCS International Conference ESM 1999 1–4 June 1999, pp. 591–598. Warsaw Poland (1999)
7. Prekopa, A.: Stochastic Programming. Kluwer Academic Publisher, Dordrecht (1995). ISBN 0-7923-3482-5
8. Najgebauer, A., Nowicki, T., Rulka, J.: The method of construction and learning of local combat generator. In: Modelling & Simulation Group Conference on Future Modelling and Simulation Challenges, NATO Modelling & Simulation Group Conference in The Netherlands, Breda, 2001, November (2001)
9. Najgebauer, A., et al.: Technical Project of Simulation System for Supporting Operational Training, volume I–VI, Faculty of Cybernetics, MUT, Warsaw 2002. (Najgebauer A. – principal designer in the phase of research and development and PM in the phase of deployment process)
10. Najgebauer, A. (PM) et al.: 2007. Technical reports on Automated decision support tools - expert system – “GURU”. Military University of Technology, Warsaw 2005–2007
11. Antkiewicz, R., Kulas, W., Najgebauer, A., Pierzchała, D., Rulka, J., Tarapata, Z., Wantoch-Rekowski, R.: The automation of combat decision processes in the simulation based operational training support system. In: Proceedings of the 2007 IEEE Symposium on Computational Intelligence in Security and Defense Applications (CISDA) 2007, ISBN 1-4244-0698-6, Honolulu (Hawaii, USA), 1–5 April 2007
12. Najgebauer, A.: Polish initiatives in M&S and training. simulation based operational training support system (SBOTSS) Zlocien. In: Proceedings of the ITEC 2004, London, UK, 20–22 April 2004

13. Ramirez, C. (red) Advances in Knowledge Representation, INTECH, Rijeka, Croatia 2012. In: Antkiewicz, R., Chmielewski, M., Drozdowski, T., Najgebauer, A., Rulka, J., Tarapata, Z., Wantoch-Rekowski, R., Pierzchała, D. (eds.) Knowledge-Based Approach for Military Mission Planning and Simulation
14. Antkiewicz, R., Gąsecki, A., Najgebauer, A., Pierzchała, D., Tarapata, Z.: Stochastic PERT and CAST logic approach for computer support of complex operation planning. In: redakcja, Al-Begain, K., Fiems, D., Knottenbelt, W. (Eds.) ASMTA 2010, LNCS 6148, pp. 159–173. Springer, Heidelberg (2010)
15. Antkiewicz, R., Najgebauer, A., Rulka, J., Tarapata, Z., Wantoch-Rekowski, R.: Knowledge-Based Pattern Recognition Method and Tool to Support Mission Planning and Simulation. 478–487. https://doi.org/10.1007/978-3-642-23935-9_47. In: Proceedings, Part I Chapter from book Computational Collective Intelligence. Technologies and Applications: Third International Conference, ICCCI 2011, pp. 478–487. Gdynia, Poland, 21–23 September 2011
16. Najgebauer, A., Antkiewicz, R., Pierzchała, D., Rulka, J.: Quantitative methods of strategic planning support: defending the front line in Europe. In: Świątek, J., Borzemski, L., Wilimowska, Z. (eds) Information Systems Architecture and Technology: Proceedings of 38th International Conference on Information Systems Architecture and Technology – ISAT 2017. ISAT 2017. Advances in Intelligent Systems and Computing, vol. 656. Springer, Cham (2018)
17. DeGregorio, E., Janssen, A., Wagenhals, W., Messier, R.: Integrating effects-based and attrition-based modeling. In: 2004 Command and Control Research and Technology Symposium the Power of Information Age Concepts and Technologies. 14–16 September Copenhagen, Denmark (2004)
18. Haider, S., Levis, A.: Effective course-of-action determination to achieve desired effects. IEEE Trans. Syst. Man Cybern. Part A Syst. Hum. **37**(6), 1140–1150 (2007)
19. The Joint Doctrine & Concept Centre Ministry of Defence UK: Joint Operations Planning (2004)
20. Falzon, L.: Using bayesian network analysis to support centre of gravity analysis in military planning. Eur. J. Oper. Res. **170**(2), 629–643 (2006)
21. Rosen, J.A., Smith, W.L.: Influence net modeling with causal strengths: an evolutionary approach. In: Command and Control Research and Technology Symposium. Naval Post Graduate School, Monterey (USA), (1996)
22. Chang, K.C., Lehner, P.E., Levis, A.H., Zaidi, A.K., Zhao, X.: On causal influence logic. Technical report, George Mason University, Center of Excellence for C31 (1994)

Sequential Function Chart to Function Block Diagram Transformation with Explicit State Representation

Maciej Hojda(✉), Grzegorz Filcek, and Grzegorz Popek

Faculty of Computer Science and Management, Wroclaw University of Science and Technology, Ignacego Lukasiewicza 5, 50-371 Wroclaw, Poland
{maciej.hojda,grzegorz.filcek,grzegorz.popek}@pwr.edu.pl

Abstract. This paper presents a method and a tool for converting Sequential Function Charts into Function Block Diagrams in a manner that models states with the use of flip-flops. Order of evaluation is enforced through the use of explicit delays. Presented approach can be used when SFC programming is not directly available but the developer wants to use advantages of SFC modeling.

Keywords: Sequential Function Chart · Function Block Diagram · Programmable Logic Controller

1 Introduction

Programmable Logic Controllers (PLCs) are microcontroller-based devices used predominantly in industrial applications. PLCs serve to execute control algorithms by providing current, either directly or indirectly, to the machines of the floor shop. Distinct advantage of PLCs over hard-wired circuits is their programmability. Control algorithms can be modified as needed and feedback from the PLC to the supervisory workstation permits on-the-fly adjustments. Compared to the other types of digital controllers, PLCs are also lauded as more suited to working in harsh environmental conditions such as high humidity, extreme temperatures and harmful vibrations. Use of PLCs has steadily grown over the past few decades as hardware solutions and programming practices developed side by side. Presently, standard PLC programming languages are formalized in the IEC 61131-3 norm [9]. Of the five languages defined therein, the Sequential Function Chart (SFC) is the one best suited for high-level design and analysis [6,11,18].

SFC retains structural similarity to its originator (through Grafcet), the Petri net. Consequently, a Petri net structure can be transformed, with some effort, into a corresponding SFC [5,15]. The inverse transformation is also possible [7]. The ability to alternate between the two representations makes it possible for SFC programmers to benefit from the methods of analysis and design common

J. Świątek et al. (Eds.): ISAT 2019, AISC 1051, pp. 115–124, 2020.
https://doi.org/10.1007/978-3-030-30604-5_10

to Petri nets. Additional modeling facilities are available through the application of Timed Automata [13], Symbolic Model Verifiers [3] and others [2].

Typically, SFC is used in conjunction with supplementary languages defined in IEC 61131-3 such as Structured Text (ST) or Function Block Diagram (FBD). The more complex the supplementary constructs are, the less descriptive power does the SFC itself have. To efficiently use the design and analysis tools limited to SFCs, it is crucial to keep the charts as pure as possible. To accomplish that, one should use SFC as the primary programming methodology and apply transformations to other languages as necessary. However, direct SFC programming might be unavailable on the given platform. A method for conversion to other languages is then invaluable. This paper introduces a method of transforming SFC into FBD and a software implementing it. The main idea of the method is to represent states defined by SFC with the use of flip-flops and to enforce the order of block evaluation with the use of explicit delays. Specific conversion methods are provided for all the common structures defined within the SFC language.

The paper is divided into five sections. This introduction is followed by a description of ST, FBD and SFC in the second section. Related works and motivation are presented shortly after. The transformation method is described in the third section and the software in the fourth one. The paper is then concluded in the final section.

2 PLC Programming – IEC 61131-3

In 1993, the International Electrotechnical Commission developed a set of guidelines for PLC programming [8,11]. This document, named IEC 61131-3, has been steadily updated (version 3 is now available [9]) and is a de-facto standard of PLC programming that the majority of PLC producing companies adhere to, in some measure at least.

The document defines, amongst others, programming languages available to a PLC programmer. They are: Instruction List (IL), Structured Text, Ladder Diagram (LD), Function Block Diagram and Sequential Function Chart. Of the five, the SFC is the most general, providing tools for organizing large portions of code into smaller, possibly nested parts. The FBD provides an input-output based control logic. Finally, ST provides convenient textual representation of common operations such as variable modification or expression evaluation. Three of the five languages – ST, FBD and SFC – are briefly explained.

2.1 Structured Text

Typical Structured Text program consists of statements such as variable assignments or variable evaluations (Fig. 1). The statements also include common programming structures including function calls, selections, loops and logic or arithmetic expressions.

```
V := A + (B * C) MOD 5;
W := A_FUNCTION(1, 2, 3) + V;
```

Fig. 1. Structured Text example – evaluation of an arithmetic expression is assigned to variable `V` then added to the result of a function and assigned to `W`

2.2 Function Block Diagram

As a legacy of signal processing, Function Block Diagrams are best suited to represent input-output operations where signals, either binary or analog, traverse across the FBD structure (Fig. 2). The elements of an FBD diagram are input-output blocks and connections between them.

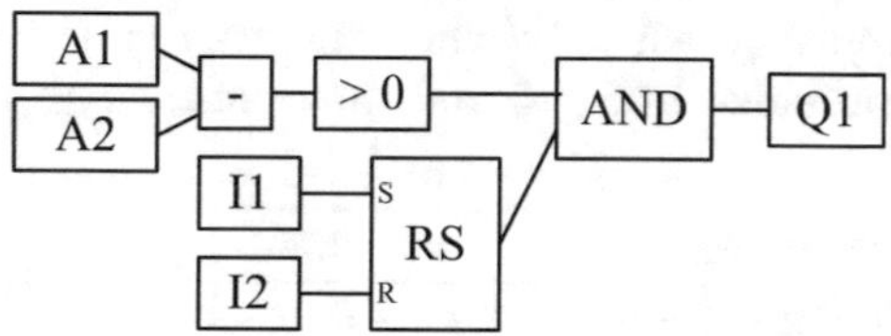

Fig. 2. Function Block Diagram example – binary output `Q1` is set if `A1 > A2` and an RS flip-flop is set (input `I1` sets, `I2` resets)

2.3 Sequential Function Chart

The concept behind Sequential Function Charts is to permit simplifying complex programs through decomposition into smaller parts. SFC contains five main types of elements: states, transitions, actions and parallel branches (Fig. 3). A state is given by its name and connected to some (or all, or none) other states with the use of transitions. A single state is selected as the starting state. Some states are connected to underlying actions. Actions can range from setting outputs, through making calculations, to executing large portions of the program. Transitions connecting states have conditions assigned to them. The conditions have the form of expressions combining logical values of inputs, outputs or structures in another PLC language.

Each action has a qualifier that determines the conditions of its execution. By default, a Non-stored (N) action guarantees continuous execution of the action, as long as the state is active. Other qualifiers affect the execution time or the permanency of the action, ex. Pulse (P) executes the action once, while Delay (D#t10s) delays the activation (here by 10 s). Finally, parallel indicators mark concurrent execution.

Execution of an SFC application can be simulated by placing a single token in the starting state. Tokens represent the states which are active, i.e. actions connected to the states are under execution. When a condition of a transition leaving an active state is satisfied, the state becomes inactive and another state, one indicated by this transition, becomes active instead. Token is then moved to the new active state. Parallel branches multiply tokens and parallel convergences merge them. If one enters a parallel block, all the states following the block are executed concurrently.

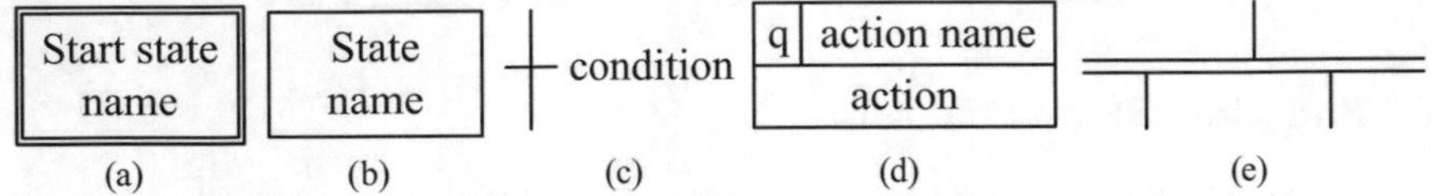

Fig. 3. Sequential Function Chart elements: (a) initial state, (b) state, (c) transition, (d) action (q – qualifier), (e) parallel branch

In Fig. 4 we have a sample SFC with two states and three actions. Signal `Execute.X` is the step flag, set only when the step is active. All the common SFC structures are provided in Fig. 5 and they are the subject of conversion.

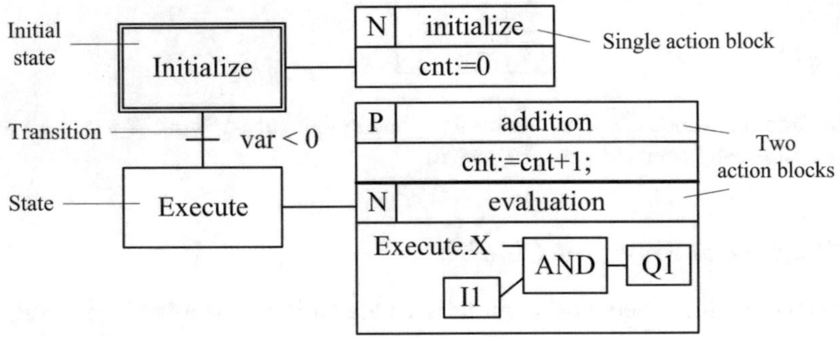

Fig. 4. Sequential Function Chart example

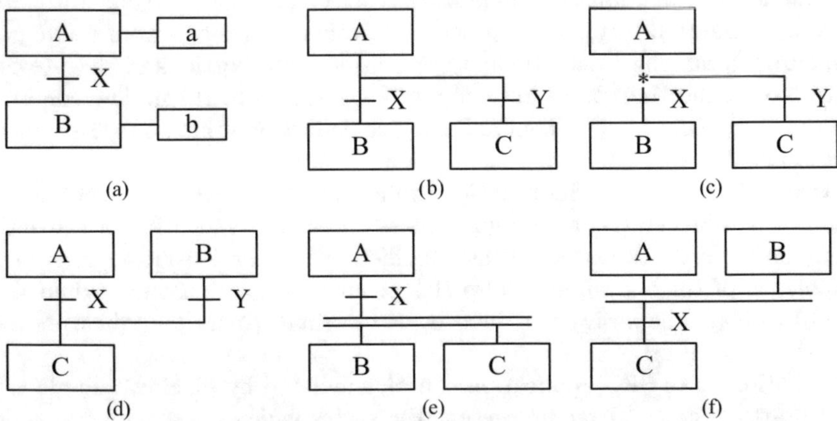

Fig. 5. Sequential Function Chart structures: (a) sequence, (b) divergence with mutual exclusion, (c) divergence with priority, (d) convergence of sequences, (e) simultaneous divergence, (f) simultaneous convergence

A sequence represents a situation in which activation of transition `X` changes the active state from `A` to `B`. While in state `A`, actions `a` are performed (`b` in state `B`). In case of divergence with mutual exclusion, if state `A` is active, then transition `X` enables state `B` while transition `Y` enables state `C`. SFC designer is responsible for ensuring that `X` and `Y` are mutually exclusive. Divergence with priority does not have this requirement and in the case both transitions are active, the leftmost one is executed. Opposite of divergence is the convergence of sequences, where `C` is the resulting state no matter if `A` or `B` was the originally active state.

Last two structures provide means of concurrent execution. For the simultaneous divergence, if state `A` is active, then transition `X` activates both `B` and `C`. In case of convergence, to activate state `C`, both `A` and `C` have to be active, and transition `X` has to be enabled. All of the structures can be extended with new states, i.e. sequence can be longer, divergences can split into more than two states, convergences can join more than two states.

2.4 Related Work and Motivation

For many PLCs, all of the languages described in IEC 61131-3 are available. However, there exists a numerous class of controllers for which that is not the case. Software for the so-called Programmable Logic Relays (PLRs) – a variant of PLCs with a small number of inputs and outputs – rarely permit direct use of SFC diagrams, opting for the more basic choices such as FBD or LD. Examples of such PLRs (also called nano/pico-PLCs) and their programming software include: Siemens LOGO! – LOGO! SoftComfort, Akytec PR200 – Akytec ALP and Eaton easyE4 – EasySoft. Although simpler than many of their PLC counterparts, PLRs see widespread use in domestic automation [10], industrial automation [17] and other control tasks [1].

To facilitate the use of languages that are not directly available, various methods of conversion were developed. Transformation between SFC and LD has been documented in [19], whereas an algorithm for the inverse transformation can be found in [14]. Multiple transformations: SFC to ST, FBD to IL and LD to IL are presented in [4]. Other conversion methods, including those for languages outside of the scope of IEC 61131-3, are also available [12].

A method converting SFC to FBD was presented by Wciślik in [20]. The authors use flip-flops to represent SFC states in FBD diagrams. Change of state is performed by setting and resetting selected flip-flops. The result is a single-network FBD representing the original SFC. While this method can be applied to many SFC diagrams, it does not specify the order of block evaluation of the resulting FBD. This can lead to issues that require additional work to ensure the FBD executes as desired.

The conversion method presented in this paper is closely related to the one presented by Wciślik. To represent states, we also use S-dominant flip-flops. Our approach forces the order of evaluation through the use of delays. Namely, if a state is active and an outgoing transition is enabled, then the change of state will occur after a short, typically a single-cycle, delay.

Without the delay, ambiguity can occur in situations when a block is simultaneously set and reset such as is the case when a convergence of sequences occurs for two different branches of a simultaneous divergence (Fig. 6a). Explicit delays are also necessary in some of the PLC programming software such as LOGO! SoftComfort (single-cycle delaying flags needed for recursion).

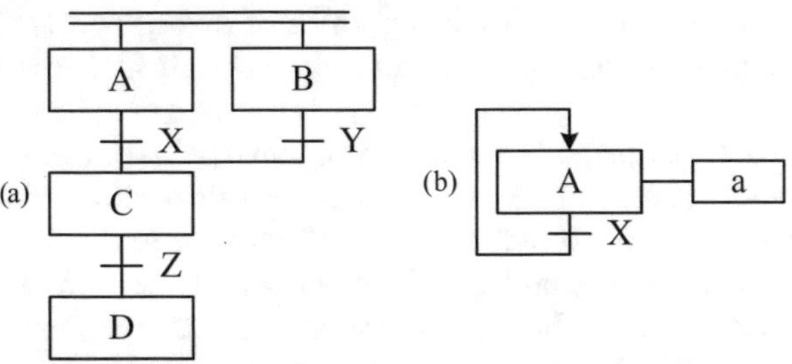

Fig. 6. SFC artifacts: (a) convergence of sequences in a simultaneous divergence, (b) self-loop

IEC 61131-3 (2013) states that "clearing time of a transition [...] cannot be zero". To reflect this ambiguity, in our transformations we also provide two methods of dealing with self-loops (Fig. 6b): with or without enforcing resetting of action controller timers. Finally, our approach removes the need for a synchronization block. This takes out the need to modify the SFC diagram in simultaneous convergence (esp. in simultaneous convergence with a sequence selection) and permits the use of any outgoing transition condition.

3 SFC to FBD Conversion Rules

The general idea of the method of conversion is to represent states as S-dominant flip-flops. When the state becomes active, a set operation is performed on the flip-flop. When the state becomes inactive, a reset operation follows. Transformation of actions is done with the use of a standard action controller (Fig. 7). We assume that all the actions that are not already in the FBD form, can be transformed into FBD. In the upcoming examples we give actions in ST.

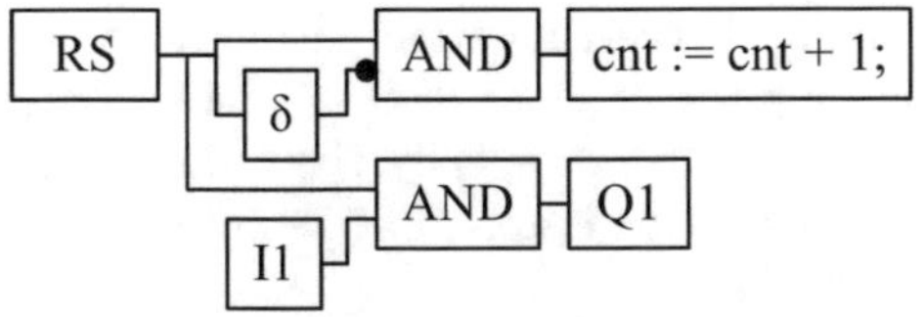

Fig. 7. Action controller and the corresponding flip-flop for the `Execute` state in Fig. 4

Furthermore, we consider a single-network FBD where recurrence order is enforced with the use of explicit delays (such as flags in LOGO!). Those blocks serve as a memory between each evaluation – they delay their output (preferably by one cycle). Transition conditions and actions are given symbolically. Flip-flops (RS) are named after the state they represent. Delay is given by δ. Transformations of structures in Fig. 5 are shown in Fig. 8.

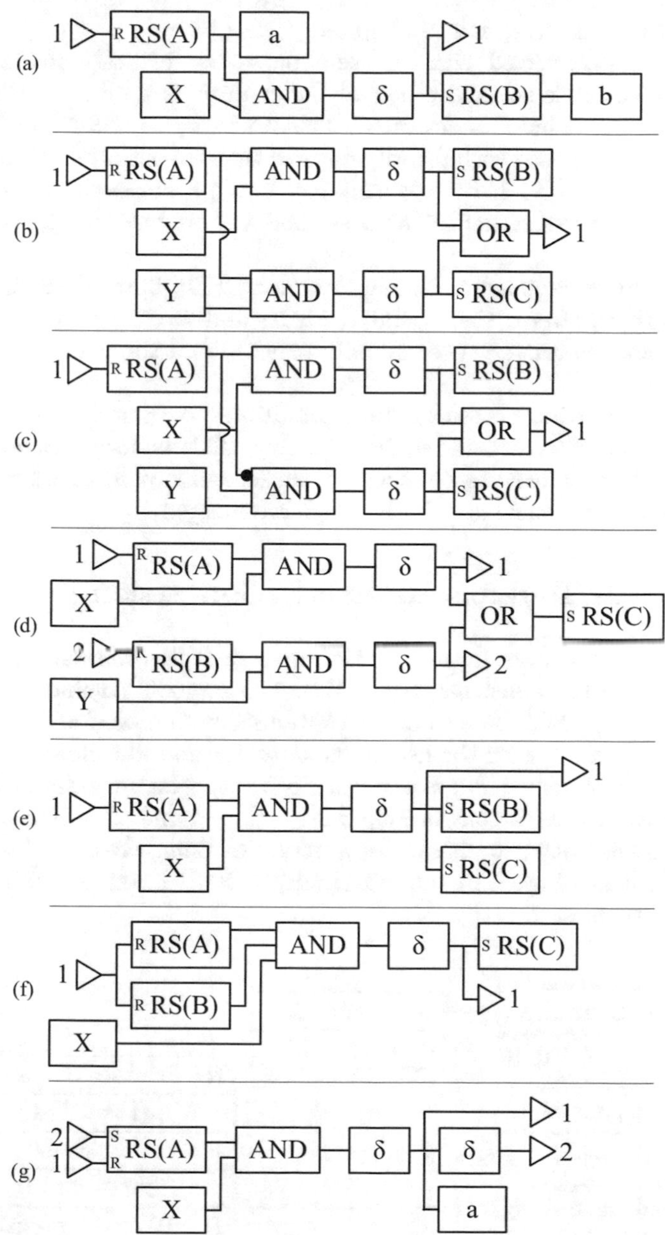

Fig. 8. FBD after conversion from SFC: (a) sequence, (b) divergence with mutual exclusion, (c) divergence with priority, (d) convergence of sequences, (e) simultaneous divergence, (f) simultaneous convergence, (g) self-loop with a timer reset

In the case of sequence, for each SFC state A and B a single flip-flop is created RS(A) and RS(B) respectively. Action controllers a and b are connected directly to flip-flops. Change of state occurs when both, RS(A) and X evaluate to truth, then the delaying block δ is activated. After the delay, RS(A) is reset and RS(B) is set. This concludes the change of state.

Divergence is converted with the use of one delay for every possible destination state. If RS(A) is set and either of the conditions X or Y is satisfied, then the corresponding delay δ is activated. This, in turn, sets RS(B) or RS(C) and resets RS(A). Divergence with priority ensures that transition to RS(B) executes over transition to RS(C) if both transitions, X and Y are enabled. Convergence to state C occurs when either RS(A) is set and X is enabled, or RS(B) is set and Y is enabled.

Simultaneous divergence results in two enabled flip-flops. If RS(A) is set and X is enabled then, after a short delay δ, RS(B) and RS(C) are set and RS(A) is reset. Simultaneous convergence can only happen if all three: RS(A), RS(B) and X are set.

Finally, we provide transformation conditions for a self-loop from Fig. 6b when action controller timers require resetting. This is accomplished by using two delays δ. First is used to reset RS(A) the second is used to set it again. For a self-loop without timer reset, both delays are merged.

4 S2FC – A Tool for Automatic Conversion

Prototype of the software for automatic conversion uses textual representation of the SFC and FBD diagrams (see [9,11]). The input file defines steps, transitions and actions, while the output file contains function blocks. The software is written in Python 3.6 with the use of standard Python libraries.

An example of conversion is given for a 3-step system presented in Fig. 9. The system represents a control loop with a fixed number of iterations. After initialization, the output q1 is set, for a period of time, whenever the user activates the input i1. Figure 10 shows the output FBD obtained with the use of transformation rules.

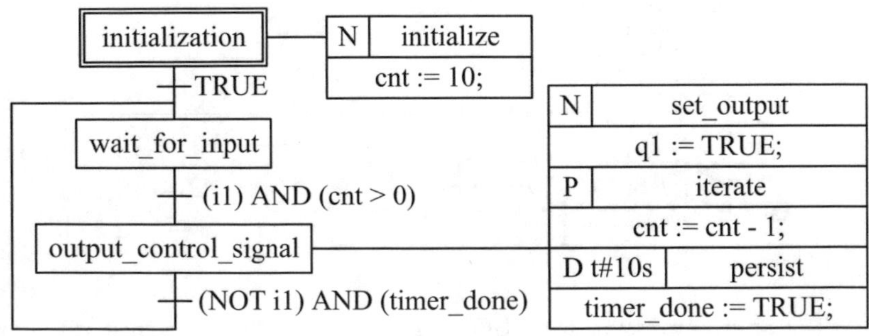

Fig. 9. Control loop example SFC

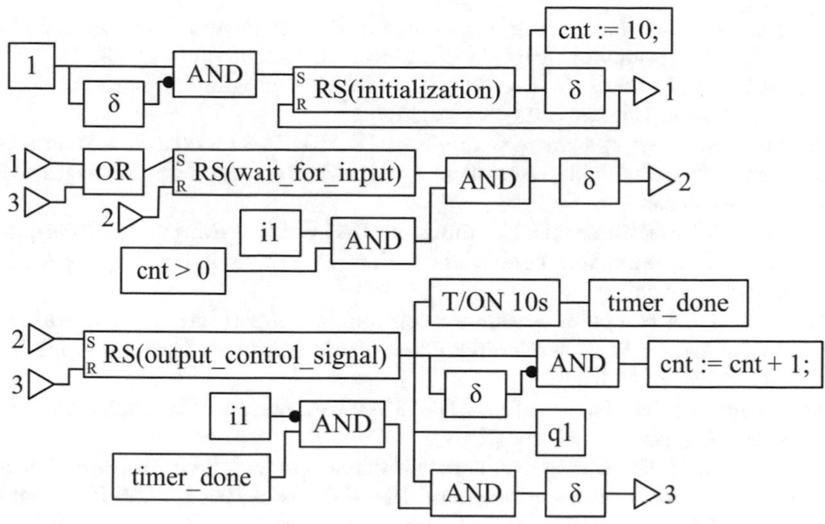

Fig. 10. Control loop example FBD

5 Conclusions

Ability to quickly perform conversions of SFC into FBD allows for more oversight when programming PLCs. Presented transformation methodology allows for easy flow control, and analysis of a program under execution. On-the-fly adjustments can be performed in order to modify the controller operation or to eliminate design mistakes. Presented methodology is accessible to future extensions such as multi-network FBDs. Included example illustrates the ease of use of presented conversion rules. Further works will focus on including other PLC programming languages in the conversion methodology.

References

1. Alnaib, A., Altaee O., Al-jawady N.: PLC controlled multiple stepper motor using various excitation methods. In: Proceedings of ICETA, 2018 International Conference on Engineering Technologies and their Applications, pp. 54–59 (2018)
2. Bauer, N., et al.: Verification of PLC programs given as sequential function charts. In: Ehrig, H., et al. (eds.) Integration of Software Specification Techniques for Applications in Engineering. Lecture Notes in Computer Science, vol. 3147, pp. 517–540. Springer, Heidelberg (2004)
3. Bornot, S., et al.: Verification of sequential function charts using SMV. In: proceedings of PDPTA, 2000 International Conference on Parallel and Distributed Processing Techniques and Applications, pp. 2987–2993 (2000)
4. Darvas, D., Majzik, I., Viñuela, E.: PLC program translation for verification purposes. Periodica Polytech. Electr. Eng. Comput. Sci. **62**(2), 151–165 (2017)
5. Dideban, A., Mohsen, K., Alla, H.: Implementation of petri nets based controller using SFC. Control Eng. Appl. Inf. **13**(4), 82–92 (2011)

6. Fengyun, H., Hao, P., Ruifeng, G.: Design of PLC sequential function chart based on IEC61131-3 standard. Appl. Mech. Mater. **325**, 1130–1134 (2013)
7. Fujino, K., et al.: Design and verification of the SFC program for sequential control. Comput. Chem. Eng. **24**, 303–308 (2000)
8. International Electrotechnical Commision: IEC 61131-3 First edition. International Standard. Programmable controllers – Part 3: Programming languages (1993). http://www.iec.ch
9. International Electrotechnical Commision: IEC 61131-3 Edition 3.0. International Standard. Programmable controllers – Part 3: Programming languages (2013). http://www.iec.ch
10. Jarmuda, T.: A computer system for controlling temperature in a two-state mode and by means of a PI controller in an "intelligent building". Comput. Appl. Electr. Eng. **10**, 372–385 (2012)
11. Karl-Heinz, J., Tiegelkamp, M.: IEC 61131–3: Programming Industrial Automation Systems. Springer, Heidelberg (2001)
12. Kim, H., Kwon, W., Chang, N.: Translation method for ladder diagram with application to a manufacturing process. In: Proceedings of ICRA, 1999 IEEE International Conference on Robotics and Automation, vol. 1, pp. 793–798 (1999)
13. L'Her, P., et al.: Proving Sequential Function Chart Programs Using Automata. LNCS, vol. 1660, pp. 149–163 (1998)
14. Lopes, V., Sousa M.: Algorithm and tool for LD to SFC conversion with state-space method. In: Proceedings of INDIN, 2017 IEEE 15th International Conference on Industrial Informatics, pp. 565–570 (2017)
15. Mello, A., et al.: A transcription tool from Petri net to CLP programming languages. In: ABCM Symposium Series in Mechatronics, vol. 5, pp. 781–790 (2012)
16. Peng, S., Zhou, M.: Design and analysis of sequential function charts using sensor-based stage petri nets. In: Proceedings of SMC, 2003 IEEE International Conference on Systems, Man and Cybernetics, Conference Theme - System Security and Assurance, pp. 4748–4753 (2003)
17. Sarac, V.: Application of PLC programming in cost efficient industrial process. Int. J. Inf. Technol. Secur. **1**, 69–78 (2016)
18. Tsukamoto, T., Takahashi, K.: Modeling of elevator control logic based on mark flow graph and its implementation on programmable logic controller. In: Proceedings of GCCE, 2014 IEEE 3rd Global Conference on Consumer Electronics, pp. 599–600 (2014)
19. Wciślik, M.: Programming of sequential system in ladder diagram language. IFAC Proc. Vol. **36**(1), 37–40 (2003)
20. Wciślik, M., Suchenia, K., Łaskawski, M.: Programming of sequential control systems using functional block diagram language. IFAC-PapersOnLine **48**(4), 330–335 (2015)

Online Environment for Prototyping and Testing Graph and Network Algorithms

Kamil Banach(✉) and Rafał Kasprzyk

Faculty of Cybernetics, Military University of Technology,
2 Kaliskiego Street, 00-908 Warsaw, Poland
kamil.banach@wat.edu.pl

Abstract. In this paper there is presented an online, high-available software environment for prototyping and testing graph & network algorithms. The environment is divided into two components: an algorithms' code editor and a graph & network visual editor. The aim of the tool was to provide an easily accessible, extendable environment for prototyping and testing algorithms with real-time interactive visualization. Developed environment can be also successfully used for educational purposes in graph and network related university courses.

Keywords: Graph and network theory · Algorithms prototyping · Algorithms visualization

1 Introduction

Last years have seen a huge interest in network systems. The number of interdisciplinary researches undertaken in this field is affected by the strategic importance of network systems [3, 4, 6, 7]. The need to collect information about genuine network structures revealed in the late twentieth century that these networks have a number of specific features, which have not been known so far. It turned out that although we have been surrounded by exhaustively examined networks, their topology and principles of evolution may still be enigmatic. Analyses carried out on the actual networks proved the existence of their specific characteristics [5]. In particular, these included a relatively small number of edges (sparse graph), a relatively short diameter of the graph (shortest longest path) and a surprisingly short average path length, while high clustering coefficient value is given. Another extremely interesting feature of most genuine networks, is a power law distribution of node's degree. The above-mentioned features have contributed to creation of a wide range of models which describe genuine networks. A solution approach presented in the research will allow to explore by prototyping and testing any kind of graph and network algorithms (in particular models which describe genuine network formation) using a unique software environment. The environment enables interactive visualization of algorithms implemented therein, which in turn allows quick verification of the algorithms results as well as its correctness. The developed tool can also be successfully used for educational purposes.

J. Świątek et al. (Eds.): ISAT 2019, AISC 1051, pp. 125–134, 2020.
https://doi.org/10.1007/978-3-030-30604-5_11

2 Definition and Notation

Graph is an abstract representation of the structure of any system. Formally, graph can be defined as follows [1, 8]:

$$G = \langle V, B, I \rangle \tag{1}$$

where:

V - a set of graph vertices

B - a set of graph branches;

I - an incident relation ($I \subset VxBxV$), which meets two conditions:

1. $\forall b \in B \quad \exists x, y \in V \quad \langle x, b, y \rangle \in I$
2. $\forall b \in B \; \forall x, y, v, z \in V$
   $\langle x, b, y \rangle \in I \wedge \langle v, b, z \rangle \in I \Rightarrow (x = v \wedge y = z) \vee (x = z \wedge y = v)$

Based on the incident relation I, three types of branches can be specified:

$\tilde{B}$- a set of edges, which meets a condition

$$\langle x, b, y \rangle \in I \wedge \langle y, b, x \rangle \in I \wedge x \neq y$$

$\vec{B}$- a set of arcs, which meets a condition

$$\langle x, b, y \rangle \in I \wedge \langle y, b, x \rangle \notin I \wedge x \neq y$$

$\dot{B}$- a set of loops, which meets a condition

$$\langle x, b, x \rangle \in I$$

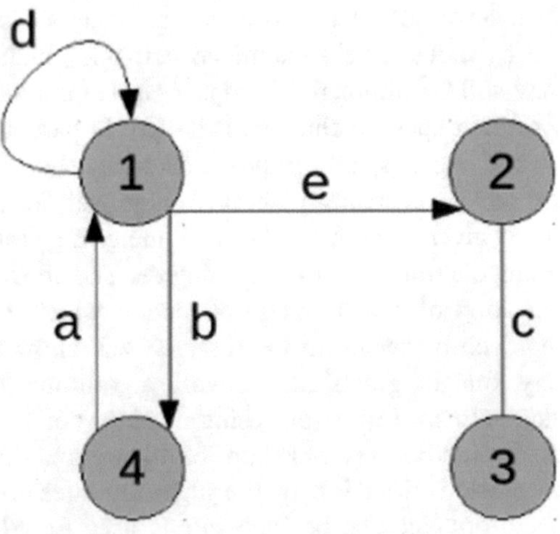

Fig. 1. Graph example: vertices – 1, 2, 3, 4; edge – c; arcs – a, b, e; loop – d. (source [1])

In the Fig. 1 there is an example graph - numbers and letters on vertices and branches are so called labels. So, presented graph contains four vertices (labeled *1*, *2*, *3* and *4*), one edge (labeled *c*), three arcs (labeled *a*, *b*, *e*) and one loop (labeled *d*). Labels are introduced only for identification purpose. It means that labels are not the description of the systems modeled using graphs. To describe elements of graph a concept called network is introduced.

Let's now define the network as follows:

$$N = \left\langle G, \{f_i(v)\}_{\substack{i \in \{1,\ldots,NF\} \\ v \in V}}, \{h_j(b)\}_{\substack{j \in \{1,\ldots,NH\} \\ b \in B}} \right\rangle \quad (2)$$

where:

G - is the graph defined by (1);

$f_i : V \rightarrow ValV_i$ - the i-th function on the graph vertices, $i = 1,..,NF$, (NF - number of vertex functions), $ValV_i$ - is a set of f_i values;

$h_j : B \rightarrow ValB_j$ - the j-th function on the graph branches, $j = 1,..,NH$, (NH - number of branch functions), $ValB_j$ - is a set of h_j values.

Results of functions on vertices and branches are commonly called weights. Because of that in the literature we can find that networks are commonly referred as weighted graphs.

In the Fig. 2 we can see an example of a social network – there are four different people, modeled as vertices (that are described with their names as weights) and relationships between them, modeled as the branches. There are different types of weights – text and numeric.

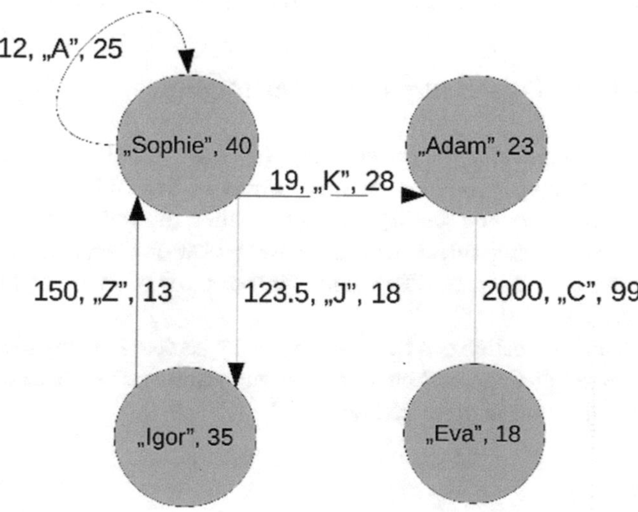

Fig. 2. Network example: two functions described on vertices and three functions described on branches (source: [2])

Moreover, there are a lot of other concepts that are used in graph and network theory. Worth mentioning is a directed graph, called also digraph, which is a graph containing only arcs and loops (set of edges is empty). An undirected graph is a graph that contains only edges and loops (set of arcs is empty). Additionally, we can divide graphs by the number of branches with the given type. A graph that can have multiple branches of the same type between the same pair of vertices, we will call multigraph. As opposite, a graph which allows exactly one branch of the specific type between the same pair of vertices, we will call unigraph.

We can notice different names for vertices and branches in literature and software. Common name for vertices is nodes, edges are called lines and arcs are called directed lines or directed edges.

3 Idea of the Environment

During researches, we used many different software solutions. Every one of them demanded from us a lot of workloads – learn about architecture, what libraries are used and so on. There were a lot of actions to be done to simply write and test algorithms.

For the reasons given above, we wanted to create a simple environment that will fulfil our needs – allow us to quickly type algorithm code and see results. In addition, we asked ourselves – when almost every user is using a web browser why almost all current solutions require downloading and installation of some additional software? Because of that, we decided to build the environment as a web application that is fully featured in any web browser.

The created environment allows a user to type algorithm in *JavaScript*, run it and see results instantly. We choose *JavaScript*, as a language used to define algorithms because it is one of the most common programming languages. Also, its syntax is like commonly used languages like *C#* or *Java* (sometimes called *C-like* languages).

4 Architecture, Deployment and Technologies

The environment, as stated in the previous chapter, was created as the web application. That means the main part of it is devoted to run in the end-user browser. Unlike previous iteration, described in [2], current solution doesn't contain any stateful backend tier (used to communicate with database) – now environment can be deployed to any web server (like *Tomcat*, *JBoss,* and more) or *CDN* (*Content Delivery Network*) as a set of static files.

That means an application can be served by many servers in many places. In case of *CDN* code of the application will be served by the nearest server to reduce load times. This type of architecture is presented on Fig. 3.

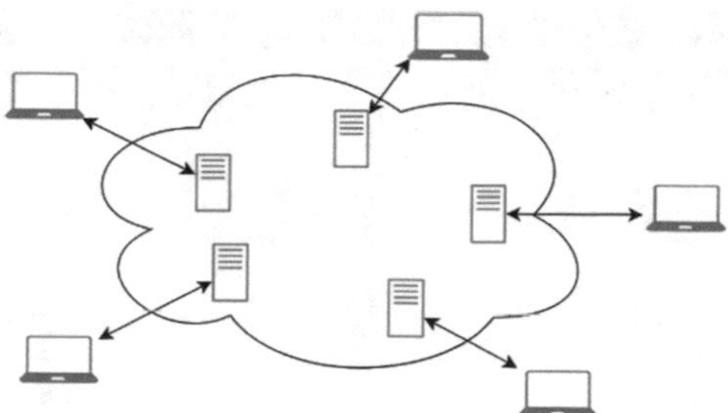

Fig. 3. Architecture of Content Delivery Network – many clients asks *"cloud"* for files and those files are distributed by nearest server.

Environment was developed by using modern technologies and *Open Source* libraries:

- *ParcelJS* [9] as an application bundler which combines all files of one type (JS, CSS and so on) into one bundle that can be easily loaded by the browser,
- *JS-Interpreter* [10] as a foundation of algorithms interpreter,
- *CodeMirror* [11] as a text editor with syntax coloring,
- *D3.js* [12] as a graph and network visualization library,
- *Bulma* [13] as a set of *Cascading Style Sheets (CSS)* for a nice look and feel.

Packages and programs mentioned above are stable and commonly used tools by developers around the world.

5 GUI of Software Environment

The software environment for prototyping graph and network algorithms contains two main components – an algorithm's code editor and a visual graph editor.

The former one allows the user to create and modify graph/network. It allows setting various, both text and number, properties of created vertices and branches.

The latter one allows typing algorithm code, checking its correctness and executing it in an isolated environment (*sandbox*). Underneath there is an interpreter which parses the code into *Abstract Syntax Tree*. By default, it allows executing it by walking over the tree – this is quite unintuitive for end-users. Because of that, we created a more user-friendly solution that allows tracking algorithm line-by-line. Moreover, the user can "run" algorithm – line-by-line until paused (by clicking pause button or when `pause();` instruction is invoked) or finished.

In the Fig. 4 we can see a main screen of the software environment.

Both components are connected by a shared graph object. The effects of the execution of the algorithm is presented in the real-time on graph/network that is drawn in the visual editor.

The user can edit every object in graph visualization – e.g. change the color of node/edge or modify other properties of graph elements. A node edit screen is presented in Fig. 5.

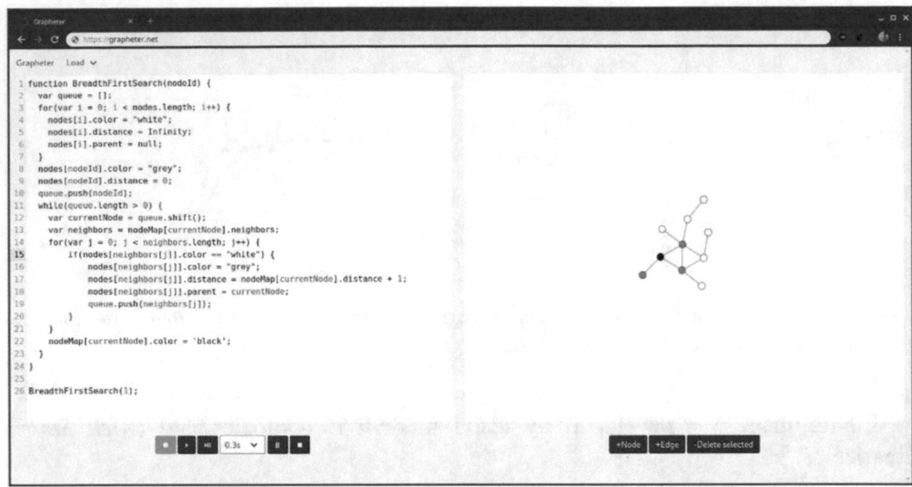

Fig. 4. The main screen of the environment. On the left we can see algorithm code (*Breadth First Search*), on the right – graph.

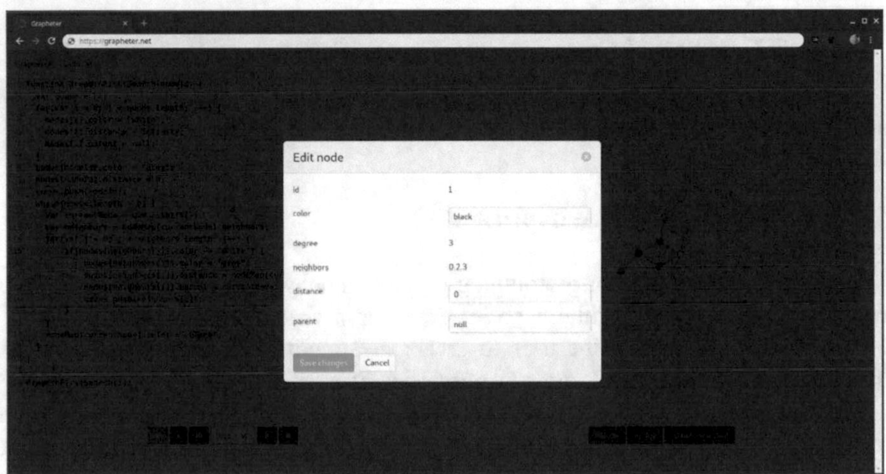

Fig. 5. Edit node screen

The application provides a predefined set of algorithms and graphs that can be loaded and used by the user – list contains e.g. *Breadth First Search*, *Dijkstra's* algorithms and example graphs like *Les Misérables* graph.

6 Case Study

The described environment can be used for both research and educational cases.

6.1 Educational Example

At the Military University of Technology, there is an introductory course to Graph and Network Theory. During the semester students learn about graphs, networks, and algorithms that use those data structures.

A basic algorithm, which is the cornerstone for other ones, is *Breadth First Search (BFS)*. A source code of it, written in JavaScript, we can see on Fig. 6. Notice additional `pause();` instruction – reaching that instruction will force the interpreter to stop executing the code. Execution of code can be continued by clicking run or step button. As a result, the end-user can quickly execute code to the selected line and proceed with the line-by-line approach if necessary.

```
function BreadthFirstSearch(nodeId) {
  var queue = [];
  for(var i = 0; i < nodes.length; i++) {
    nodes[i].color = "white";
    nodes[i].distance = Infinity;
    nodes[i].parent = null;
  }
  pause();
  nodes[nodeId].color = "grey";
  nodes[nodeId].distance = 0;
  queue.push(nodeId);
  while(queue.length > 0) {
    var currentNode = queue.shift();
    var neighbors = nodeMap[currentNode].neighbors;
    for(var j = 0; j < neighbors.length; j++) {
        if(nodes[neighbors[j]].color == "white") {
            pause();
            nodes[neighbors[j]].color = "grey";
            nodes[neighbors[j]].distance = nodeMap[currentNode].distance + 1;
            nodes[neighbors[j]].parent = currentNode;
            queue.push(neighbors[j]);
        }
    }
    pause();
    nodeMap[currentNode].color = 'black';
  }
}

BreadthFirstSearch(1);
```

Fig. 6. Screenshot from environment code editor containing *Breadth First Search* algorithm that can be run step-by-step (because of *pause();*) instruction.

The main value of the environment for students is that it provides an easy way to learn how algorithms work step-by-step. Also, they can check correctness of their own implementations – they can run both built-in and own algorithm and compare results.

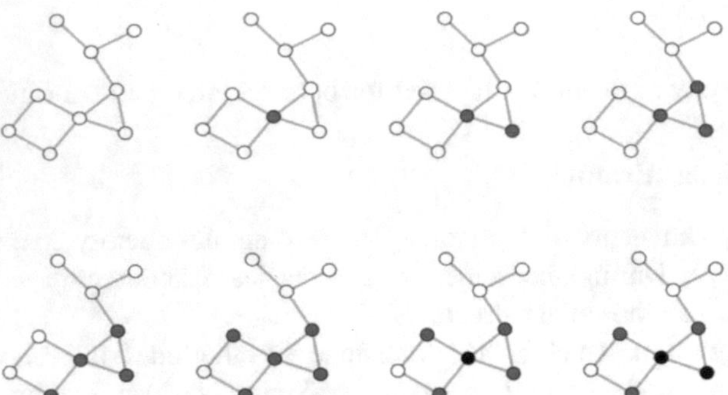

Fig. 7. First eight steps of *BFS* algorithm – screenshot from the environment.

Example of *BFS* step-by-step run results (first eight steps) is presented in the Fig. 7.

6.2 Research Example

As the research case study, we will show how to implement a simple network evolution model using a copying mechanism. Our algorithm will start with full graph generation with *m*0 parameter – the number of vertices. Next, we will add new vertices, one at a time, and apply the copying mechanism to create edges. The algorithm will end when there are *N* vertices in the graph.

The copying mechanism will work as follows for every added vertex *i*:

1. Randomly select existing vertex *j*, excluding newly added *i*,
2. For every arc from *j* to some vertex *k*:
   a. with a probability *p* create the arc from *i* to some vertex *k*,
   b. with the probability $1 - p$ create the arc from *i* to randomly selected vertex, if the arc already exists – do nothing.

In the Fig. 8 we can see bootstrap source code – function `createFullGraph` is omitted.

```
function copying(m0, N, p) {
  createFullGraph(m0);
  for(var i = m0; i < N; i++) {
    // 1. add node
    // 2. select node to copy
    // 3. create edges
  }
}
```

Fig. 8. Starting source code for simple copying network evolution model.

To add a new vertex, we will use a built-in function `addNode`, which returns the id of the created vertex. Then we will choose vertex uniformly at random from previously existing (excluding newly created one) – as we can assume that there are n vertices with id from 0 to $n - 1$ we will select random integer from that set. Next, we will implement the mechanism mentioned above.

The final source code of our model is presented in Fig. 9 – the copying mechanism is implemented in lines 6–15.

```
1 function copying(m0, N, p) {
2   createFullGraph(m0);
3   for(var i = m0; i < N; i++) {
4     var nodeId = addNode();
5     var selectedNode = getRandomInt(nodeId - 1);
6     var neighbors = nodes[selectedNode].neighbors;
7     for(var i = 0; i < neighbors.length; i++) {
8       var random = Math.random();
9       if (random < p) {
10        addLink(nodeId, getRandomInt(nodeId - 1));
11      } else {
12        addLink(nodeId, neighbors[i]);
13      }
14    }
15  }
16 }
```

Fig. 9. Source code of the network evolution model with copying mechanism.

7 Summary

The presented environment is a modern solution for easy and quick prototyping of graph and network algorithms. It allows real-time presentation of the results of algorithms run in either line-by-line or continuous mode. Moreover, it is easily accessible and can be used in any modern web browser. A core feature of the presented application is the ability to run it without any downloading and installing additional software (like *Java Runtime Environment* or *.NET Framework*) and it is an operating system independent.

Presented software can be successfully used for prototyping and testing algorithms or simply for educational purposes at university courses like graph and network theory or parallel and distributed computations. We decided to come forth with our online environment because we have not discovered any solutions which satisfy our requirements and we strongly believe that researchers and lectures are gasping for this kind of software. It is worth to mention that our software environment smoothly process and visualise not only tens-vertex graphs or even hundreds-vertex graphs for demonstration purposes but also thousands-vertex graphs for use in real-life or applied cases. Obviously, the environment is able to visualise the different steps of algorithms and their results for a case of a large graph, still algorithms with exponential time complexity can be prototype and test only for small size of a task.

References

1. Korzan, B.: Elementy teorii grafów i sieci: metody i zastosowania. WNT (1978)
2. Banach, K., Kasprzyk, R.: Software environment for rapid prototyping of graph and network algorithms. Comput. Sci. Math. Model. **5**, 11–16 (2017)
3. Tarapata, Z., Kasprzyk, R.: An application of multicriteria weighted graph similarity method to social networks analyzing. In: Proceedings of the 2009 International Conference on Advances in Social Network Analysis and Mining, Athens (Greece), pp. 366–368. IEEE Computer Society (2009)
4. Tarapata, Z., Kasprzyk, R.: Graph-based optimization method for information diffusion and attack durability in networks. In: Szczuka, M., Kryszkiewicz, M., Ramanna, S., Jensen, R., Hu, Q. (eds.) RSCTC'2010. Lecture Notes in Artificial Intelligence, vol. 6086, pp. 698–709. Springer, Heidelberg (2010)
5. Bartosiak, C., Kasprzyk, R., Tarapata, Z.: Application of graphs and networks similarity measures for analyzing complex networks. Biuletyn Instytutu Systemów Informatycznych **7**, 1–7 (2011)
6. Kasprzyk, R.: Diffusion in networks. J. Telecommun. Inf. Technol **2**, 99–106 (2012). ISSN 1509-4553
7. Tarapata, Z., Kasprzyk, R., Banach, K.: Graph-network models and methods used to detect financial crimes with IAFEC graphs IT Tool. In: MATEC Web Conference, vol. 210 (2018)
8. Diestel, R.: Graph Theory. Springer-Verlag, Heidelberg (2010)
9. ParcelJS project. https://parceljs.org. Accessed 19 May 2019
10. JS-Interpreter project source code: https://github.com/NeilFraser/JS-Interpreter. Accessed 19 May 2019
11. CodeMirror project. https://codemirror.net. Accessed 19 May 2019
12. D3.js project: https://d3js.org/. Accessed 19 May 2019
13. Bulma project. https://bulma.io/. Accessed 19 May 2019

Evolutionary Strategies of Intelligent Agent Training

Assel Akzhalova[1(✉)], Atsushi Inoue[2(✉)], and Dmitry Mukharsky[3(✉)]

[1] Kazakh-British Technical University, Almaty, Kazakhstan
assel.akzhalova@gmail.com
[2] Eastern Washington University, Cheney, USA
inoueatsushij@gmail.com
[3] Kazakh National University named after al-Farabi, Almaty, Kazakhstan
amiddd@rambler.ru

Abstract. Groups of interacting agents are able to solve complex tasks in a dynamic environment. Robots in a group can have a simpler device than single stand-alone robots. Each agent in the group has the ability to accumulate interaction experience with the environment and share it with other members of the group. In many cases, the group behavior is not deduced from any properties of its parts. The paper proposes an approach to modeling the mobile agent group behavior that is busy with a common goal. The main purpose of the agents is to study the greatest territory at minimal time. The agents interact with the environment. A control of each agent is carried out by a modified neural network with restrictions imposed on it. Weights of the neural network are chosen by a genetic evolution method. The agents compete among themselves for obtaining the greatest reward from the environment. An efficiency of the proposed model is confirmed by some convergence speed tests by computer simulation. The proposed model can be applied to a group of robots that perform search tasks in a real physical space.

Keywords: Intelligent agents · Reinforcement learning · Genetic algorithms · Neural network · Hybrid algorithm

1 Introduction

The main factor of efficiency during rescue operations in places of ecological and man-made catastrophes is time. The less time it takes to find then more efficient a rescue operation will be. Possibilities for people work are limited or impossible in catastrophes that are associated with radiation leakage, toxic substances and so on. Use of intelligent robotic-agents groups for these purposes is an effective solution to a problem of rapid survey of entire disaster area [1, 2]. Each intelligent robot-agent in the group is focused to performing the search task. A large number of identical robot-agents are able to cover the entire area of the catastrophe in the shortest possible time.

Managing a group of robot-agents that are united by a common goal is a difficult task [3]. The robot-agents can be controlled from outside. External control assumes presence of a qualitative, wide and stable communication channel between an operator

J. Świątek et al. (Eds.): ISAT 2019, AISC 1051, pp. 135–145, 2020.
https://doi.org/10.1007/978-3-030-30604-5_12

and the robot-agent. Remote control of a large group of the robot-agents without possibility of their autonomous movement is confronted with purely technical problems and problems of qualified operators' lack.

There is a need for a clear and coordinated interaction of the robot-agents for the most rapid investigation of the disaster zone [4, 5]. The robot-agents cannot build a trajectory in advance. They should plan the trajectory in real time. In a monograph [6] study on a grant № 14-19-01533 considers planning of flat trajectories. In the paper quality estimation criteria of offered methods are formulated.

The problem of planning the optimal route is discussed in many sources. The most of the proposed approaches do not show significant results. The subtask of finding optimal trajectories for the task of minimizing the search time requires paying attention to a individual robot-agent control system. Rigid programs for bypassing obstacles and targeting radiation sources are not able to effectively control the robot-agents.

We have taken as a basis for the intellectual control model that we developed in the article a biological nervous system organization [7].

2 A Problem Statement

2.1 Physical Statement of the Problem

There is a distorted surface here. We will use a term physical space or an environment to the distorted surface at our article. There may be obstacles that prevent free movement in the physical space. Localization of the obstacles is not known in advance. Sources of radiation or radiation pollution are located in arbitrary points of the physical space. Localization of the radiation sources, their number and physical parameters are not known in advance. There is a task to detect in the shortest possible time all or as many as possible the sources of radiation.

A group of robotic agents is looking for the radiation sources or the radiation pollution into the physical space. Each of the robot-agent from the group is placed in an arbitrary point of the physical space. The search can begin with a limited area of physical space which we will call the base. The robot-agents have a decision-making system, a sensor system and effectors system.

The environment is partly observable and stochastic for each of the robot-agent. The given model is based by real tasks of search of the radiation sources by robots in places that are unfit for people's work [8].

2.2 Formal Problem Statement

We have a two-dimensional physical space E. There are many physical sources of radiation $T = \{T_r, r \in [1 : T_{max}]\}$, T_{max} is a number of the radiation sources in the space E. Each source T_r has many physical properties. We highlight only an intensity of radiation I_r which is essential for the source characteristic in our model. A physical size of the source is not a significant property. The sources are represented as points. Thus, each source has two characteristics the intensity I_r and a localization point X_r^T. The intensity of the source falls back proportionally to square of distance from it.

The intensity of radiation in the arbitrary point X of the physical space is superposition intensity $I(X) = \sum I_r(X_r^T)$ for all T_{max} sources.

The obstacles $\breve{O}$ occupy arbitrary areas of the space. Points of the physical space $X^{\breve{O}}$ that are occupied by the obstacles, are not available to accommodate other objects in them. We accept that the areas $\breve{O}$ do not distort the intensity field $I(X)$ of the sources in our model.

The mobile robot-agents $A = (A_1, A_2, \ldots, A_N)$, N is the number of the robot-agents, are place in the physical space at the initial moment of time. We will use a term "agent" to designate a model that combines only properties that are essential for the formal description of a real robot-agent that functions in the real physical space [9].

Mobility of the agents is a freedom to move in the two dimensions physical space E in any direction at a rate that does not exceed the maximum velocity V_{max}. The each agent has three degrees of freedom (two coordinates in the two dimensions physical space and an angle of orientation of the velocity vector). All agents have a same architecture.

A state of the entire system can be described with help four elements $\Sigma(t_0) = \left\langle E, T, \breve{O}, A(t_0) \right\rangle$ at the initial point of time t_0. An environmental impacts and internal system parameters (agent state) are independent (exogenous) variables. An output characteristics of the system are dependent (endogenous) variables. The process of system functioning in time Σ can be described by an operator F_Σ which converts the exogenous variables into the endogenous variables and translates the system into a new state.

$$\Sigma(t) = F_\Sigma(\Sigma(t-1)) \tag{1}$$

A sequence $\Sigma(t_0) \rightarrow \Sigma(t_1) \rightarrow \Sigma(t_2) \rightarrow \cdots$ forms a system evolution trajectory. An ultimate goal of the evolution is to achieve any of many absorbing states $\Sigma_p = \{\Sigma_p^{'}, \Sigma_p^{''}, \ldots\}$. The absorbing state is achievement by all agents of such points of the physical space E which are concentrated in a small vicinity of ε at least one of the many sources T.

$$\forall l, \exists r : \left\| X_l^A - X_r^T \right\|_E \leq \varepsilon \tag{2}$$

where, X_l^A are the agents coordinates, X_r^T are the radiation sources coordinates, ε is a number that has a dimension length and an order of an agent radius R^A, $\left\| X_l^A - X_r^T \right\|_E$ is the Euclidean's distance between vectors. Thus the each agent moves on an own events trajectory and aspires to achievement the own absorbing condition.

We are interested in finding an optimal the system evolution trajectory which minimizes a time of reaching the absorbing state $t_p \rightarrow min$.

3 Building an Intelligent Agent Architecture

Each agent from the set $A = \{A_l, l \in [1 : N]\}$ has an internal structure. The each agent state l at the current time t is described by a vector:

$$\boldsymbol{A}_l(t) = (a_1(t), a_2(t), \ldots, a_m(t)), \tag{3}$$

where, $a_i(t)$ is an agent parameter value at the time t, m is an agent internal parameters number. Parameters can include various characteristics of the agent such as: coordinates in the space, velocity and acceleration vectors, angles of orientation in the space, energy reserves, CPU power of the control, a memory capacity, etc. The agents parameters can be divided into static parameters for example a radius of the agent R^A, and dynamic parameters for example the position in the physical space. The agents operate in an environment with a discrete time $= 0, 1, 2, \ldots$.

Any agent l can be in one of many possible dynamic states $S_l(t) = \{S_p, p \in [1 : S_{lmax}]\}$, S_{lmax} is a possible states number that are composing of sensor and sensor states $In = (f, \ldots, r_k, \ldots, b, \ldots, l_k, \ldots, D_1, \ldots, D_n, \ldots, D_{2n-1})$, the agent coordinates in the physical space X_l^A and the agent velocity vectors V_l^A. The agent state term will indicate the agent dynamic state in a next article part.

We will characterize state of the group of agents using a vector-function:

$$\mathfrak{H}(t) = f_R(\boldsymbol{S}_1(t), \boldsymbol{S}_2(t), \ldots, \boldsymbol{S}_N(t)). \tag{4}$$

3.1 A Sensor System Architecture

The agent gets knowledge about the environment using a sensor system. The scheme of the location of the agent sensors is shown (see Fig. 1a).

The purpose of sensors f, r_k, b, l_k is to measure the intensity of radiation at the front, on the right, on the far side and on the left of the agent. These intensity sensors measure the intensity $I(X)$ directly at a point where they are located. Therefore, the maximum distance between the intensity sensors cannot exceed the agent radius. The set of intensity sensors forms at the agent images of intensity distribution of radiation of sources in the limited area of the physical space E and allows developing specific reactions to these images.

Here is the reaction of the intensity sensors on the adjacent radiation source (see Fig. 1b).

The color density encodes the corresponding intensity sensor activity. The closest to the radiation source intensity sensors have the greatest signal. The agent must interpret the image that is obtained on the intensity sensors aggregate as a recommendation to change its state at a next step.

The intensity sensors location that we reviewed above is used in azimuth instrument AMI. The AMI contains 12 counters (SI29BG) which provide a circular overview. The intensity chart which gives a recorded radiation distribution that measures the device by its maximum determines a direction to the radiation source.

Sensors $D_1 - D_{2n-1}$ estimate a distance from the agent to the obstacle. Their main task is to create images of objects of different nature that surround agent. If there is no any obstacle within the sensor range then the corresponding sensor returns a zero value. The distance sensors have a limited range of action. The resolution of the sensors decreases over long distances. The sensors can give incorrect information at the limit of range (see Fig. 1c).

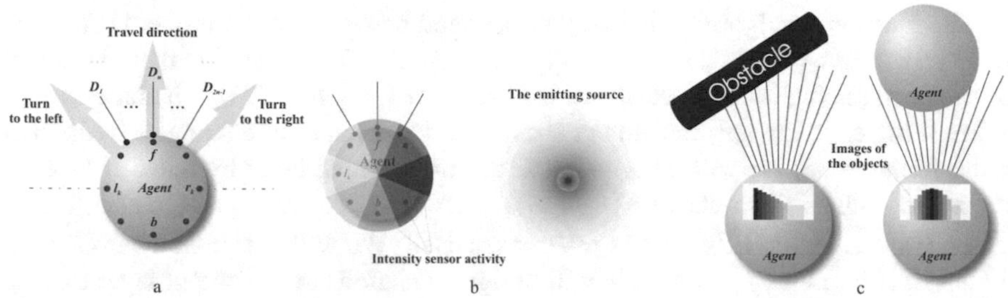

Fig. 1. (a) The sensors location scheme on the agent body. Following designations are used: D are sensors of distance to obstacles, f, r, b, l are intensity sensors; (b) Intensity sensor activity as reaction to radiation source; (c) Formation of images by combination of the distance sensors D

We will use a term base set for minimum set of sensors on the agent body. The base set consists of four intensity sensors which are localized in the corners of the square and one the distance sensor which is directed along the agent velocity vector. The basic set of sensors is the minimum set that is required for the agent to still be able to perform the primary task. The basic set idea is important for an issue of system fault tolerance.

3.2 An Intellectual Subsystem Architecture

Here there is a task to select a trajectory which minimizes time for which will be reached absorbing state for given initial condition among many possible system trajectories $\Sigma(t_0) \rightarrow \Sigma(t_1) \rightarrow \Sigma(t_2) \rightarrow \cdots$.

The agent state $\boldsymbol{S}_p(t)$ is composed of all sensors states $\boldsymbol{In}_l = (f, \ldots, r_k, \ldots, b, \ldots, l_k, \ldots, D_1, \ldots, D_n, \ldots, D_{2n-1})$, from agent coordinates in the space $\boldsymbol{X}_l^A$ and from its speed $\boldsymbol{V}_l^A$. A path function into a phase space of generalized coordinates $\boldsymbol{X}_l^A$ and $\boldsymbol{V}_l^A = d\boldsymbol{X}_l^A/dt = \boldsymbol{X}_l^A$ can describe the agent movement in the physical space. We will designate as symbol $\boldsymbol{Q}_l$ a set of coordinates and velocities of the agent l, which form the generalized coordinates. We set the task to build some function of displaying an exogenous variables space in an endogenous variables space.

$$\pi_l : \boldsymbol{In}_l \rightarrow \boldsymbol{Q}_l, \tag{5}$$

that if applied it to the each agent in the group will minimize the time to reach the condition (2). We will call this function π_l as a control function or a target function.

In our model we have broken the agent intelligent subsystem into two closely related parts $\pi_l = \pi_l^{'}, \pi_l^{''}$, each of which implements own target function and is built on own functioning algorithm.

The target function $\pi_l^{'}$ implements the agent orientation to the radiation source. Decisions are made on the basis of indications of the sensors f, r_k, b, l_k. The target function $\pi_l^{''}$ implements the bypass of the obstacles $\breve{O}$ based the distance sensors $D_1 - D_{2n-1}$.

The functioning algorithm is based on a neural network for our model [10]. Sensor activity vector $\boldsymbol{In} = (f, \ldots, r_k, \ldots, b, \ldots, l_k, \ldots, D_1, \ldots, D_n, \ldots, D_{2n-1})$ there is an input for the neural network. The set of actions $\boldsymbol{Out} = \{q_R, \ldots, q_F, \ldots, q_L\}$ includes turns at different angles to the left, movement straight and turns at different angles to the right. In the further work we will designate neurons of the input layer by index i. Indexes i can accept values from set $i = \{f, \ldots, r_k, \ldots, b, \ldots, l_k, \ldots, D_1, \ldots, D_n, \ldots, D_{2n-1}\}$ or values $i = \overline{0\ldots 2k+1+2n}$, unless otherwise stated. We will designate neurons of the hidden layer by index $j, j \in \mathcal{N}$. We will designate neurons of the output layer by index k. The index takes values from a set $k = \{q_R, \ldots, q_F, \ldots, q_L\}$ or numerical values $k \in \mathcal{N}$.

Weights of relationship matrix between neurons we will designate by symbol $w_{\beta\alpha}$, where the letter α denotes some neuron on which the connection begins and the letter β denotes some neuron on which the connection ends.

The neural network architecture includes two types of connections between neurons in our model. A first kind of connection is the excitation link. The excitation connection weight has a plus sign and amplifies a resulting signal. A second type of connection is inhibition connection. The inhibition connection has a minus sign and decreases the resulting signal. The types of neural network connections must be set initially and cannot be changed during network operations [11, 12].

This is an example of a complete network that is built for the intensity sensors basic set and for five distance sensors (see Fig. 2). The inhibitory negative connections are highlighted in black color and the excitatory positive connections do in red color. All connections in the first subnet are represented by solid lines and all connections in the second subnet are dotted lines.

The neural network which is given in figure has been used by us in part of numerical experiments and gives a good account of oneself as a network which gives enough detailed information about the surrounding space and has simple architecture for acceptable learning speed.

We will call a *reflex network* the architecture which was built in this section.

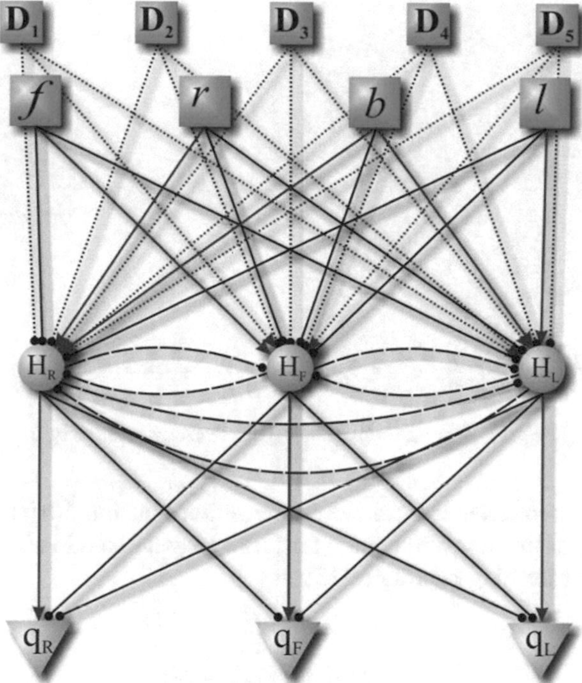

Fig. 2. A complete neural network scheme for the intensity sensors basic set and for five distance sensors. Solid lines belong to the first subnet, and dotted lines correspond to the second subnet. A red color marks the excitatory connections, and a black color marks the inhibitory connections.

4 An Intelligent Agent Training Model Building

Each agent in the group is operated by its own neural network. All neural networks have the same architecture which is described above and has different sets of weights. Application of operator (5) many times for the each agent generates a system trajectory $\Sigma(t_0) \rightarrow \Sigma(t_1) \rightarrow \Sigma(t_2) \rightarrow \ldots$. There is a possibility that the trajectory will lead to one of the many absorbing states. We build a correction model of the function π_l by the method of genetic selection in next part of our paper.

The general scheme of the training model is shown in Fig. 3.

The agents receive information about their state in the environment at each iteration. The digit I denotes this step of the algorithm in the figure. The sensory neurons activity vector moves through the neural network layers and is transformed into the output neurons activity vector. A digit II denotes this step. The output vector is transformed into an agent action which is rotate to some angle and movement into physical space. It is stage III.

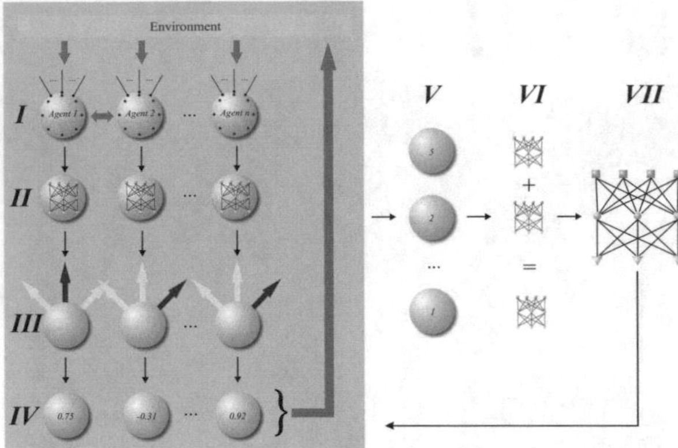

Fig. 3. It is an intelligent agents learning architecture.

Between any two states is introduced a transition function $\delta(\boldsymbol{S}(t), \boldsymbol{S}(t+1))$ which has the meaning of this transition value for further agent evolution in accordance with the reinforcement learning paradigm [13].

$$r_l = \delta_l(\boldsymbol{S}(t), \boldsymbol{S}(t+1)) \\ = \begin{cases} 0.27 \text{ } if \text{ } I_l(t+1) > I_l(t) \\ -0.3 \text{ } if \text{ } I_l(t+1) < I_l(t) \\ -0.35 \text{ } if \left(\exists \boldsymbol{X}^{\breve{O}} : \left\| \boldsymbol{X}_l^A - \boldsymbol{X}^{\breve{O}} \right\| \leq R_l^A \right) \vee \left(\exists e : \left\| \boldsymbol{X}_l^A - \boldsymbol{X}_e^A \right\| \leq R_l^A + R_e^A \right) \\ -0,27 \text{ } if \text{ } \dot{\boldsymbol{X}}_l^A = \boldsymbol{V}_l^A = 0 \\ -0.001 \cdot \frac{\sum D}{2n-1} \text{ } if \text{ } \sum D > 10 \cdot (2n-1) \\ 1 \text{ } if \text{ } \exists r : \left\| \boldsymbol{X}_l^A - \boldsymbol{X}_r^T \right\| \leq \varepsilon \end{cases}, \tag{6}$$

$$I_l = \left(I_f + I_b + \sum\nolimits_{k=1}^{\frac{m-2}{2}} \left(I_{r_k} + I_{l_k} \right) \right) / m, \tag{7}$$

$$\sum D = \sum_{k=1}^{2n-1} D_k, \tag{8}$$

where, $\boldsymbol{X}^{\breve{O}}$ are the obstacles coordinates, r_l is reward of the actin, R_l^A is the agent radius, n is the distance sensors number, m is the radiation intensity sensors number.

Point IV designates the reward. The accumulated reward is basis for calculating an agent rating which is a measure of the agent target function effectiveness. We subtract a fractional part from the total accumulated reward. The integer which remains we will call the rating of the agent for the previous period.

$$R = r_\Sigma - \{r_\Sigma\}, \tag{9}$$

where, r_Σ is the total reward for all time modeling.

All control matrices are encoded in a genotype $\boldsymbol{\Gamma} = (w_{\beta\alpha})$. Single weights are alleles. Each genotype can be represented by a point in the multidimensional phase space of the weights $w_{\beta\alpha}$. The ranked genotypes series is exposed to genetic operators of a crossing, a mutation and a weighted averaging weights operator which is specially developed for our model.

$$w^a_{\beta\alpha} = \frac{w^b_{\beta\alpha} \cdot R^b + w^c_{\beta\alpha} \cdot R^c}{R^b + R^c} + q_{\beta\alpha} \cdot sgnw^a_{\beta\alpha}, \tag{10}$$

where, $w^a_{\beta\alpha}$ is the resulting weight (allele) of neural connection from the neuron α to the neuron β, $w^b_{\beta\alpha}, w^c_{\beta\alpha}$ are the initial weights (alleles) of neural connections from the neuron α to the neuron β, R^b, R^c are the agents ratings that are samples to create a new neural network, $q_{\beta\alpha}$ is a little random amendment. The random value of $q_{\beta\alpha}$ is intended to prevent premature the process convergence. Its addition brings the network out of balance and allows not lingering in irregularities of the errors function.

Application of the genetic operators to the population is marked by digits V, VI and VII (see Fig. 3).

The updated genotypes are decoded in phenotypes for each agent. The converted agent group starts a new state-action-reward cycle.

5 Numerical Experiments

A test of the described model was performed on a specially designed simulator.

Dependence plots of the average reward which are received by the agents from the iteration number are used to assess a progress that the agents reach in the evolution process by the reinforcement learning method. Monotone increase of the plot tells that the agents accumulate positive experience and improve their behavior over time.

We conducted tests with training of the agents group on the proposed model. We used a neural network which is built on the architecture discussed above in a first series of experiments to manage agents. We used a classic perceptron where is not fixed a priori connection weights signs in a second series of experiments. Experiments were conducted with identical groups of the agents into same environment in each series. An initial agents location was accidental. An agents quantity in the group gradually changed from 1 agent to 60 agents. 60 experiments were carried out with each type of control neural network. Next we summarized the reward that proportional to the iterations number for each group and a result divided by 60. As a result we received averaged by the number of agents in the group curves of the rewards that proportional to the number of iterations for the classic perceptron and for the reflex network. Resulting plots for 1000 iterations are shown in Fig. 4.

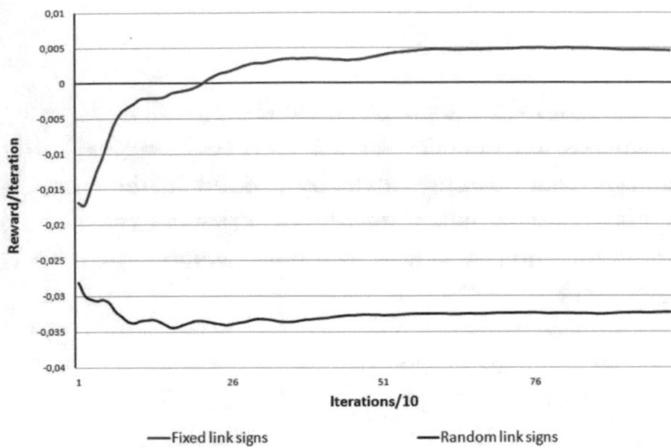

Fig. 4. Group-averaged rewards proportional to the number of iterations for period at 1000 iterations.

We tested for 5000 iterations for each group. After 500 iterations, the curves stabilize and change little in the future shows plots for first 1000 iterations only. Stabilizing curves means small fluctuations in the rewards values that agents earn at the each iteration.

The curves in figure are a basic confirmation of working capacity, efficiency and effectiveness of the model which is built in the article.

6 Conclusion

Swarm intelligence is used in many fields of industry and science. Use of large groups of simple agents is justified in situations that involve danger to life and health of people. Examples of such situations are radioactive hotbeds and chemical contamination, high temperature zones and high pressure zones. Intelligent agents have been actively studied since middle of last century. Active study of swarm intellect began relatively recently. This is due to large computational difficulties in a collective intelligence study. This article purpose is to build the hybrid algorithm to learning the intelligent agents group in a continuous, non-deterministic, partially observable environment.

The intellectual agent structure is considered in work [14]. The agents are described by a reactive model based on the neural network. The neural network architecture was originally built with the aim of finding the radiation sources and bypassing the obstacles. During execution of the task, the neural network is learned by the genetic selection method. The target function optimization model for the neural network is based on the reinforcement learning algorithm. A system of rewards has been developed to inform the neural network about learning results.

References

1. Maza, I., Caballero, F., Capitan, J., Martinez-de-Dios, J.M., Ollero, A.: A distributed architecture for a robotic platform with aerial sensor transportation and self-deployment capabilities. J. Field Robot. **28**(3), 303–328 (2011)
2. De Greeff, J., Hindriks, K., Neerincx, M.A., Kruijff-Korbayova, I.: Human-robot teamwork in USAR environments: the TRADR project. In: Proceedings of the Tenth Annual ACM/IEEE International Conference on Human-Robot Interaction Extended Abstracts, pp. 151–152 (2015)
3. Nelson, E., Micah, C., Nathan, M.: Environment model adaptation for mobile robot exploration. Auton. Robots **42**(2), 257–272 (2018)
4. Couceiro, M.S., Rocha, R.P., Ferreira, N.M.F.: Fault-tolerance assessment of a darwinian swarm exploration algorithm under communication constraints. In: 2013 IEEE International Conference on Robotics and Automation, pp. 2008–2013 (2013)
5. Couceiro, M.S., Figueiredo, C.M., Rocha, R.P., Ferreira, N.M.: Darwinian swarm exploration under communication constraints: initial deployment and fault-tolerance assessment. Robot. Auton. Syst. **62**(4), 528–544 (2014)
6. Pshikhopov, V.: Path Planning for Vehicles Operating in Uncertain 2D Environments. Butterworth-Heinemann, Oxford (2017)
7. Izquierdo, E.J., Beer, R.D.: The whole worm: brain–body–environment models of C. elegans. Curr. Opin. Neurobiol. **40**, 23–30 (2016)
8. Krishnanand, K.N., Amruth, P., Guruprasad, M.H., Bidargaddi, S.V., Ghose, D.: Glowworm-inspired robot swarm for simultaneous taxis towards multiple radiation sources. In: Proceedings 2006 IEEE International Conference on Robotics and Automation, ICRA 2006, pp. 958–963 (2006)
9. Pshikhopov, V., Medvedev, M., Gaiduk, A., Neydorf, R., Belyaev, V., Fedorenko, R., Krukhmalev, V.: Mathematical model of robot on base of airship. In: 52nd IEEE Conference on Decision and Control, pp. 959–964 (2013)
10. Finn, C., Tan, X.Y., Duan, Y., Darrell, T., Levine, S., Abbeel, P.: Learning visual feature spaces for robotic manipulation with deep spatial autoencoders. arXiv preprint. arXiv:1509.06113 (2015)
11. Petrushin, A., Ferrara, L., Blau, A.: The Si elegans project at the interface of experimental and computational Caenorhabditis elegans neurobiology and behavior. J. Neural Eng. **13**(6), 065001 (2016)
12. Sarma, G.P., Lee, C.W., Portegys, T., Ghayoomie, V., Jacobs, T., Alicea, B., Cantarelli, M., Currie, M., Gerkin, R.C., Gingell, S., Gleeson, P.: OpenWorm: overview and recent advances in integrative biological simulation of Caenorhabditis elegans. Philos. Trans. Royal Soc. B **373**(1758), 20170382 (2018)
13. Bojarski, M., Del Testa, D., Dworakowski, D., Firner, B., Flepp, B., Goyal, P., Jackel, L.D., Monfort, M., Muller, U., Zhang, J., Zhang, X.: End to end learning for self-driving cars. arXiv preprint. arXiv:1604.07316 (2016)
14. Akzhalova, A., Inoue, A., Mukharsky, D.: Intelligent mobile agents for disaster response: survivor search and simple communication support. In: AROB 2014 International Symposium on Artificial Life and Robotics, pp. 254–259 (2014)

An Optimization Algorithm Based on Multi-free Dynamic Schema of Chromosomes

Radhwan Al-Jawadi[1](✉), Marcin Studniarski[2], and Aisha Younus[2]

[1] Engineering Technical College of Mosul, Northern Technical University, Mosul, Iraq
radwanyousif@yahoo.com

[2] Faculty of Mathematics and Computer Science, University of Łódź, Banacha 22, 90-238 Łódź, Poland
marcin.studniarski@wmii.uni.lodz.pl, azeezzena74@yahoo.com

Abstract. In this work, continuing the line of research from our previous papers [1–4], we further explore the notion of a schema in evolutionary algorithms and its role in finding global optima in numerical optimization problems. We present another optimization algorithm called Multi-Free Dynamic Schema (MFDS) which differs from the Free Dynamic Schema (FDS) algorithm introduced in [6] because in the MFDS the free dynamic schema operator is applied six times to different groups of chromosomes. The results of numerical experiments show that this change speeds up the search for a global optimum for most of the test problems.

Keywords: Dynamic schema · Dissimilarity and similarity of chromosomes · Double population · Free dynamic schema

1 Introduction

The idea of double population in evolutionary algorithms was used in our previous papers [1–4] to improve the search for optimal solution by increasing the diversity of a population. Also, in [5] the authors have used a double population with Swarm Optimization Algorithm, while in [6] a dual-population genetic algorithm was presented, where the main population was used to find a good solution, and the second population was used to evolve and provide controlled diversity to the main population. In this paper, a new evolutionary algorithm for solving numerical optimization problems called Multi-Free Dynamic Schema (MFDS) is presented. This algorithm is similar to the Free Dynamic Schema (FDS) algorithm introduced in [4]. The main difference between these two algorithms is that in the FDS the free dynamic schema operator is applied two times, while in the MFDS it is applied six times to six different parts of the population. The other operators used in the MFDS are dynamic dissimilarity, similarity, dissimilarity, dynamic schema, and random generation of chromosomes. These five operators have already been used in our previous algorithms: MDSDSC [3] and FDS [4].

J. Świątek et al. (Eds.): ISAT 2019, AISC 1051, pp. 146–156, 2020.
https://doi.org/10.1007/978-3-030-30604-5_13

The free dynamic schema operator aims at finding the optimal solution by fixing the highest bits of a chromosome (i.e., fixing the highest bits of all variables $(x_1, \ldots, x_n)$ which are contained in the chromosome) and changing the lower bits at the same time, thus the algorithm focuses on the searching for optimal solution in a small area.

After noticing that the FDS [4] algorithm was more effective than the DSC [1], DSDSC [2] and MDSDSC [3] algorithms in terms of speed in finding the best solution, we now propose here another way of using the same principle as in FDS, but now a larger number of free schema types (six) are selected at random from the first quarter of the sorted generation.

We now briefly explain the principle of multi-free schema. Suppose f is a one-dimensional function with domain [0, 1], as shown in Fig. 1, This domain is discretized by using a binary representation consisting of four bits, $(0000, 0001, \ldots, 1111)$, which means that the interval [0, 1] is divided into 15 segments. Suppose there are 6 randomly chosen good solutions $(S_1, \ldots S_6)$, and the free dynamic schema operator (see Sect. 2.1) is applied 6 times, with $R_1(S_1) = 3$, $R_1(S_2) = 3, \ldots, R_1(S_6) = 1$, where $R_1(S_i)$ is a randomly chosen number from the set $\{3, \ldots, \mathrm{m_i}/2\}$ selected for solution S_i. Now we discover a multi-free schema: (101*) for $R_1(S_1)$, (111*) for $R_1(S_2)$, and so on. Then we randomly put 0 or 1 in positions having *s. Here the multi-free schema will cover all the subspaces colored with red, as shown in Fig. 1.

An example, the first three bits are fixed, $R_1(S_1) = 3$.

Bits: 1234
S_1: 1010
Schema : 101*
Sol.1: 1011
Sol.2: 1010

Here another example, the first three bits are fixed, $R_1(S_2) = 3$.

Bits: 1234
S_2: 1110
Schema : 111*
Sol.1: 1111
Sol.2: 1110

……

Here another example, the first bit is fixed, $R_1(S_6) = 1$.

Bits: 1234
S_6: 0100
Schema : 0***
Sol.1: 0000
Sol.2: 0001
Sol.3: 0010
….
Sol.8: 0111

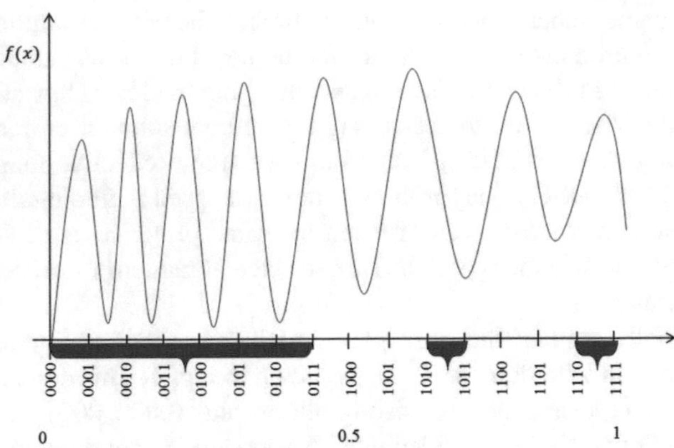

Fig. 1. Multi-free dynamic schema

2 Methodology

The MFDS algorithm starts with a random population (P0) of *M* elements representing a number of solutions to the problem. This population is sorted, then a new population (P1) is formed whose first 40% of chromosomes are copied from a part of (P0). Population (P0) is divided into four equal groups ($\boldsymbol{G}_1, \boldsymbol{G}_2, \boldsymbol{G}_3, \boldsymbol{G}_5$), then population (P1) is divided into 8 not equal groups ($\boldsymbol{G}_5, \boldsymbol{G}_6, \ldots, \boldsymbol{G}_{12}$). Then we apply different operators to these groups (see Table 1).

Table 1. Populations (P0) and (P1) and the twelve groups of chromosomes.

<table>
<tr><th colspan="2">Original groups of chromosomes (P0)</th><th colspan="2">Copy groups of chromosomes (P1)</th></tr>
<tr><td>Ch_1</td><td rowspan="5">$\boldsymbol{G}_1$: To the first group the dynamic dissimilarity operator is applied</td><td rowspan="4">20% of population</td><td rowspan="4">$\boldsymbol{G}_5$: To the fifth group the dissimilarity operator is applied</td></tr>
<tr><td>Ch_2</td></tr>
<tr><td>Ch. …</td></tr>
<tr><td>Ch. …</td></tr>
<tr><td>$Ch_{M/4}$</td><td rowspan="4">20% of population</td><td rowspan="4">$\boldsymbol{G}_6$: To the six group the dynamic dissimilarity operator is applied.</td></tr>
<tr><td>$Ch_{M/4+1}$</td><td rowspan="5">$\boldsymbol{G}_2$: To the second group the similarity operator is applied</td></tr>
<tr><td>$Ch_{M/4+2}$</td></tr>
<tr><td>Ch. …</td></tr>
<tr><td>Ch. …</td><td rowspan="2">10% of population</td><td rowspan="2">$\boldsymbol{G}_7$: To this group the FDS operator is applied</td></tr>
<tr><td>$Ch_{M/2}$</td></tr>
<tr><td>$Ch_{M/2+1}$</td><td rowspan="5">$\boldsymbol{G}_3$: To the third group the dynamic schema operator is applied</td><td rowspan="2">10% of population</td><td rowspan="2">$\boldsymbol{G}_8$: To this group the FDS operator is applied</td></tr>
<tr><td>$Ch_{M/2+2}$</td></tr>
<tr><td>Ch. …</td><td rowspan="2">10% of population</td><td rowspan="2">$\boldsymbol{G}_9$: To this group the FDS operator is applied</td></tr>
<tr><td>Ch. …</td></tr>
<tr><td>$Ch_{M/2+ M/4}$</td><td rowspan="2">10% of population</td><td rowspan="2">$\boldsymbol{G}_{10}$: To this group the FDS operator is applied</td></tr>
<tr><td>$Ch_{M/2+ M/4+1}$</td><td rowspan="5">$\boldsymbol{G}_4$: The fourth group is generated randomly</td></tr>
<tr><td>$Ch_{M/2+ M/4+2}$</td><td rowspan="2">10% of population</td><td rowspan="2">$\boldsymbol{G}_{11}$: To this group the FDS operator is applied</td></tr>
<tr><td>Ch. …</td></tr>
<tr><td>Ch. …</td><td rowspan="2">10% of population</td><td rowspan="2">$\boldsymbol{G}_{12}$: To this group the FDS operator is applied</td></tr>
<tr><td>Ch_M</td></tr>
</table>

To groups (G_1, G_2, G_3) of population (P0), the dynamic dissimilarity, similarity and dynamic schema operators are applied, respectively, and in (G_4), random chromosomes are generated. To groups (G_5, G_6) of population (P1), the dissimilarity and dynamic dissimilarity operators are applied respectively, where each of (G_5, G_6) represents 20% of population (P1). For the next groups ($G_7, G_8, \ldots, G_{12}$), where each group represents 10% of population (P1), six types of free dynamic schema are applied, where chromosomes are randomly chosen from the first quarter of sorted population (P0), see Table 1. Each free dynamic schema represents a group of solutions, these solutions are close to the area of best solutions because they are chosen form the first quarter in the sorted population (P0).

The descriptions of the dissimilarity, similarity, dynamic schema and dynamic dissimilarity operators were presented in our previously published papers [1–3]. The free dynamic schema operator is described in the next subsection.

2.1 Free Dynamic Schema Operator

The free dynamic schema operator finds schemata from selected chromosomes in the first quarter of population (P0). It fixes the higher bits of each x_i and changes the remaining bits of each x_i. This operator works as follows (see Table 2):

1- First, from the first quarter (G_1) of sorted population (P0) choose 6 chromosomes randomly.
2- Then, divide the chosen chromosomes into n parts corresponding to variables $(x_1, \ldots, x_n)$, the i-th part having length m_i, where m_i is the number of bits for x_i. Next, for each variable x_i, generate a random integer R_i from the set $\{3, \ldots, m_i/2\}$.

Table 2. The free dynamic schema operator. This operator applies to groups $G_i, i = 8, ..12$. Before change: an example for finding schema from the first chromosome. Here shadow bits are not destroyed.

Ch. No.	m_1						m_2			
	R_1		$m_1 - R_1$				R_2	$m_2 - R_2$		
Ch_i	1	1	0	0	1	0	1	0	1	0
Schema	1	1	*	*	*	*	1	*	*	*
After finding the schema: put it in $M/10$ positions in group G_i.										
G_i (1)	1	1	*	*	*	*	1	*	*	*
G_i (2)	1	1	*	*	*	*	1	*	*	*
G_i (...)	1	1	*	*	*	*	1	*	*	*
G_i (...)	1	1	*	*	*	*	1	*	*	*
G_i (M/10)	1	1	*	*	*	*	1	*	*	*
After change: put randomly 0 or 1 in (*) bits.										
G_i (1)	1	1	1	1	1	0	1	0	0	1
G_i (2)	1	1	1	0	0	0	1	1	1	1
G_i (...)	1	1	0	1	1	0	1	1	1	0
G_i (...)	1	1	0	1	0	1	1	0	0	0
G_i (M/10)	1	1	1	0	1	1	1	0	1	1

Define the "gray" part of $\boldsymbol{x_i}$ as the first segment of length $\boldsymbol{R_i}$ of the string corresponding to $\boldsymbol{x_i}$. Define the "white" part of $\boldsymbol{x_i}$ as the second segment of length $\boldsymbol{m_i - R_i}$ of the same string.

3- For the "white" parts of chromosomes, put a star (*) in the schema. After finding the schema, copy it $K = M/10$ times and put it in group ($\boldsymbol{G_7}$), then put randomly 0 or 1 in the positions having (*). The positions marked in "gray" are kept unchanged.

4- Repeat step 2, 3 for the groups $\boldsymbol{G_8}, \ldots \boldsymbol{G_{12}}$.

3 The MFDS Algorithm

The following optimization problem is considered:

$$f : \mathbb{R}^n \to \mathbb{R}$$
$$\text{minimize}|\text{maximize} f(x_1, \ldots, x_n) \text{ subject to}$$
$$x_i \in [a_i, b_i], i = 1, \ldots, n$$

where $f : \mathbb{R}^n \to \mathbb{R}$ is a given function.

In the algorithm described below, we use a standard encoding of chromosomes as in the book of Michalewicz [7]. In particular, we use the following formula to decode a real number $x_i \in [a_i, b_i]$:

$$x_i = a_i + \text{decimal}(1001..001) * \frac{b_i - a_i}{2^{m_i} - 1}$$

where m_i is the length of a binary string and "decimal" represents the decimal value of this string. The value of m_i for each variable depends on the length of the interval $[a_i, b_i]$. To encode a point $(x_1, \ldots, x_n)$, a decimal string of length $m = \sum_{i=1}^{n} m_i$ is used.

Let M be a positive integer divisible by 8. The MFDS algorithm consists of the following steps:

1. Generate $2M$ chromosomes, each chromosome representing a point $(\mathrm{x}_1, \ldots, \mathrm{x}_\mathrm{n})$. Divide the chromosomes into two populations (P0) and (P1), where (P0) consists of four groups (G_1, G_2, G_3, G_4), and (P1) consists of eight groups (G_5, $G_6, \ldots, G_{12}$), each group in (P0) having M/4 chromosomes, but in (P1) the size is equal to 20% of population for (G_5, G_6) and 10% for $(G_7, \ldots, G_{12})$.
2. Compute the values of the fitness function f for each chromosome in the population $(G_1, \ldots, G_{12})$.
3. Sort the chromosomes according to the descending (for maximization) or ascending (for minimization) values of the fitness function.
4. Copy the first 40% from (P0) onto (G_5, G_6), replacing the original chromosomes.
5. Copy C times the first chromosome and put it in C randomly chosen positions in the first half of population (P0), replacing the original chromosomes, where C = M/8.

6. Apply the dynamic schema operator for chromosomes A = Ch_1 and B = $Ch_{M/4}$ from populations (P0), (that is, the chromosomes on positions 1 and M/4, respectively). Copy this schema M/4 times and put it in (G_3).
7. Apply the dynamic dissimilarity and similarity operators to groups (G_1) and (G_2) respectively. Apply the dissimilarity and dynamic dissimilarity operators to group (G_5) and (G_6) respectively.
8. Apply the free dynamic schema operator 6 times to generate six groups ($G_7, \ldots, G_{12}$). To generate each group, a chromosome is chosen randomly from the first quarter of solutions in (P0). Then put 0 or 1 randomly in positions having *s in each group.
9. All the chromosomes created in Steps 4 to 8 replace the original ones in positions from 2 to 2M in populations (P0) and (P1). Then randomly generate chromosomes for group (G_4).
10. Go to Step 2 and repeat until the stopping criterion is reached.

Notes:

- The stopping criterion for the algorithm depends on the example being considered, see Sect. 4.
- An injection technique to the population was suggested in [8] to preserve the diversity of the population. They use fixpoint injection, which means that they introduce new randomly generated chromosomes to the population for certain numbers of generations. A similar strategy has been applied in the DSDSC, MDSDSC, FDS, MFDS algorithms by generating the group (G_4) randomly in each iteration.

4 Experimental Results

In this section, we report on computational testing of the MFDS algorithm on 18 functions of 2 variables, 5 functions of 10 variables. After each test, the result of MFDS has been compared with the known global optimum and with the result of a CGA taken from our experimental result (see Table 3), also, in Table 4 a comparison of the mean number of function evaluations and success rate of CMA-ES, DE and FDS algorithms is presented. All 22 tested functions with optimal solutions are mentioned in our previous researches (see [1–4]). We have applied the algorithm with 80 chromosomes (P0) with the stopping criterion that the difference between our best solution and the known optimal solution is less than or equal to a given threshold. The threshold was equal to 0.001 for most two-dimensional functions, only for Shubert the threshold was 0.01, for Michalewicz it was 0.04 and for 10-dimensional functions it was 0.1.

The MFDS algorithm has found optimum solutions for some optimization problems (like Easom, Booth's, Schaffer, Schwefel's,) that the classical genetic algorithm cannot solve, as shown in Table 3. Column nine shows 0% success rate with bit string as it has been mentioned in [9, 10]. For our algorithm all success rates are 100% with 80 chromosomes in (P0) for all problems.

Table 3. The best value of functions for 50 runs of the MFDS algorithm (140 chromosomes).

Function name	Min no. of iter.	Max no. of iter.	Mean no. of iter. for all successful runs	Mean of the best solution fitness from all runs	Success rate of MFDS	Success rate of DSC	Success rate of GA – Double Vector	Success rate of GA – Bit string
Easom	4	291	62	−0.99934	100%	100%	100%	0%
Matyas	2	10	5	0.000484	100%	100%	100%	90%
Beale's	2	74	16	0.000497	100%	100%	70%	0%
Booth's	2	48	17	0.000519	100%	100%	100%	0%
Goldstein–Price	2	62	20	3.000497	100%	100%	100%	0%
Schaffer N. 2	2	39	14	0.000354	100%	100%	70%	0%
Schwefel's	8	253	65	0.000684	100%	92%	0%	0%
Branins's rcos	2	103	9	0.398415	100%	100%	100%	0%
Six-hump camel back	2	61	8	−1.03112	100%	100%	100%	0%
Shubert	2	169	33	−186.714	100%	100%	100% DV	0%
Martin and Gaddy	2	11	6	0.000445	100%	100%	40%	0%
Zbigniew Michalewicz	2	546	71	38.81849	100%	100%	80%	20%
Holder table	4	87	24	−19.208	100%	100%	80%	0%
Drop-wave	6	189	44	−0.99959	100%	100%	0%	100%
Levy N. 13	4	47	19	0.000527	100%	100%	0%	100%
Rastrigin's	8	133	38	0.000418	100%	100%	100%	100%
Rosenbrock	3	102	24	0.0005517	100%	100%	0%	100%
Sum Squares d = 10	38	251	128	0.072773	100%	25%		100%
Sphere function d = 10	14	38	23	0.073044	100%	100%		100%
Sum of different powers d = 10	1	7	4	0.029555	100%	100%		100%
Zakharov d = 10	107	700	373	0.09010	54%			100%
Rastrigin d = 10	157	456	294	0.07169	100%[a]			100%

[a]For this function we change the R_i in dynamic schema and free dynamic schema to be a random number from $\{0, 1, \ldots, m_i\}$.

The MFDS algorithm keeps the best solution from each iteration at the first position until it is replaced by a better one.

In Table 5 we compare the average number of iterations for 2-dimensional among the DSC, DSDSC, MDSDSC, FDS, MFDS, CMA-ES, DE, and GA algorithms. It is clear that the MFDS algorithm has the best values for most tested functions. Here it is possible to note the effect of multi free dynamic schema by decreasing the average number of iterations for most functions comparing with previous algorithms.

In Table 6 we compare the average number of iterations for 10-dimensional functions by using 160 chromosomes for DSC, DSDSC, MDSDSC, FDS, MFDS algorithms, It is in the MFDS has lowest iterations.

Table 4. Comparing the mean number of function evaluations and success rate of CMA-ES, DE and MFDS algorithms (50 runs, max 2500 iterations, 80 chromosomes).

Function name	CMA-ES success rate	Function evaluations of CMA-ES	DE success rate	Function evaluations of DE	MFDS success rate	Function evaluations of MFDS
Easom	70%	17053	100%	3240	100%	8120
Matyas	100%	500	100%	2700	100%	700
Beale	100%	460	100%	3060	100%	980
Booth's	100%	492	100%	2820	100%	1400
Goldstein–Price	100%	1812	100%	1620	100%	2940
Schaffer N. 2	90%	6726	100%	5016	100%	1540
Schwefel's	0%	–	0%	–	100%	4060
Branins's rcos	100%	6876	100%	840	100%	980
Six-hump camel	100%	780	100%	2160	100%	700
Shubert	90%	2220	100%	8160	100%	3640
Martin and Gaddy	100%	1660	100%	2400	100%	700
Michalewicz	100%	1848	0%	–	100%	4200
Drop-wave	50%	26470	94%	9048	100%	1680
Levy N. 13	100%	606	100%	1958	100%	4200
Rastrigin's	80%	13134	100%	2388	100%	1540
Sphere	100%	720	100%	1800	100%	5740
Rosenbrock's	100%	1644	100%	4560	100%	25760

Table 5. Comparing the average number of iterations for 2-dimensional functions for all algorithms.

Function name	DSC	DSDSC	DDS	FDS	MFDS	CMA-ES	DE	GA DV
Easom	88	51	62	89	58	384	41	124
Matyas	31	11	5	6	5	9	34	125
Beale's	93	49	16	8	7	8	38	204
Booth's	151	20	17	12	10	8	35	75
Goldstein–Price	134	34	20	35	21	31	21	82
Schaffer N.2	278	71	14	16	11	112	63	93
Schwefel's	561	41	65	39	29	*	*	*
Branins's rcos	86	28	9	11	7	115	11	68
Six-hump camel back	39	18	8	7	5	13	27	75
Shubert	198	19	33	45	26	37	102	64
Martin and Gaddy	36	15	6	5	5	28	30	320
Michalewicz	207	67	71	55	30	31	500	72
Holder Table	47	12	24	19	12	*	113	240
Drop-wave	201	48	44	45	30	441	25	*
Levy N. 13	290	45	19	19	11	10	20	*
Rastrigin's	71	25	38	58	41	219	23	51
Sphere	75	7	4	7	5	12	44	63
Rosenbrock's	101	115	24	18	13	27	41	*

Table 6. Comparing the average number of iterations for 10-dimensional functions with 160 chromosomes for all algorithms

Function name	DSC	DSDSC	DDS	FDS	MFDS
Sum squares d = 10	1936	145	128	320	245
Sphere d = 10	746	31	23	51	46
Sum of different powers d = 10	14	3	4	4	3
Zakharov d = 10	1808	217	468	581	373
Rastrigin d = 10	***	1045	***	1159	294

*** No success for this function.

Table 7. Comparing the average number of function evaluations for 2-dimensional functions of DSC, DSDSC, MDSDSC, FDS, MFDS algorithms with CMA-ES and DE algorithms

Function name	DSC	DSDSC	MDSDSC	FDS	MFDS	CMA_ES	DE
Easom	7040	4080	8680	12460	8120	17053	3240
Matyas	2480	880	700	840	700	500	2700
Beale's	7440	3920	2240	1120	980	460	3060
Booth's	12080	1600	2380	1680	1400	492	2820
Goldstein–Price	10720	2720	2800	4900	2940	1812	1620
Schaffer N. 2	22240	5680	1960	2240	1540	6726	5016
Schwefel's	44880	3280	9100	5460	4060	–	–
Branins's rcos	6880	2240	1260	1540	980	6876	840
Six-hump camel	3120	1440	1120	980	700	780	2160
Shubert	15840	1520	4620	6300	3640	2220	8160
Martin and Gaddy	2880	1200	840	700	700	1660	2400
Michalewicz	16560	5360	9940	7700	4200	1848	–
Holder Table	3760	960	3360	2660	1680	–	–
Drop-wave	16080	3840	6160	6300	4200	26470	9048
Levy N. 13	23200	3600	2660	2660	1540	606	1958
Rastrigin's	5680	2000	5220	8120	5740	13134	2388
Sphere	75	560	560	980	700	720	1800
Rosenbrock's	101	9200	3360	2520	1820	1644	4560

The results are presented in Table 7 of comparing function evaluations for 2-dimensional functions of DSC, DSDSC, MDSDSC, FDS, MFDS algorithms with CMA-ES and DE algorithms, in most cases the MFDS has the best value.

We have observed in [9] that the success rate and number of functions evolution of Bees Algorithm (BA), Particle Swarm Optimization (PSO), these results are compared with the MFDS algorithms for two dimensions functions in terms of average number of functions evaluations and success rate as shown in Table 8.

Table 8. Comparison of BA, PSO and MFDS algorithms in terms of average number of functions evaluations and success rate.

Function name	BA	Fun. eval. of BA	PSO	Fun. eval. of PSO	MFDS	Fun. eval. of MFDS
Easom	72%	5868	100%	2094	100%	8120
Shubert	0%	–	100%	3046	100%	3640
Schwefel's	85%	5385	86%	3622	100%	4060
Goldstein–Price	7%	9628	100%	1465	100%	2940
Martin and Gaddy	100%	1448	3%	9707	100%	700
Rosenbrock	46%	7197	100%	1407	100%	1820

5 Conclusion

MFDS is a new multi-population evolutionary algorithm that uses two populations. This algorithm uses different operators to find the optimal solution, where each application of the free dynamic schema operator generates a group of solutions ($G_7, \ldots, G_{12}$), these solutions are close to the area of best solutions because they are chosen form the first quarter in the sorted population (P0), also, the dynamic schema operator the algorithm obtains the best area of solutions and searches within that area in each iteration in (G_3), as it detects the schema from the best solution in the population. While the dynamic dissimilarity operator performs searching in a wide range of solutions in (G_6), where the high bits are kept without change and the lower bits are changed. The dissimilarity and similarity operators possess the ability of searching in the whole search space because every bit of a chromosome can be changed by them. The fifth operator generates chromosomes randomly in (G_4) to help increasing the diversity and not to stick in a local optimum solution.

We have applied the GA, DSC, DSDSC, MDSDSC, FDS, MFDS algorithms on 22 tested function with 2 and 10 dimensions. The results show the MFDS algorithm is superior on the others algorithms for most functions.

Acknowledgments. The first author I would like to thank is the Ministry of Higher Education and Scientific Research (MOHESR), Iraq.

References

1. Al-Jawadi, R., Studniarski, M., Younus, A.: New genetic algorithm based on dissimilaries and similarities. Comput. Sci. J. AGH Univ. Sci. Technol. Pol. **19**(1), 19 (2018)
2. Al-Jawadi, R.: An optimization algorithm based on dynamic schema with dissimilarities and similarities of chromosomes. Int. J. Comput. Electr. Autom. Control Inf. Eng. **7**(8), 1278–1285 (2016)
3. Al-Jawadi, R., Studniarski, M.: An optimization algorithm based on multi-dynamic schema of chromosomes. In: International Conference on Artificial Intelligence and Soft Computing, vol. 10841, pp. 279–289. Springer, Cham (2018)

4. Al-Jawadi, R.Y., Studniarski, M., Younus, A.A.: New optimization algorithm based on free dynamic schema. In: Submitted to ICCCI 2019, Hendaye, France (2019)
5. Wu, Y., Sun, G., Su, K., Liu, L., Zhang, H., Chen, B., Li, M.: Dynamic self-adaptive double population particle swarm optimization algorithm based on lorenz equation. J. Comput. Commun. **5**(13), 9–20 (2017)
6. Park, T., Ryu, K.R.: A dual-population genetic algorithm for adaptive diversity control. IEEE Trans. Evol. Comput. **14**(6), 865–884 (2010)
7. Michalewicz, Z.: Genetic Algorithms + Data Structures = Evolution Programs. Artificial Intelligence, 3rd edn. Springer, Heidelberg (1996)
8. Sultan, A.B.M., Mahmod, R., Sukaiman, M.N., Abu Bakar, M.R.: Maintaining diversity for genetic algorithm: a case of timetabling problem. J. Teknol. Malaysia **44**(D), 123–130 (2006)
9. Eesa, A.S., Brifcani, A.M.A., Orman, Z.: A new tool for global optimization problems-cuttlefish algorithm. Int. J. Comput. Electr. Autom. Control Inf. Eng. **8**(9), 1198–1202 (2014)
10. Ritthipakdee, A., Thammano, A., Premasathian, N., Uyyanonvara, B.: An improved firefly algorithm for optimization problems. In: ADCONP, Hiroshima, no. 2, pp. 159–164 (2014)

Knowledge Discovery and Data Mining

Scientists' Contribution to Science and Methods of Its Visualization

Oleksandr Sokolov[1], Veslava Osińska[2], Aleksandra Mrela[3(✉)], Włodzisław Duch[1], and Marcin Burak[1]

[1] Faculty of Physics, Astronomy and Informatics, Nicolaus Copernicus University, Toruń, Poland
{osokolov,286270}@fizyka.umk.pl, wduch@is.umk.pl
[2] Faculty of Humanities, Nicolaus Copernicus University, Toruń, Poland
wieo@umk.pl
[3] Faculty of Technology, Kujawy and Pomorze University, Bydgoszcz, Poland
a.mrela@kpsw.edu.pl

Abstract. In the paper, fuzzy sets technology is used to find out the levels of the scientific contribution of a scientist or a team of scientists into scientific fields or discipline. Namely, fuzzy logic is applied to describe the fuzzy input and output relations between scientific fields, disciplines, journals and researchers, afterward, the compositions of these relations and the proposed optimistic fuzzy aggregation norm over fuzzy variables are used. Two techniques are applied to show the contribution to the science of the team of researchers graphically: one graphical method uses the radar graphs, and the other one is the application of the t-SNE technique based on the Euclidean metrics in the space of the scientists' contribution into the different scientific fields. Moreover, the proposed methods are the foundation for the web-software "ScienceFuzz" developed for calculations levels of contribution to the sciences by an individual researcher or by a team of them.

Keywords: Scientometrics · Contribution to science · Fuzzy relation · Visualization method · t-SNE

1 Introduction

All the time, the scientific achievements of scientists and scientific units, where they are employed, are evaluated and parametrized. For scientists, this estimation is essential during the process of their promotion to be a doctor or professor. For researchers and the teams of them, it is valid during the application for grants. Moreover, for scientific units, the evaluation of their staff achievements is crucial when they are trying to get the highest levels in parametrization and so on. Evaluation of professional output of research teams can be valuable information for scientific policymakers in deciding how to fund particular domains, and which ones are progressive or declining. For the managers of scientific units (institutes, colleges, universities and so on), it is essential to

J. Świątek et al. (Eds.): ISAT 2019, AISC 1051, pp. 159–168, 2020.
https://doi.org/10.1007/978-3-030-30604-5_14

find the method of useful presentation of their staff scientific achievements to decide about applying for grants or start the new curriculum for students.

Scientists, of course, try to find out the best way of scientific estimation and presentation of the results. The problem is not easy because there are many disciplines collected in scientific domains, a lot of journals where researchers can publish their papers, and there is a considerable number of scientists. The fields are not disjoint (for example, IT sciences can be a part of exact and technical sciences). Many people work not in the inside discipline but on the borders using the methods of the other discipline or disciplines. From time to time, new disciplines are created, and so on.

2 Scientific Contribution of Scientists

In this section, the method of estimation of the scientist's scientific contribution is proposed. All scientists publish their scientific achievements writing papers in journals and proceedings, monographs, chapters in books, and so on. These publications can be coupled with one discipline or sometimes more disciplines via the coupling of the journal with the discipline or the disciplines.

Let us consider four sets:

- $\boldsymbol{F}$ – the set of scientific fields and $\boldsymbol{F} = \{F_i, i = 1, 2, \ldots, I\}$, in Poland, at present, $I = 8$ [1],
- $\boldsymbol{D}$ – the set of disciplines; $\boldsymbol{D} = \{D_k, k = 1, 2, \ldots, K\}$,
- $\boldsymbol{J}$ – the set of journals; $\boldsymbol{J} = \{J_l, l = 1, 2, \ldots, L\}$,
- $\boldsymbol{S}$ – the set of scientists; $\boldsymbol{S} = \{S_n, n = 1, 2, \ldots, N\}$.

Upon the Regulations of the Ministry of Science and Higher Education [1], Table 1 is prepared.

Table 1. Scientific domains adopted in the Regulations [1]

Symbol	Scientific fields	Symbol	Scientific fields
F_1	Social Sciences	F_5	Humanities
F_2	Agricultural Sciences	F_6	Engineering and Technical Sciences
F_3	Exact and Natural Sciences	F_7	Theological Sciences
F_4	Medical and Health Sciences	F_8	Arts

Based on these four sets, four fuzzy input relations are built:

- $R_1 \subseteq \boldsymbol{F} \times \boldsymbol{D}$ – the fuzzy relation between scientific fields and disciplines;
- $R_2 \subseteq \boldsymbol{D} \times \boldsymbol{J}$ – the fuzzy relation between disciplines and journals;
- $R_3 \subseteq \boldsymbol{J} \times \boldsymbol{S}$ – the fuzzy relation between journals and scientists.

In the Regulations [1], there are also defined scientific disciplines and $K = 47$. Some disciplines are related to one scientific field, and some are related to more. For example, History is associated with Humanities, and Computer Science is related to two scientific fields F_3 - Exact and Natural Sciences and F_6 – Engineering and Technical Sciences.

Table 2 presents a portion of the relation R_1 between scientific fields, and disciplines and 1 denotes that the discipline is considered as s part of the scientific field and 0 otherwise.

Table 2. The portion of relation R_1 based on the Regulations [1]

Disciplines	Scientific fields						
	F_1	F_2	F_3	F_4	F_5	F_6	F_7
D_1 - History	0	0	0	0	1	0	0
D_2 - Mathematics	0	0	1	0	0	0	0
D_3 - Computer Science	0	0	1	0	0	1	0
D_4 - Law	1	0	0	0	0	0	0
D_5 - Medical Sciences	0	0	0	1	0	0	0
D_6 - Veterinary	0	1	0	0	0	0	0
D_7 – Architecture and Urban Planning	0	0	0	0	0	1	0
D_8 - Chemistry	0	0	1	0	0	0	0

Table 3 shows the part of relation R_2 between disciplines and journals.

Table 3. The part of relation R_2 based on journals descriptions

Journals	Disciplines						
	D_1	D_2	D_3	D_4	D_5	D_6	D_7
J_1 - Palamedes a journal of ancient history	1	0	0	0	0	0	0
J_2 - Acta Applicandae Mathematicae	0	1	0	0	0	0	0
J_3 - Acta Neurologica Belgica	0	0	0.8	0	0.8	0	0
J_4 - European Journal of International Law	0	0	0	1	0	0	0
J_5 - Family Practice	0	0	0	0	0.8	0	0
J_6 - Acta Biologica Cracoviensia	0	0	0	0	0.2	0.4	0
J_7 - Acta Universitatis Lodziensis	0.3	0	0	0.3	0	0	0.7
J_8 - Advanced Materials	0	0	0	0	0	0	1

The third relation R_3 can present the number of articles published by the scientists in all considered journals. Because the values of this relation can be larger than 1, so they have to be normalized.

Let us introduce the concept of the unit of contribution to science or scientific fields or discipline. If the considered scientist publishes one article in the scientific journal, their contribution to science (scientific field or discipline) is equal to a (the unit of contribution to science), which is a small number, for example, $a = 0.01$. Table 4 shows the part of relation R_3.

Table 4. The contribution to science - the part of relation R_3

Scientists	Journals					
	J_1	J_2	J_3	J_4	J_5	J_6
S_1	0.01	0	0	0	0	0
S_1	0	0	0.01	0	0	0
S_1	0	0	0.01	0	0	0
S_2	0	0	0	0.01	0	0
S_2	0	0	0	0	0.01	0
S_3	0	0.01	0	0	0	0
S_3	0	0	0	0	0	0.01
S_3	0	0,01	0	0	0	0

Since very often the scientists publish more than one article in one journal, especially in the case of long time, then some aggregation norm must be applied to calculate the value of the contribution of one scientist into one discipline by one journal.

On the foundation of these input relations, one fuzzy output is built:

$R_4 \subseteq \boldsymbol{F} \times \boldsymbol{S}$ – the fuzzy relation between scientific fields and scientists.

3 Operations of Fuzzy Logic

Zadeh introduced the definitions of fuzzy sets [2] and fuzzy relations [3]. A fuzzy set $A \subseteq X$ is a set of pairs $\{(x, \mu_X), x \in X, \mu_X : X \to [0,1]\}$, where μ_X is a membership function which shows the level of membership of element x to set A. Let X and Y be two spaces. If $R \subseteq X \times Y$ is a fuzzy set, then it is called a fuzzy relation.

Let $P_1 \subseteq X \times Y$ and $P_2 \subseteq Y \times Z$ be two fuzzy relations. Then $P_3 \subseteq X \times Z$ is called a S-T composition of relations P_1 and P_2 (see [4]) if for each $(x, z) \in X \times Z$,

$$\mu_3(x, z) = S_{y \in Y} T(\mu_1(x, y), \mu_2(y, z))$$

where T is a T-norm and S is a S-conorm (comp. [4]). In this paper, the algebraic norm and conorm are applied:

$$T(x, y) = xy \text{ and } S(x, y) = x + y - xy \text{ for } x, y.$$

To aggregate values of fuzzy relation R_3, the optimistic fuzzy aggregation norm [5] is used because the “optimism” denotes that the result of applying this norm is not less than the calculated values. Therefore for each $a, b \in [0, 1]$, the following conditions should be fulfilled:

$$S_o(0,0) = 0,\ S_o(a,b) = S_o(b,a),\ S_o(a,0) \geq a,\ S_o(a,b) \geq max\{a,b\} \tag{1}$$

In the paper, the following optimistic fuzzy aggregation norm is applied, for $a, b \in [0, 1]$ let

$$S_o(a, b) = a + b - ab \tag{2}$$

These conditions show that the if for the given scientist who has not published any paper and in the given time they have not published any articles, so their contribution to science/scientific field/discipline is still 0. Of course, the result of calculating the contribution is not depended on the order of numbers. Moreover, for the given period, if the researcher has not published any paper, their contribution to science is not diminished. Finally, the contribution to science after publishing the paper is higher than before.

4 Application of Fuzzy Logic Operations

In the beginning, we use the optimistic fuzzy aggregation norm S_o to reduce the number of rows of fuzzy relation R_3 in such a way that one row is devoted to contributing to science/scientific field or discipline of one scientist by one journal by (2).

Table 5. The contribution to science - the part of relation R_3

Scientists	Journals					
	J_1	J_2	J_3	J_4	J_5	J_6
S_1	0.01	0	0.0199	0	0	0
S_2	0	0	0	0.01	0.01	0
S_3	0	0.0199	0	0	0	0.01

Values of the relations R_1 and R_2 have to be prepared by experts. In Poland, the relations between scientific fields and disciplines are described by the Ministry of Science and Higher Education [1] and based on this information, the crisp relation R_1 can be built, or after the fuzzification, the fuzzy relation can be developed (Table 2).

The next relation, R_2, can be built by experts – bibliometricians, who describe the levels of the relation between disciplines and journals based on the description of the journal and papers published in this journal. Estimating values of this relation is not easy because some journals declare that they are related to many disciplines and some are devoted to only one discipline (Table 3).

The following relation, R_3, is built on the basis of the number of papers published by the discussed researchers (Table 4) and then aggregated with the application of the optimistic fuzzy aggregation norm (2), and presented in Table 5.

The next set is calculated values of the fuzzy relation R_4 based on the values of the fuzzy relations R_1, R_2, and R_3 (after aggregation) applying the S-T-composition of these relations (1). Table 6 presents the values of fuzzy relation R_4 prepared based on ten researchers.

Table 6. The contribution to science of some scientists

Scientific fields	Scientists									
	S_1	S_2	S_3	S_4	S_5	S_6	S_7	S_8	S_9	S_{10}
F_1	0.053	0.126	0.01	0	0.004	0.243	0.020	0.142	0.070	0.032
F_2	0.163	0.01	0	0	0.002	0	0	0	0	0
F_3	0.997	0	0.458	0.423	0.121	0	0.356	0.044	0.597	0.012
F_4	0.253	0	0.005	0.067	0.010	0.002	0.377	0.533	0.040	0.012
F_5	0	0.154	0	0	0.221	0.039	0	0.028	0.020	0
F_6	0.475	0.025	0.030	0.012	0	0	0.020	0.012	0.019	0.05
F_7	0	0	0	0	0	0.01	0	0	0	0
F_8	0	0.115	0.01	0.238	0	0	0.059	0	0	0

Based on the values of relation R_4, the managers of universities and scientific units can see the contribution of each considered scientist into different scientific fields. On the foundation of this relation, the ranking of scientists concerning to their contribution to the chosen scientific fields can be prepared. Moreover, the graph of these results can be prepared to help users see the contribution of the group.

Based on the results presented in Table 6, Fig. 1 visualizes the contribution to the scientific fields.

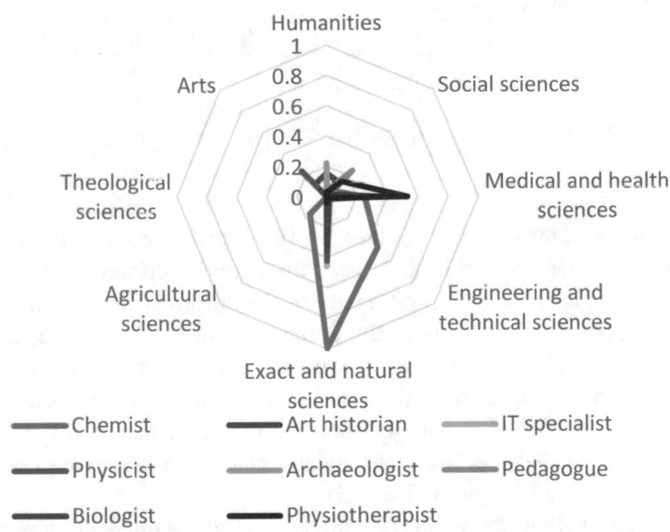

Fig. 1. The contribution of scientists into scientific fields

It can be noticed that S_1's (chemist's) contribution into the Exact and Natural Sciences is the highest and moreover, there is some (smaller) contribution into Engineering and Technical Sciences and also Medical and Health Sciences. S_1's contribution into Exact and Natural Sciences is an outlier of this discussed data set (comp. Table 6), so comparing it with the contributions of all other scientists' contributions is high and causes the difficulty of the presentation of the group's contribution.

One of the problems with this presentation method of scientific contribution is challenging to compare the contribution of scientists visually if there are one or two scientists with great scientific achievements. To solve this problem, the considered group of scientists can be divided into groups with similar achievements, for example, professors, adjuncts, and assistants. The next issue is presenting a scientific contribution of a large group of scientists like the staff of a university.

5 Visualization Based on the T-SNE Method

In 2008, Laurens van der Maaten and Geoffrey Hinton [6] presented the new method of visualization of a significant amount of multidimensional data. This technique. called "t-SNE" (t-distribution Stochastic Embedding Neighbors), transforms each point of the given multidimensional space to 2D or 3D space. This technique is a development of Stochastic Embedding Neighbors method prepared by Hinton and Roweis [7].

Let X be the eight-dimensional space and each point of this space is a vector, which each component presents the scientific contribution of given scientist into chosen scientific field. For example,

$$S_1 = x_1 = (0.053 \quad 0.163 \quad 0.997 \quad 0.253 \quad 0 \quad 0.475 \quad 0 \quad 0)^T.$$

In the paper, the technique t-SNE transforms this eight-dimensional space of scientists' contribution of eight scientific fields into two-dimensional space, which is easy to analyze visually. The basic formula used in the t-SNE technique is the conditional probability $p_{j|i}$ of the event that point x_i be a neighbor of x_j whenever they were chosen as neighbors in a proportion of their probability density of a Gaussian centered at x_i, then [6]

$$p_{j|i} = \frac{exp\left(-\|x_i - x_j\|^2/2\sigma_i^2\right)}{\sum_{k\neq i} exp\left(-\|x_i - x_k\|^2/2\sigma_i^2\right)} \tag{3}$$

where $\|a - b\|$ denotes the distance in the eight-dimensional space and σ_i^2 is the variance of Gaussian probability density centered at x_i.

Figure 2 presents the graph – the visualization of the group of scientists concerning their contribution into eight scientific fields.

As it can be noticed, the t-SNE technique let place the scientist on the 2D space preserving the similarity and the structure (local and global) of the data. Since the data presented in Table 6 and then in Fig. 2 is dispersed, and there are only a few objects (10 scientists), it is difficult to see clusters. However, it can be seen that in the case of

(a)

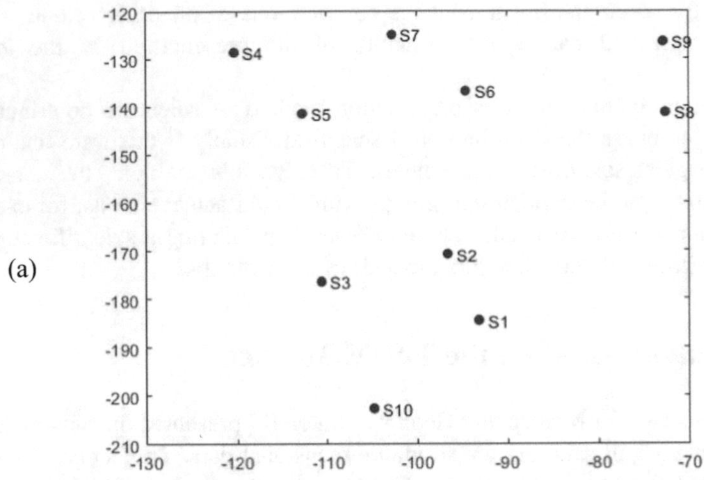

(b)

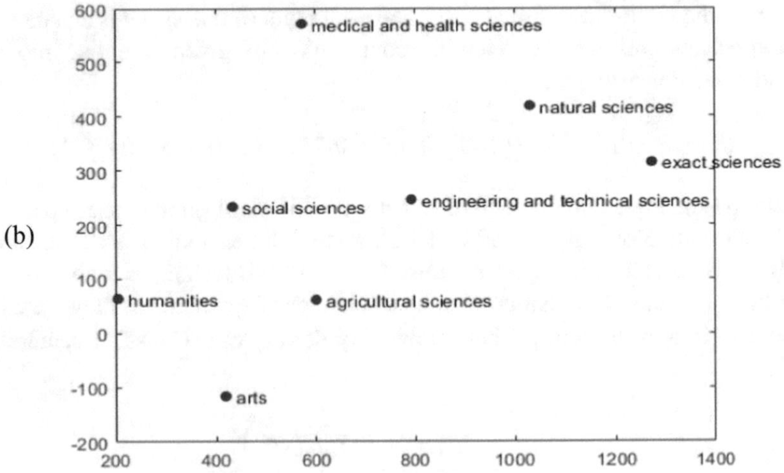

Fig. 2. Visualization of (a) the chosen group of scientists considering their contribution into the scientific fields, (b) contribution to the scientific fields based on these researchers

Fig. 2(a) there are two clusters (one in the upper part of the figure, and the other one in the bottom). In the case, the data is too dispersed to see clusters. However, in the case of recognizing some clusters of researchers, there will be considered larger groups, or their contribution to sciences will be more similar.

For managers, it is convenient to recognize groups of their staff concerning their scientific contribution to discussed scientific fields or disciplines.

6 Software for Calculating and Visualizing the Contribution to Sciences

The application, called ScienceFuzz, Application for visualizing scientific achievements [8], is developed by Authors, is a software which enables visualization of scientific achievements. This system stores data about publications submitted by scientists, journals where the papers are published and a list of scientific disciplines and fields. The main component of the application is a calculator, which allows to input data and determines how the program will calculate the level of contribution for each science field of a given researcher. The calculator analyzes both its internal and external data and calculates the contribution of the scientist to each scientific discipline and domain. The calculations carried out by the program are based on optimistic fuzzy aggregation norms.

The sample graph prepared by this software for a large group of researchers is presented in Fig. 3. As it can be noticed there are clusters of scientists because there are groups of them with similar contributions to the sciences.

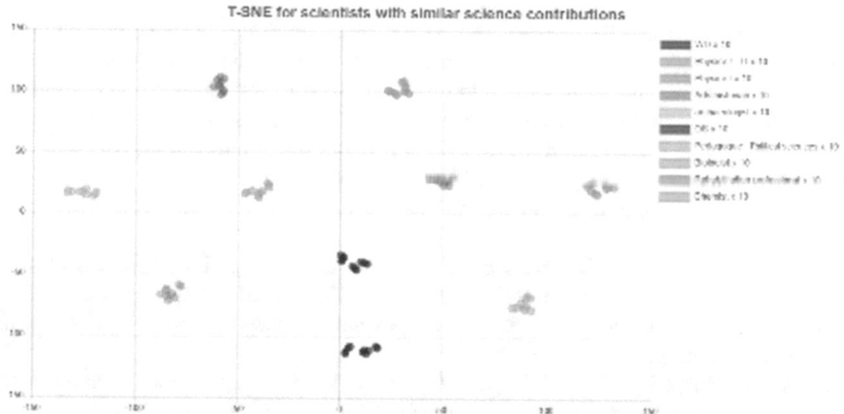

Fig. 3. ScienceFuzz layout to visualize the group of scientists concerning their contribution to sciences

7 Conclusions

Scientists look for methods of estimating the scientific achievements of a group of scientists to forecast and manage the scientific development of the staff of universities or scientific units. The contribution to sciences is the concept which helps to solve this problem. The method using fuzzy relations and optimistic fuzzy aggregation norms let calculate the contribution to science individual researchers and teams of them. Finally, the two methods of visualization are proposed, one is based on the radar style of graphs, and the other one applies the t-SNE technique which was developed by Laurens van der Maaten and Geoffrey Hinton to visualize data of many-dimensional space to two or three-dimensional space. Visualization of scientific achievements, in subject literature called maps of science [9] can facilitate scientometric analyses at different levels, including the parametrization of an individual's output in a more visual portfolio.

This method founds the practical application in the software ScienceFuzz developed for calculating the levels of the scientific contribution of scientists and the scientific units automatically.

References

1. Regulation of the Minister of Science and Higher Education of 8 August 2011 on the areas of knowledge, fields of science and art as well as scientific and artistic disciplines, Official Journal Nr 1790, item 1065 (2018). http://prawo.sejm.gov.pl/isap.nsf/download.xsp/WDU20111791065/O/D20111065.pdf. (Polish)
2. Zadeh, L.A.: Fuzzy sets. Inf. Control **8**, 338–353 (1965)
3. Zadeh, L.A.: The concept of a linguistic variable and its application to approximate reasoning–1. Inf. Sci. **8**, 199–249 (1975)
4. Rutkowski, L.: Methods and techniques of artificial intelligence, PWN, Warsaw (2009). (in Polish)
5. Sokolov, O., Osińska, W., Mreła, A., Duch, W.: Modeling of scientific publications disciplinary collocation based on optimistic fuzzy aggregation norms. In: Advances in Intelligent Systems and Computing. Information Systems Architecture and Technology: Proceedings of 39th International Conference on Information Systems Architecture and Technology - ISAT 2018, Part II, vol. 853, pp. 145–156. Springer, Switzerland (2018). https://link.springer.com/chapter/10.1007/978-3-319-99996-8_14. ISBN 978-3-319-99995-1
6. van der Maaten, L., Hinton, G.: Visualizing data using t-SNE. J. Mach. Learn. Res. **9**, 2579–2605 (2008)
7. Hinton, G.E., Roweis, S.T.: Stochastic neighbor embedding. In: Advances in Neural Information Processing Systems, vol. 15, pp. 833–840. The MIT Press, Cambridge (2002)
8. ScienceFuzz: Application for visualizing scientific achievements. https://sciencefuzz.azurewebsites.net. Accessed 20 May 2019
9. Waltman, L., Boyack, K.W., Colavizza, G., van Eck, N.J.: A principled methodology for comparing relatedness measures for clustering publications (2009). https://www.researchgate.net/publication/330553654_A_principled_methodology_for_comparing_relatedness_measures_for_clustering_publications. Accessed 20 May 2019

Towards a Pragmatic Model of an Artificial Conversational Partner: Opening the Blackbox

Baptiste Jacquet[1,2(✉)] and Jean Baratgin[1,2]

[1] P-A-R-I-S Association, Paris, France
[2] Laboratoire CHArt (EPHE & Université Paris VIII), Paris, France
baptiste.jacquet28@gmail.com

Abstract. In this paper, we suggest a new model of chatbot aiming to enhance the pragmatic aspects of language processing. We believe that dissociating the sentence processing aspect and the information processing aspect to be of the utmost importance in any system aiming to generate relevant utterances in response to the user's input. Therefore, we suggest a cognitively plausible model of a conversation processing system, containing: (1) a sentence decoder, responsible for translating a sentence into a purely semantic information; (2) an information processor, responsible for choosing the appropriate information to send back to the user; (3) a sentence encoder, responsible for generating a sentence able to convey the information chosen by the information processor in a relevant way; (4) a world model, containing all the information the system has learnt about the world; and (5) a user's mental state model, containing all the information the system assumes the user is or is not aware of. We review the literature giving rise to this model and attempt to detail it.

Keywords: Chatbots · Pragmatics · Human-machine interaction · Conversations

1 Introduction

The interest in chatbots has been on the rise in the recent years [1, 2, for examples], especially in businesses [3–5, for examples]. This interest stems from the fact that customer interaction seems to be moving towards mobile systems [6]. In consequence, it is expected that users will require different interfaces more suitable to a mobile experience, including using conversational agents [5] and, more specifically, textual conversational agents like chatbots.

Indeed, having a textual conversation on a mobile device is a very common user experience, and feels quite intuitive. Many challenges remain regarding the efficiency of these systems in providing relevant information to the user and managing their expectations [7]. In particular, the frustration caused by the inability of chatbots to infer the true meaning of the user's utterance, forcing the user to adapt to it rather than having it adapt to them.

While the quality of the syntax of chatbots is now near perfect, and while their semantic capacities have also greatly improved (in particular, when answering simple questions that do not require understanding the context in which they arise), their

J. Świątek et al. (Eds.): ISAT 2019, AISC 1051, pp. 169–178, 2020.
https://doi.org/10.1007/978-3-030-30604-5_15

ability to process contextualized utterances within a conversation is still far from being close to what humans can expect of a conversational partner [8], in particular regarding the way conversational partners refer to previous information given earlier within the conversation.

For example, consider this extract from a conversation with the chatbot A.L.I.C.E from pandorabots.com:

- User: Do you like children?
- A.L.I.C.E: Yes they are the future.
- User: I have two.
- A.L.I.C.E: What are their names?

  […]
- User: Do you like computers?
- A.L.I.C.E: Yes, I live in a computer, so I must like them.
- User: I have two.
- A.L.I.C.E: What are their names?

It is clear here that A.L.I.C.E does not really take context into account, but only processes text sentence by sentence. While this example might feel a bit forced, such events occur quite often in natural conversations, as people are not used to having to keep all the contextual information required to processing their utterances within it. Consider now what happens if we include the contextual information (that we mean to say: I have two [computers]) in the utterance:

- User: I have two computers.
- A.L.I.C.E: Tell me about them.

While the answer here might be generic, it is at least believable.

In order to improve the ability of chatbots to use contextual information, we suggest in this paper a theoretical model of a textual conversation processing system considering contextual information and user expectations.

2 Literature Review

2.1 Pragmatics in Human Conversation

Humans do not use the definition of the words they know in day-to-day language. Instead, the definitions of the words we use tends to vary a lot depending on the context they are used in. This principle does not only apply for words, but also for entire expressions. Hence, making the distinction between what is *said* and what is *meant* is extremely important in order to develop our understanding of human language processing.

Cooperation. One of the first authors to suggest this distinction within conversations was Grice [9]. He suggested that all participants in a conversation expect their conversation partner to cooperate with them in order to communicate. Indeed, efficient communication is a complex task that requires the participants to cooperate if they want to make sure what they mean to say is understood.

Grice suggested that all participants in a conversation follow an implicit "Cooperative Principle" (CP), which is defined in the following terms:

> "Make your contribution such as is required, at the stage at which it occurs, by the accepted purpose or direction of the talk exchange in which you are engaged".

This general principle is subdivided into four super maxims, which can be considered as categories of expectations. These maxims are the maxim of Quality, the maxim of Quantity, the maxim of Relation and the maxim of Manner.

This implies that the speaker has an implicit model of what the listener expects to hear, either in order to conform to these expectations, or to voluntarily use these expectations to provide additional information implicitly.

Relevance. Sperber and Wilson have later attempted to generalize the different maxims suggested by Grice in a single theory: The Relevance Theory [10]. They suggest that all utterances in fact go through a process of effect/cost calculation which defines their relevance. The more effect an utterance has on the mental representations of the target, given a certain context, the more relevant it is. On the other hand, the higher the cognitive cost (effort) required to process the utterance, the less relevant it becomes.

In this theory, if an utterance contains a piece of information that the listener already knows, the sentence becomes irrelevant, for it has no effect on the mental representations of the user, and the effort required to process it was wasted.

Similarly, an utterance containing too many pieces of information that the listener does not know can become irrelevant as well, for while it would potentially have a large effect on the mental representations of the user, the effort required to process it would also greatly increase. Unless the listener is motivated to extract all the information, it is likely that he or she will only process some of it instead.

In consequence, optimal relevance is achieved when an utterance contains just the right amount of new information for the listener, while maintaining the cognitive cost required to process it as low as possible.

2.2 Pragmatics in Artificial Conversational Partners

The quality of the processing of pragmatics in conversational partners is still far from being near the level humans expect [8, for a short review], even though experimental data shows the importance of that kind of processing in the ability of the system to establish cooperative interactions with the user [11–13]. These papers show that, when enough pragmatic clues are given to the user about the entity's willingness to partake in a social interaction, they will indeed consider it to be a social entity and will follow similar "bias" to those observed in interactions between humans. In the experimental protocol used in these studies, the bias consists in being quite equalitarian in how much currency they will take for themselves and leave to the robot. This is not observable without these pragmatic cues.

Chatbots have also been found to feel more "machine like" when they do not use pragmatic cues correctly [14]. Indeed, in this paper the authors investigate the influence of violations of Grice's maxims on participants in a protocol inspired by the Turing Test, in which they needed to guess which of the two conversational partner was a human, and which one was a machine, on excerpts of conversations, found the chatbot much more easily when it violated Grice's maxim of Relation ("Be relevant") in particular, but also for some other maxims, like the second maxim of quantity ("Do not make your contribution more informative than is required"). Interestingly, when the participant did not know that an artificial intelligence was included, they still saw the difference compared to their expectations of a human behavior, as some reacted by saying things like "Are they mentally ill?".

Similar conclusions were drawn from experiences attempting to measure the cognitive cost (effort) as defined by the Relevance Theory [10] relative to the understanding of a sentence, and to the production of its response in interactive online conversations, through the recording of the participants' response times [15–17].

In these experiments, violations of Grice's maxim of relation took, on average, seconds longer to respond to than utterances without any violation, even after factoring in the utterances' length [15, 16]. This can be explained by the time required to (1) understand that the utterance was not relevant to the current topic of conversation, as this is assumed to be true by default and thus needs to be inhibited in such cases [9, 10], (2) consider the new topic of conversation, and (3) write a reply to an unexpected, surprising utterance.

Another experiment with a similar experimental design has shown that a similar thing happens when the chatbot writes too much information that is not directly necessary within their utterance (violation of Grice's second maxim of quantity) [17]. While such violations did not significantly decrease the humanness of the chatbot (certainly because the additional pieces of information could be interpreted simply as offering more diverse conversations topics, like when asking about the chatbots job, not only answering the question, but also providing the job of the chatbot's character's wife), writing unrequested pieces of information did increase the response times of participants, which can be interpreted as needing more effort to process the utterance.

2.3 Current Implementations

As of today, most chatbots use Seq2Seq models (models mapping an input sequence to output sequences) to answer the user's input sentence. This type of system directly maps a vector representation of the input sentence (a sequence) to different output vectors representing potential output sentences (other sequences), without doing any further treatment of the user's expectations regarding the use of context and in general, pragmatic cues, or at least, limited ones [18, 19].

These models traditionally use variants of neural networks to learn the mapping between input sequences and output sequences, along with learning the grammar and syntax of the language. They also often use an attention system, which significantly improves the ability to use contextual information within previous messages [20].

Some hybrid models exist, that add an intermediary within the process of mapping a sequence to another, sometimes using different technologies, like graphs, to not only map sequences to other sequences, but intents to sequences instead [21].

Although Seq2Seq models are interesting for their ability to learn the correct behavior from input data, they still fall short in considering the user's mental state. The model we suggest aims at focusing on this aspect.

3 Model

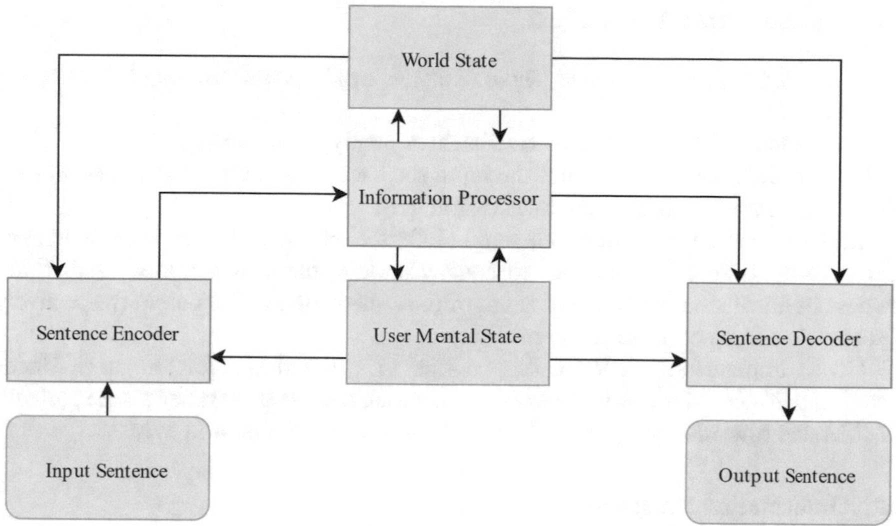

Fig. 1. Model of the conversational processing system we are suggesting. Orange boxes represent storage of inner models (World and User). Blue boxes represent processing systems, and the green boxes represent the input and output.

The model we suggest using is built around two types of components (Fig. 1). On one side we have inner models of the system. These models are divided into two systems: The World State Model (WSM), and the User Mental State Model (UMSM). Both systems contain assumptions, beliefs, the first one about the world in general, and the second one about the user the system is currently interacting with.

On the other side, we have processors. These processing units are the Sentence Encoder (SE), the Information Processor (IP) and the Sentence Decoder (SD). Each of these systems transform the input information they receive, given contextual information extracted from the models.

3.1 Sentence Encoder

The Sentence Encoder is the entry point of the system. Any sentence written by the user will be processed through this unit first in order to extract the semantic content (what was *meant*) of the sentence (what was *said*). This extraction of the semantic content is achieved thanks to the information stored in the UMSM and in the WM. This is the first pragmatic effect. This unit produces predictions of what kind of information the system expects to receive from the user's state modelled within the UMSM and from its beliefs about the world modelled within WM. If the sentence received in input is incompatible with this prediction, the UMSM is updated and the sentence is interpreted in its decontextualized state. It is also the Sentence Encoder which decides whether to process the input sentence at all. For example, if the amount of words that are recognizable is not enough, or if it is in a language that is not known. It should be tolerant to noise (such as typos, special characters, unknown words), as human reading ability is also tolerant to noise [22].

$$SE_{output} = \textit{function of}\ (\textit{input sentence},\ \textit{world model},\ \textit{user model})$$

The output of the SE would consist in a purely semantic representation of the semantic content contained within the input sentence, or at least what can be inferred given the system's expectations towards the user.

In other words, the SE does not simply infer the meaning of a sentence. It attempts to infer what a specific user tries to convey through the sentence they used. It is a process of translation, as the input is not of the same nature as the output (respectively, a sentence, and a semantic representation).

On an implementation level, this consists in a biased Natural Language Understanding unit, biased towards the system's expectations given the user's state stored in UMSM and towards the systems' beliefs about the world stored in WM.

3.2 Information Processor

This is certainly the most complex processing system within the model. It takes as input the result of the processing of SE, using the system's knowledge (or beliefs) about the world to generate a purely semantic representation of the answer that should be sent to the user. If the information given by the user is new to the system, it updates its beliefs about the world, or about the user, depending on the type of the information.

$$IP_{output} = \textit{function of}\,(SE_{output},\ \textit{world model},\ \textit{user model})$$

The output of the IP should be similar in nature to the output of the SE, except that it contains the semantic information the system means to send to the user, which is not necessarily in a human readable format.

This is most likely the most complex of the three processing systems, as it is essentially meant to be emulating different cognitive processes, including some aspects of conscious thought. This role could be taken either by an expert system or by a learning algorithm.

3.3 Sentence Decoder

This processing device is the opposite of the SE. It translates a purely semantic representation of a piece of information (what is *meant*) into a sentence (what is *said*). This sentence might not contain every single aspect of the representation. Indeed, some of them might not be relevant to the user, as they might already be quite aware of them. In consequence, the sentence produced will only contain aspects of the information that were not already known by the user.

$$SD_{output} = function\, of(IP_{output}, world\, model, user\, model)$$

The SD is, just like the SE, a translation unit, turning a purely semantic representation of a message into a human-readable sentence.

The SD does not simply translate the semantic content into a sentence, but it uses the model of the user's mental state to consider their expectations. Indeed, a user does not expect to receive in response to their question information that they are already aware of, or are simply not interested in.

On an implementation level, this processing unit is a Natural Language Generation device, producing sentences biased by the assumptions generated by the UMSM and by the WM to produce sentences containing only the relevant pieces of information.

3.4 World Model

This unit is where the model of the world inferred by the system is stored. It contains all the information that the system has learnt during its entire "life". Its contents can be updated by the IP, sending new information to it. The information stored within it can also be recollected by the IP.

It should not only contain a list of properties, but also the links between each of them.

On an implementation level, this unit should be the equivalent of a Working Memory, of a Short-Term Memory and of a Long-Term Memory combined.

3.5 User Mental State Model

This unit can be considered as a more specific extension of the WM. The main difference is that unlike WM, which is general, this model is user specific. In other words, it contains what the system assumes the user's beliefs about the world to be. In consequence, for each user, there should be one model. It should always closely match the WM of the system. For example, if the system has a certain information stored in its WM, it should be able to decide whether the user would likely have some knowledge about this specific information, or not.

The way the system would generate new models for users that it does not know could depend on general assumptions that it would learn. For example if it learns that most people know of the concept A, any new user will be assumed to know it as well, unless the conversation with them becomes incompatible with that assumption, prompting the system to ask directly about their knowledge of A, or more simply to explain A.

4 Discussion

The model we are suggesting is still far from being complete and remains in development in order to reach a progressively more detailed structure of a cognitively plausible model of a conversation processing system.

We are quite aware that the distinction between the SE and SD might seem arbitrary in a cognitively plausible model, as studies seem to show that both language comprehension and expression seem to be using the same regions of the brain to process lexical, grammatical and syntactic information [23]. Yet the model we describe here is first and foremost a functional one, independent of its implementation.

In a similar way, the distinction between WM and UMSM might also seem arbitrary to some readers. Indeed, in the brain, the UMSM would likely be included within WM, perhaps as single neuronal patterns that inhibit or activate other neuronal patterns of the WM to represent a specific user's knowledge of some information. Still, for the purpose of this first attempt at suggesting a model, we believe making that distinction clear to highlight the importance of the modeling of the user to be of great importance, as it is on the interaction with the user that most pragmatic effects occur within conversations.

It is also worth noting that the way in which new user models are created within the system would likely participate in the appearance of prejudices, just like they exist within humans. Indeed, should the system notice that certain profiles tend to know about certain concepts, it might later wrongly generalize that every profile similar will also have some knowledge about this concept. While this would in most cases allow for a more personalized experience for the user, it should not be forgotten that the system might sometimes be wrong in its assumptions. In consequence, the importance of making sure that the system can update its beliefs about the user is of utmost importance in order to counter-balance this aspect of the system's processing.

The main aim of this model is to offer a way of exploring what was previously considered to be a Blackbox: the interaction between expectations of conversational partners in the way they generate and interpret sentences.

5 Conclusion

While this type of model has not been tested yet, we suspect it could be greatly beneficial to the field of artificial intelligence applied to conversation processing. Indeed, to our knowledge, no system currently attempts to include the modeling of the user's state of mind in the process of generating and interpreting sentences to offer a personalized interaction with the chatbot. Indeed, while the pragmatics of conversation will not be thoroughly taken into consideration, chatbots will never be able to fully grasp the complexity of human conversations, leading to an increased frustration of the user preventing any efficient cooperation between the two conversational partners, human and artificial.

Acknowledgments. We would like to thank Charles Tijus, Frank Jamet, Olivier Masson and Jean-Louis Stilgenbauer for their support and advices in the development of our model.

References

1. Dahiya, M.: A tool of conversation: chatbot. Int. J. Comput. Sci. Eng. **5**(5), 158–161 (2017)
2. Qiu, M., Li, F.L., Wang, S., Gao, X., Chen, Y., Zhao, W., Chen, H., Huang, J., Chu, W.: AliMe chat: a sequence to sequence and rerank based chatbot engine. In: Proceedings of the 55th Annual Meeting of the Association for Computational Linguistics, Vancouver, vol. 2, pp. 498–503 (2017)
3. Thomas, N.T.: An e-business chatbot using AIML and LSA. In: International Conference on Advances in Computing, Communications and Informatics (ICACCI), Jaipur, pp. 2740–2742 (2016)
4. Heo, M., Lee, K.J.: Chatbot as a new business communication tool: the case of naver talktalk. Bus. Commun. Res. Pract. **1**(1), 41–45 (2018)
5. Zhou, A., Jia, M., Yao, M.: Business of bots: how to grow your company through conversation. Topbots Inc. (2017)
6. Faulds, D.J., Mangold, W.G., Raju, P.S., Valsalan, S.: The mobile shopping revolution: redefining the consumer decision process. Bus. Horiz. **61**(2), 323–338 (2018)
7. Chaves, A.P., Gerosa, M.A.: How should my chatbot interact? A survey on human-chatbot interaction design. arXiv, 1–44 (2019)
8. Jacquet, B., Masson, O., Jamet, F., Baratgin, J.: On the lack of pragmatic processing in artificial conversational agents. In: International Conference on Human Systems Engineering and Design: Future Trends and Applications, pp. 394–399. Springer, Cham (2018)
9. Grice, H.P.: Logic and conversation. In: Cole, P., Morgan, J.L. (eds.) Syntax and Semantics 3: Speech Arts, pp. 41–58 (1975)
10. Sperber, D., Wilson, D.: Relevance: Communication and Cognition, 2nd edn. Blackwell, Oxford (1995)
11. Masson, O., Baratgin, J., Jamet, F.: Nao robot and the "endowment effect". In: IEEE International Workshop on Advanced Robotics and its Social Impacts, Lyon, pp. 1–6 (2015)
12. Masson, O., Baratgin, J., Jamet, F.: Nao robot as experimenter: social cues emitter and neutralizer to bring new results in experimental psychology. In: Proceedings of the International Conference on Information and Digital Technologies, pp. 256–264 (2017)
13. Masson, O., Baratgin, J., Jamet, F.: Nao robot, transmitter of social cues: what impacts? In: Benferhat, S., Tabia, K., Ali, M. (eds.) Advances in Artificial Intelligence: From Theory to Practice. IEA/AIE 2017. Lecture Notes in Computer Science, vol. 10350, pp. 559–568. Springer, Cham (2017)
14. Saygin, A.P., Cicekli, I.: Pragmatics in human-computer conversations. J. Pragmat. **34**, 227–258 (2002)
15. Jacquet, B., Baratgin, J., Jamet, F.: The Gricean maxims of quantity and of relation in the Turing Test. In: 11th International Conference on Human System Interaction, pp. 332–338. IEEE (2018)
16. Jacquet, B., Baratgin, J., Jamet, F.: Cooperation in online conversations: the response times as a window into the cognition of natural language processing. Front. Psychol. **10**, 1–15 (2019)
17. Jacquet, B., Hullin, A., Baratgin, J., Jamet, F.: The impact of the Gricean maxims of quality, quantity and manner in chatbots. In: Proceedings of the 2019 International conference on Information and Digital Technologies, pp. 1–10 (2019)
18. Nayak, N., Hakkani-Tür, D., Walker, M.A., Heck, L.P.: To plan or not to plan? Discourse planning in slot-value informed sequence to sequence models for language generation. In: INTERSPEECH, pp. 3339–3343 (2017)

19. Cui, L., Huang, S., Wei, F., Tan, C., Duan, C., Zhou, M.: Superagent: a customer service chatbot for e-commerce websites. In: Proceedings of ACL 2017, System Demonstrations, pp. 97–102 (2017)
20. Mei, H., Bansal, M., Walter, M.R.: Coherent dialogue with attention-based language models. In: Thirty-First AAAI Conference on Artificial Intelligence, pp. 3252–3258 (2017)
21. Arora, A., Srivastava, A., Bansal, S.: Graph and neural network-based intelligent conversation system. In: Banati, H., Mehta, S., Kaur, P. (eds.) Nature-Inspired Algorithms for Big Data Frameworks, Hershey, pp. 339–357 (2019)
22. Goodman, K.S.: Reading: a psycholinguistic guessing game. J. Read. Spec. **6**(4), 126–135 (1967)
23. Menenti, L., Gierhan, S.M.E., Segaert, K., Hagoort, P.: Shared language: overlap and segregation of the neuronal infrastructure for speaking and listening revealed by functional MRI. Psychol. Sci. **22**(9), 1173–1182 (2011)

Random Forests in a Glassworks: Knowledge Discovery from Industrial Data

Galina Setlak(✉) and Lukasz Pasko

Rzeszow University of Technology,
al. Powstancow Warszawy 12, 35-959 Rzeszow, Poland
{gseltak, lpasko}@prz.edu.pl

Abstract. This paper presents the use of random forests for knowledge discovery from large data sets. The data was obtained from sensors of glassworks technological process monitoring system. Based on original and standardized data, several random forest models were created. Then they were used to establish a significance ranking of explanatory variables. The ranking presents information about explanatory variables that most affect the product defects.

Keywords: Random forests · Knowledge discovery · Decision trees

1 Introduction

The new concept of production automation Industry 4.0 provides for the creation of so-called smart enterprises and the widespread use of modern information and telecommunications technologies in the manufacturing industry [1]. Industry 4.0 involves the need to collect, integrate and process huge amounts of data in real time and use advanced methods to analyze these data. Thanks to the use of analytical tools to process of huge datasets collected from networked machines and devices, we gain insight into their work and the ability to use this information to implement intelligent procedures in decision-making processes.

In the Industry 4.0 modern information and communication technologies like cyber-physical system, big data analytics and cloud computing, will help early detection of defects and production failures, thus enabling their prevention and increasing productivity, quality, and agility benefits of the enterprise.

In connection with the above, the problem of developing and applying advanced data analytics techniques for industrial analytics of big data sets becomes more and more relevant. Currently, many research centers carry out research in the field of artificial intelligence methods and techniques such as "machines learning applications", tools for "knowledge discovery from databases" and data mining [2–5].

Knowledge Discovery in Databases is a process whose task is comprehensive data analysis, starting from a proper understanding of the problem under investigation, through data preparation, implementation of relevant models and analyzes, to their evaluation. The purpose of discovering knowledge from data is to extract information invisible to the user due to the need to process too much data. Then the identified information is transformed into knowledge that can be used, among others to support decision-making [5].

J. Świątek et al. (Eds.): ISAT 2019, AISC 1051, pp. 179–188, 2020.
https://doi.org/10.1007/978-3-030-30604-5_16

Data mining is the second, key stage in discovering knowledge in databases [7–11]. The task of data mining is to automatically discover non-trivial, hitherto unknown dependencies, similarities or trends, generally referred to as patterns, in large data resources. The methods used for this purpose are associated with statistical multidimensional analysis, artificial intelligence, machine learning and deep learning. These include, among others: discriminant analysis, neural networks, fuzzy logic, and decision trees [8]. Data mining technology is now a response to the need for advanced and automatic analysis of big data collected in an enterprise.

This paper is a continuation of scientific research in the selection of methods and tools for knowledge discovery from data sets obtained from the system for monitoring the production process of glass packaging in order to use it in the knowledge base of the developed decision support system. The results of previous research have been presented, among others in [6], where artificial neural networks were used to classification of defects in glass packaging products.

The aim of the research in this work is to assess the possibility of using random forests (RF henceforth) to knowledge discovery from large data sets obtained from sensors of the glassworks manufacturing monitoring system. Models of RF were used to establish a significance ranking of explanatory variables that most affect the product defects.

2 Decision Trees and Random Forests

Decision trees and related RF are data analysis tools belonging to machine learning methods. Decision trees are the most popular tool commonly used to classification and generate decision rules based on a data set [3, 8, 9]. They can also be used to investigate the significance of variables present in this data set. Models of decision trees and RF are created based on learning sets.

The decision tree is a structure composed of nodes and leaves connected by branches. Logical conditions consisting of an explanatory variable and a corresponding value determined by the tree-forming algorithm are placed in tree nodes. However, in the leaves there are values of the explained variable. In case the explained variable is of the numerical type (as in the presented research), the decision tree is called the regression tree.

Decision trees are distinguished by their simplicity and speed of action. The ability to reproduce unknown, non-linear dependencies is also mentioned as an advantage. In addition, the decision tree is a very transparent, easily interpretable model. Unfortunately, the trees also have disadvantages. Above all, their response is of a step nature. Another disadvantage of trees is their instability: even a small change of data (e.g. the removal of several cases from learning data) may result in a completely different tree.

To improve decision trees, retaining at least some of their advantages, while reducing their drawbacks, RF were proposed [12, 13]. The idea of RF consists in using many decision trees instead of one. Individual trees are created not based on the entire data set, but on random samples. A subset of observations is selected for each tree, which then acts as a learning set on the basis of which the tree is created. With this approach, can be ensured the right variety of trees included in the model.

RF are used in various fields. In [14] RF predicts the modification of auditor opinion. The performance of a production system and failure events are predicted using among others RF in [15]. Machine learning-based classification algorithms using RF help in diagnosing the symptoms of diabetes mellitus at early stages [16]. In [17] RF is used to recognize protein folds.

In the case of a regression problem, it is usually assumed that the pool of variables random selected before establishing each tree node consists of $m/3$ variables, where m is the number of explanatory variables. An important feature of RF models is the ability to determine the significance of individual variables. The significance of variables is calculated for each created RF and is presented in the form of variable weights. The higher the weight of a given variable, the greater its effect on the explained variable is.

3 A Research Example

The subject of the research is the data received from the technological process monitoring system in the glass industry enterprise (Glassworks). In the glass production process there are nine main activities related to the transformation of raw materials into finished products. The production lines include several dozen machines. The monitoring system of the technological process registers the operating parameters of these machines and the results of the quality control of finished products. The data from measurement points located on production lines are collected using specialized software monitoring the parameters of the technological process and product defects. From conversations with employees of the Glassworks follows that an anomaly was observed. The anomaly involves the occurrence of production periods in which an increased number of defective products is recorded. The increase in the number of defective products is above average, and its cause is unknown even to specialist of glassworks-making processes. The anomaly is primarily associated with one type of defect, referred to as EBI (Empty Bottle Inspector), therefore in the considered research problem only this type of product defects is analyzed.

In this study, the training set included all of the data collected by the sensors recorded at an interval of 10 min for a single production line. The output variables (explained) are the number of products with EBI defect on each of the three production lines. On the other hand, the input (explanatory) variables can include the parameters of the forehearths and the furnace as well as data on the cooling of moulds and atmospheric conditions. To study the causes of the increased number of defective products on the analyzed L21 line, 70 input variables can be used.

The purpose of creating RF was to obtain:

- a model that predict values of EBI depending on explanatory variables,
- the significance ranking of explanatory variables, presenting information on which explanatory variables most affect the EBI values.

3.1 Original Data

Based on original data, several RF models were created. During their creation, two parameters of the algorithm building the RF model were modified:

- the number of trees in the model and
- the number of variables drawn into the pool from which the tree node is selected.

Observing the results, particular attention was paid to the significance ranking of explanatory variables. Initially, the standard pool size was used (*m*/3 what is equal 23 variables for the L21 line), and the number of trees in the model was set to 1000.

Figure 1 shows that when creating a model, i.e. adding more trees to it, the mean square error (MSE) of the model prediction initially decreases very rapidly, and then stabilizes when the model contains about 100 trees. The MSE value for a model composed of 1000 trees is 318.9. As the MSE decreases, the value of the R^2 coefficient increases in the same way. After stabilization, its value is 0.605. However, in order to reduce the number of trees in the model, to shorten the time needed to generate it, a problem of inconsistencies in the significance of variables was noticed. To show this problem, two models were created, each containing 200 trees.

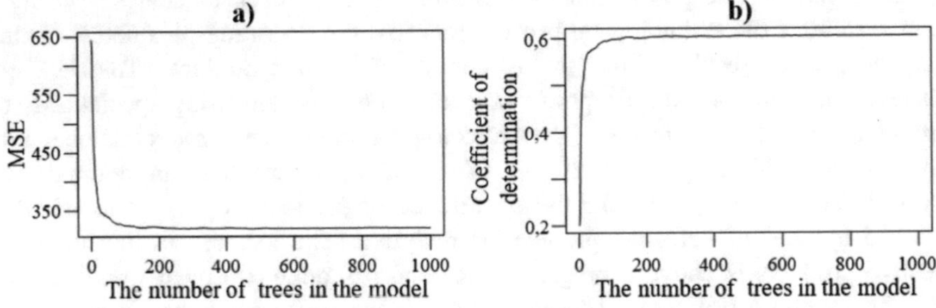

Fig. 1. Change in the mean square error of the prediction (a) and the determination coefficient (b) of the model created for the data obtained from the L21 production line

Then, for both models, information about the thirty variables that most affect the value of the EBI was generated. They are presented in Fig. 2. The value of MSE for the first and second model was 324.7 and 326.7 respectively, and the determination coefficient R^2 was equal to 0.598 and 0.596. It turns out that models with such similar values of MSE and R^2 give different results of the significance of variables. In both generated variable rankings, the first place is taken by J21.A1C.TOP.TP.TC.PV. The next few places have the same variables, but their order in both rankings is different. However, from about the 10th place the differences between the weights become less and less noticeable. To stabilize the ranking of variables (in particular on the first few positions), the number of trees in the model was increased to 1200. This value was established by creating many models with different number of trees. For models with 1200 trees, stable results were obtained in the top ten in the significance ranking of variables.

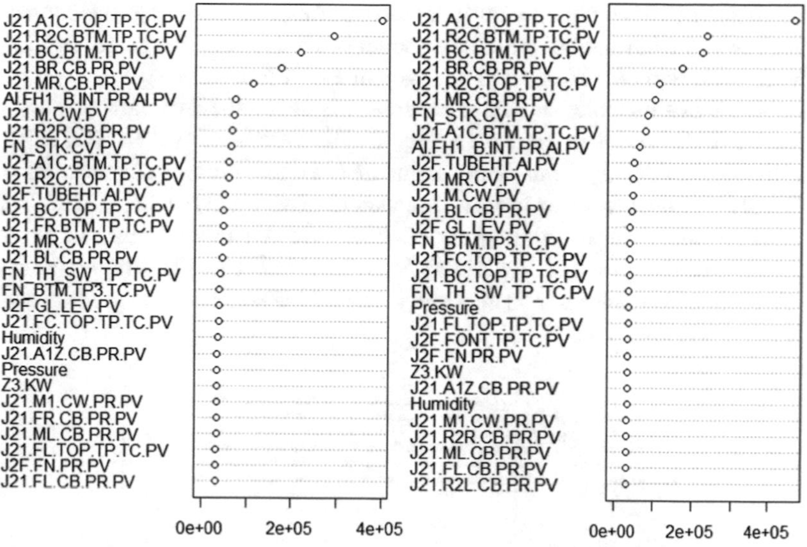

Fig. 2. Weights of thirty variables that most affect the value of the EBI: (a) model 1; (b) model 2

Thee model containing 1200 trees was marked as RF1. The MSE and R^2 values for RF1 were 318.7 and 0.605, respectively. In addition, to the above measures on the accuracy of the created model, it can be also visually checked how the model maps the course of the EBI variable. Figure 3 presents a graph of EBI values predicted using the RF1 model (a) and a graph of the actual values of the EBI variable (b).

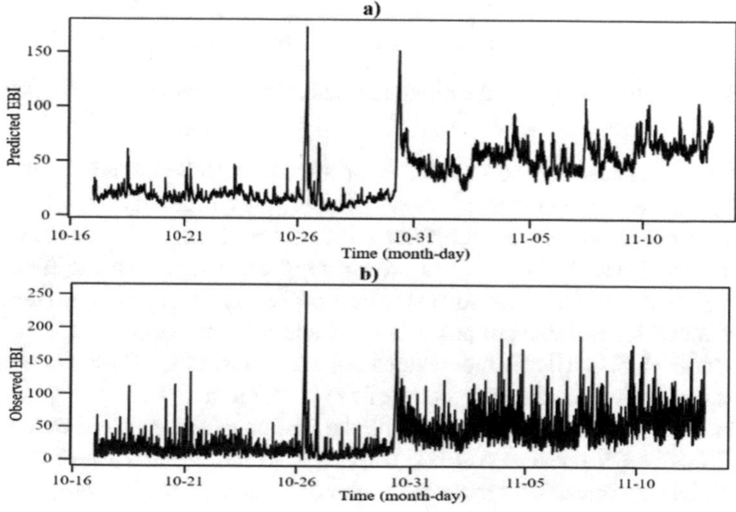

Fig. 3. EBI values: (a) predicted using the RF1 model; (b) real

The course of the predicted values does not correspond exactly to the course of the EBI. However, the advantage of the RF1 model is that it can map changes in EBI levels and the most important outliers and changes in EBI variance.

The ranking of variables that have the most impact on the EBI value created on the basis of the RF1 model is shown in Fig. 4. The variable J21.A1C.TOP.TP.TC.PV seems to be particularly significant compared to the other variables. Its weight (401133.02) is almost twice as large as the weight of variable J21.BC.BTM.TP.TC.PV occupying second place in the ranking (249120.96). A similar weight to the second variable is also given by J21.R2C.BTM.TP.TC.PV (227136.77). The weights of the three subsequent variables are in the range from 118780.92 to 161157.90, while the other variables do not exceed 80000.

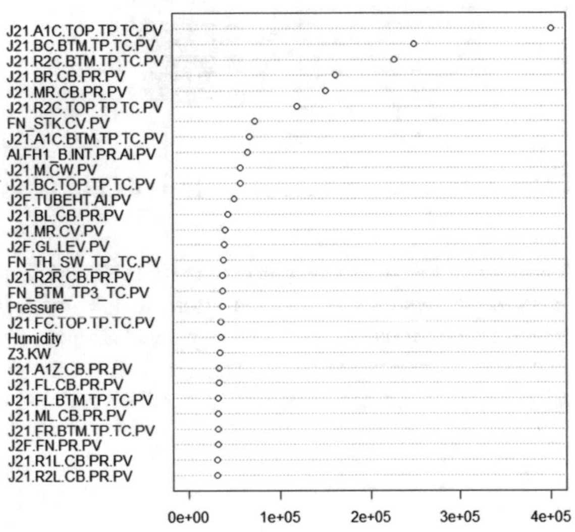

Fig. 4. The weights of thirty variables that most affect the EBI on the L21 line (RF1 model)

To better understanding the variables that turned out to be the most important, the following plots show the course of these variables over time together (Table 1). The correlation results of the ten most important variables in the RF1 model and the EBI variable indicated that these variables are strongly correlated with the EBI, with the exception of one variable. The lowest value of Pearson's correlation coefficient of −0.66 was noted for variables in positions 1, 2 and 6 in the ranking. In turn, variables taking places 3, 4, 5, 8, 10 obtained values not exceeding −0.6. These were one of the lowest results obtained during the study of the correlation.

Two variables positively correlated with the EBI take the top places of the analyzed ranking: FN_STK.CV.PV and AI.FH1_B.INT.PR.AI.PV. In the case of the first one, one of the highest correlation results was recorded (0.63), however, the correlation of the second variable was only 0.27. Comparing its plot with the EBI, there are clear differences in the course of both variables, hence the low correlation between these variables. Figure 4 shows that the significance of only the first six places in the ranking is clearly greater than the others. The remaining variables do not show significant differences in weights, hence the presence of a variable with such a low correlation with the EBI on the 9th position of the ranking.

Table 1. The plots of the course over ten variables that most affect the EBI value on the L21 production line according to the RF1 model

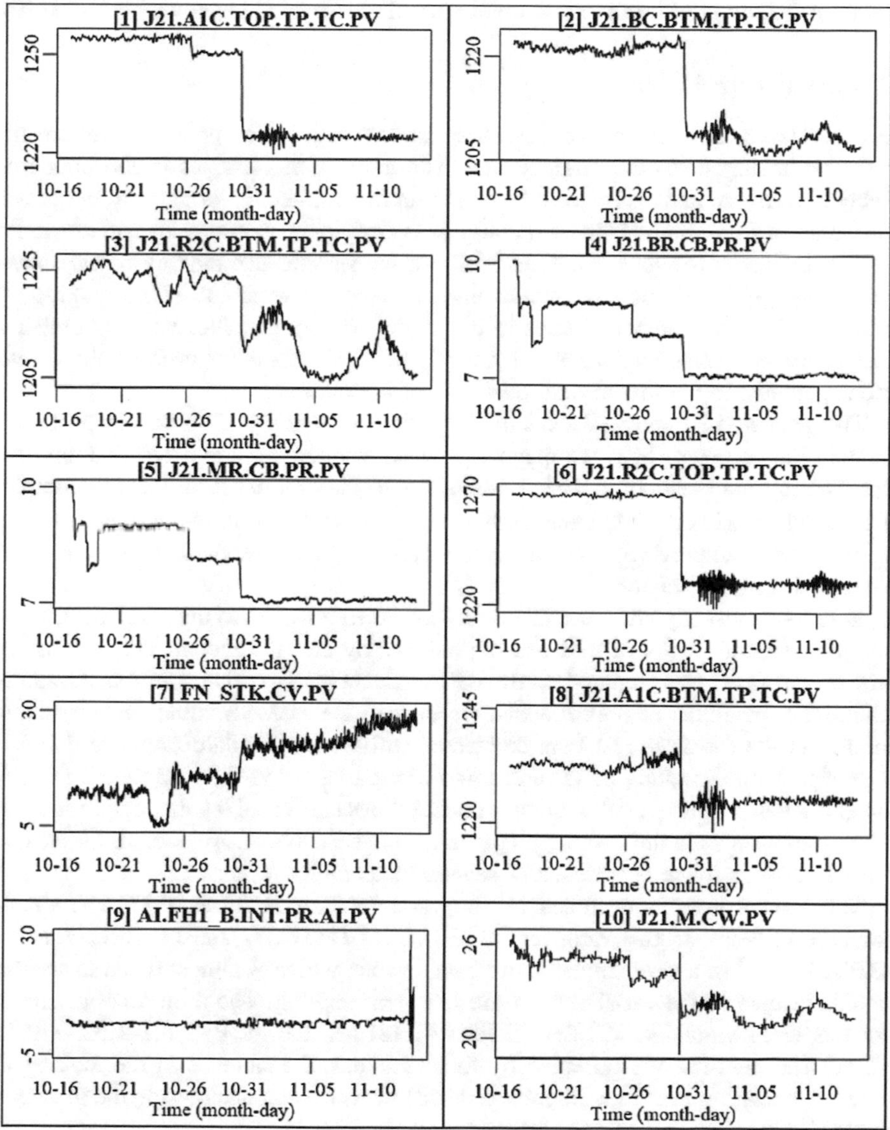

To compare the studied explanatory variables and find those whose course is similar in time, hierarchical grouping was performed. The Euclidean distance was used as a measure of data difference. The grouping results showed that the highest similarity occurs between variables in positions 4th and 5th in the significance ranking. Variables 1st and 6th are similar too. Moreover, variables in positions 2nd, 8th and 3rd are similar each other. Among the negatively correlated variables from the EBI, the most distant in the sense of Euclidean distance is 10th variable. Variables positively correlated with the EBI do not show great similarity to other variables correlated negatively with the EBI.

3.2 Standardized Data

RF models were also created for data after standardization. The procedure for creating them was analogous to the original data. The number of variables in the pool from which a given tree node is selected has been set to *m*/3, i.e. 23 variables in the case of data obtained from the L21 line. Similarly to the original data, the adoption of too few trees in the model resulted in the instability of the significance ranking of variables – each of the generated models gave a different order of variables in the ranking. To counteract this, the number of trees in the model was set to 1200. Such a number of trees guaranteed obtaining the same ranking of significance for each of the created models, at least in the first several positions of the ranking.

The generated RF model for standardized data, containing 1200 trees, was marked as RF2. The course of both the mean square error value and the determination coefficient during the creation of RF2 is analogous to the original data. For the complete RF2 model containing 1200 trees, MSE and R^2 amounted to 0.39 and 0.605, respectively. Much smaller MSE error values are caused by data standardization, which has scaled their original values.

In order to visually checking the ability of the RF2 model to map EBI data, Fig. 5 was prepared with a plot of EBI values predicted by the RF2 model. The plot has the same course as the plot prepared for the RF1 model (Fig. 3), which confirms the similar predictive capabilities of both models. As in the case of MSE values, much smaller amounts of the forecasted EBI are connected with earlier standardization of data.

A significance ranking of variables was created based on the RF2 model (Fig. 6). The six initial positions of the ranking occupy the same variables that occurred at the same positions also in the ranking prepared for the RF1 model. Another similar feature of both rankings is the occurrence of several “materiality levels”.

The highest level is occupied by the variable J21.A1C.TOP.TP.TC.PV. On the second level there are two variables: J21.BC.BTM.TP.TC.PV and J21.R2C.BTM.TP.TC.PV. Both have a very similar significance value, which is almost twice lower than the significance of the variable from the first item ranking. The third level of significance is three variables: J21.BR.CB.PR.PV, J21.MR.CB.PR.PV, J21.R2C.TOP.TP.TC.PV. The next level is occupied by four variables, the same as in the ranking for RF1, but appearing here in a different order. However, variables outside the top ten of the ranking do not show strong differences in significance.

Clear similarities in significance rankings for the RF1 and RF2 models confirm earlier observations about the very similar capabilities of both models to the EBI data representation.

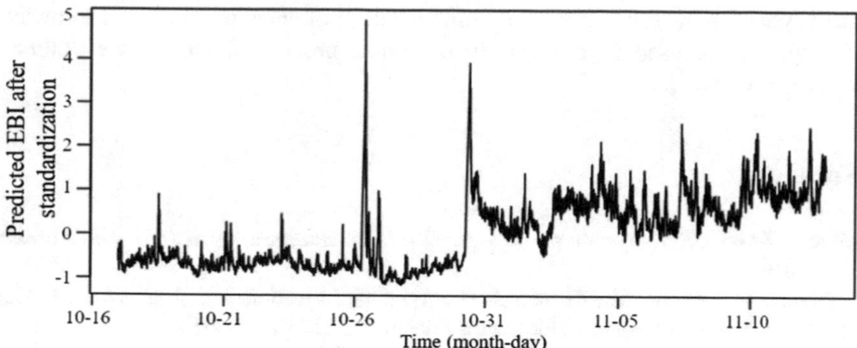

Fig. 5. EBI values predicted using the RF2 model

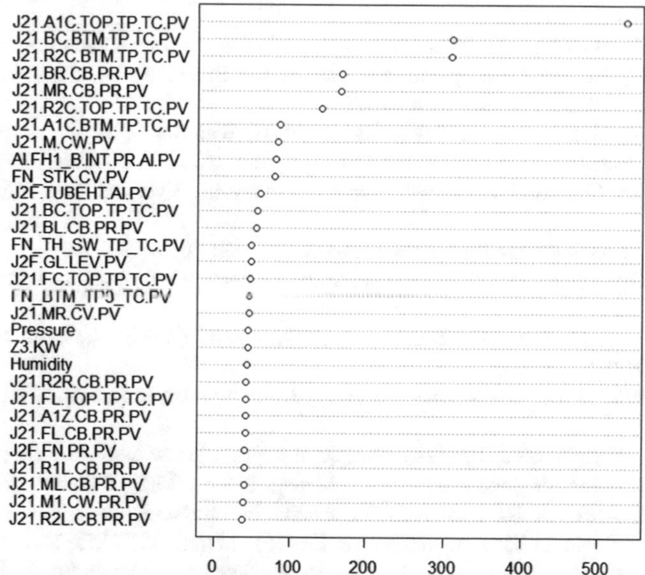

Fig. 6. The weights of thirty variables that most affect the value of a standardized EBI on the L21 line (RF2 model)

4 Conclusions

The paper presents applications of RF for knowledge discovery in large industrial datasets obtained from sensors of a technological process monitoring system and measurement points in the process of quality control of products in a glassworks. Based on original data, several RF models were created, with the help of which it is possible to determine the significance of individual explanatory variables. Acquired knowledge as a result of this research will be used in the knowledge base of the intelligent decision support system in the scope of streamlining the technological process. This knowledge, above all, can be used in the quality assurance process to reduce the number of defective products.

References

1. Piatek, Z.: http://przemysl-40.pl/index.php/2017/03/22/czym-jest-przemysl-4-0/. Accessed 9 Jan 2019
2. Harding, J., Shahbaz, M., Kusiak, A.: Data mining in manufacturing: a review. J. Manuf. Sci. Eng. **128**(4), 969–976 (2006). https://doi.org/10.1115/1.2194554
3. Cichosz, P.: Systemy uczace sie. WNT, Warszawa (2000)
4. Flach, P.: Machine Learning: The Art and Science of Algorithms That Make Sense of Data. Cambridge University Press, Cambridge (2012)
5. Gama, J.: Knowledge Discovery from Data Streams, 1st edn. Chapman & Hall/CRC, Boca Raton (2010)
6. Setlak, G., Pirog-Mazur, M., Pasko, L.: Intelligent analysis of manufacturing data. In: Setlak, G., Markov, K. (eds.) Computational Models for Business and Engineering Domains. ITHEA, pp. 109–122. Rzeszow-Sofia (2014)
7. Larose, D.T.: Odkrywanie wiedzy z danych. PWN, Warszawa (2006)
8. Morzy, T.: Eksploracja danych. Metody i algorytmy. PWN, Warszawa (2013)
9. Osowski, S.: Metody i narzędzia eksploracji danych. Wydawnictwo BTC, Legionowo (2013)
10. Sayad, S.: http://www.saedsayad.com/. Accessed 9 Jan 2019
11. Saad, H.: The application of data mining in the production processes. Ind. Eng. **2**(1), 26–33 (2018)
12. Breiman, L.: Random forests. Mach. Learn. **45**(1), 5–32 (2001). https://doi.org/10.1023/A:1010933404324
13. Liaw, A., Wiener, M.: Classification and regression by random forest. R News **2**(3), 18–22 (2002)
14. Gorska, R., Staszkiewicz, P.: Zastosowanie algorytmu lasow losowych do prognozowania modyfikacji opinii bieglego rewidenta. J. Manag. Financ. **15**(3), 339–348 (2017)
15. Accorsi, R., Manzini, R., Pascarella, P., Patella, M., Sassi, S.: Data mining and machine learning for condition-based maintenance. Procedia Manuf. **11**, 1153–1161 (2017)
16. Komal Kumar, N., Vigneswari, D., Vamsi Krishna, M., Phanindra Reddy, G.V.: An optimized random forest classifier for diabetes mellitus. In: Abraham, A., Dutta, P., Mandal, J., Bhattacharya, A., Dutta, S. (eds.) Emerging Technologies in Data Mining and Information Security. Advances in Intelligent Systems and Computing, vol. 813. Springer, Singapore (2019)
17. Jo, T., Cheng, J.: Improving protein fold recognition by random forest. BMC Bioinform. **15**, S14 (2014)

Predicting Suitable Time for Sending Marketing Emails

Ján Paralič[1(✉)], Tomáš Kaszoni[1], and Jakub Mačina[2]

[1] Department of Cybernetics and AI, Technical University Košice, Letná 9, 04200 Košice, Slovakia
jan.paralic@tuke.sk

[2] Exponea Ltd., City Business Center I, Karadžičova 8/7244, Bratislava, Slovakia

Abstract. E-mail marketing is one of the main channels of communication with existing and potential customers. The Open Rate metric is as one of the primary indicators of email campaign success. There are many features of e-mail communication affecting the behavior of individual recipients. Understanding and properly setting these features imply the success of email marketing campaigns. One of these features is the time to send the e-mail. In this paper, we present a methodology for predicting suitable email sending time. Analyzing the available data collected from e-mail communications between companies and customers creates a space for applying data mining methods to data. The proposed methodology for the generation of prediction models to determine the optimal time to send an email has been implemented and evaluated on a real dataset with very promising results.

Keywords: Data mining · E-mail marketing · CRISP-DM · Classification · Data mining methodology

1 Introduction

Imagine a medium that allows traders to understand and respond to individual customer behavior like never before. The medium that helps merchants identify products and services based on customer interests and orders. It sounds like a merchant's dream, but today, it's all possible through the use of e-mail communication [2].

We live in a time when Internet technologies have revolutionized the marketing industry [6]. Use of the Internet by companies for presentation and branding become a common, even necessary phenomenon in the field of e-commerce [1, 5]. The related electronic marketing (e-marketing) refers to the use of digital media such as the web, e-mail communications [8], wireless media but also includes electronic customer relationship management as well as digital customer data management. E-mail marketing [4] as part of e-marketing is used to increase web traffic and sales support. It is a targeted sending of commercial and non-commercial messages to a list of recipient e-mail addresses. E-mail marketing enables traders to gain a customer-driven advantage for decades [7].

J. Świątek et al. (Eds.): ISAT 2019, AISC 1051, pp. 189–196, 2020.
https://doi.org/10.1007/978-3-030-30604-5_17

Nevertheless, many marketing workers still cannot use the wealth of customer data. Merchants who use this data not only increase revenue but deepen their customer relationships. However, such advancements have created an interesting paradox: e-mail has created such effectiveness that traders have lost their motivation to invest enough resources in using this media properly. Because e-mail is so cheap, many marketing workers continue to use most of their budget for traditional marketing media, neglecting many features that affect the e-mail marketing result.

The work presented in this paper provides a methodology to determine the optimal time to send an e-mail, depending on the type of customer and the marketing campaign. The ability to set the optimal time to send an e-mail message to a client should help increase the amount of opened e-mails and direct traffic to the desired websites.

We suggested methodology based on a cascade of suitable classification models learned by data mining methods. Based on these models, one can classify individual customers and specify the best time to send an e-mail. As a result, the open-rate on the sent emails significantly increases.

2 State of the Art

2.1 Related Work

There is a lot of data from various studies which try to determine the optimal time for sending e-mails [1, 3, 4, 9, 10]. The problem is that the conclusions of some of these studies are in agreement, while others are completely contradictory.

Many of the existing studies in the result describe that determining the optimal e-mail sending time depends on the recipient list because each e-mail list consists of a different group of people with different behavior and preferences [4, 9, 10]. We, therefore, decided to propose a general methodology on how to approach the problem of suitable sending time for any group of customers for which suitable historical data is available.

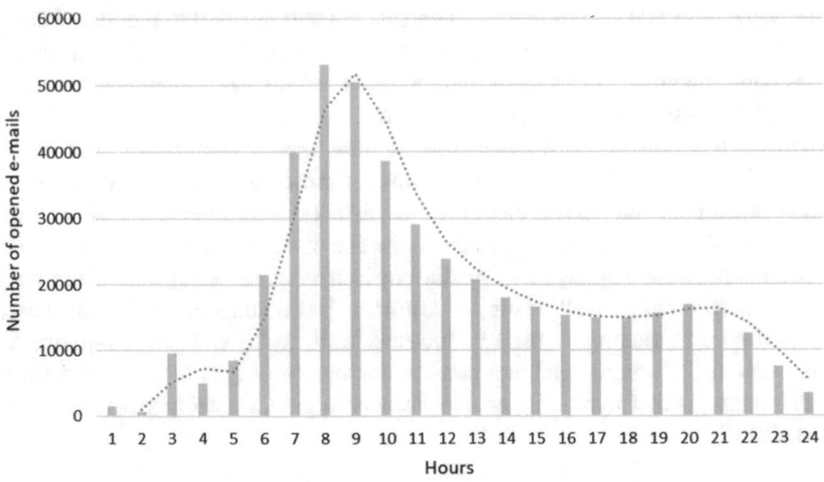

Fig. 1. The number of emails opened by the hour of a day.

2.2 Available Dataset

By analyzing the available data set, we focused on a subset of cases in which the customer responded positively. Subsequent projection of these data into a graph can show that most users open their e-mails in the morning (see Fig. 1). Interpretation of this phenomenon may describe the behavior of users who check their mailboxes when they arrive at work, assuming that most customers start working in the morning.

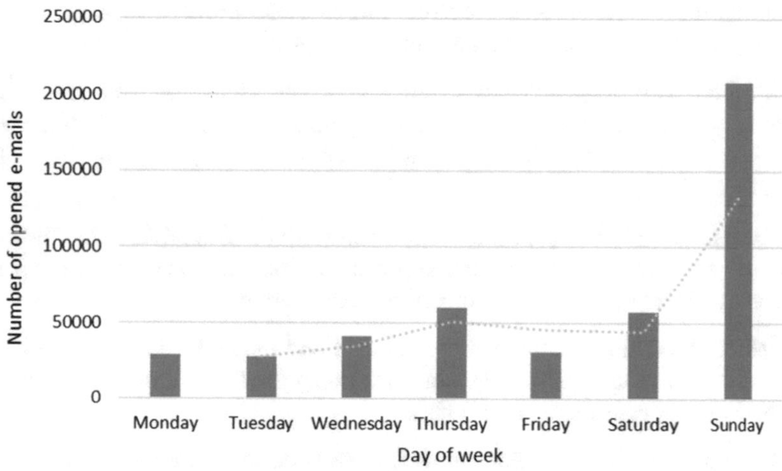

Fig. 2. The number of emails opened by the day of the week.

One can see the distribution of email openings describing a positive response to a delivered message as a graph showing the number of opened messages in particular days of the week on Fig. 2. As we can see from the graph, customers opened most of the messages at the weekends. This phenomenon may be caused by the fact that most customers have the opportunity to process their private orders over the weekend.

3 Methodology

Today's widely available e-mail communication meets many of the company's communication needs with its customers. It stimulates the creation of huge amounts of data. More than 90% of this information is unstructured, which means that the data does not have a predefined model and structure. In general, unstructured data is unnecessary unless data and knowledge mining methods or data extraction techniques are used [14]. This data only makes sense to the owner if one can process and understand the data. Otherwise, they become useless. Although a small part of this vast whole is structured (logs) or semi-structured (email, web pages), it is difficult to process and manage this data without advanced data analysis techniques [11, 13].

3.1 CRISP-DM

Cross-Industry Standard Process for Data Mining [12], known as CRISP-DM, is an open standard process model that describes common approaches used by data mining experts. It is one of the most used analytical models. CRISP-DM consists of a series of phases that are usually involved in the data mining process.

The CRISP-DM methodology is composed of these six major phases:

1. Business Understanding. Focuses on understanding the project objectives and requirements from a business perspective, and then converting this knowledge into a data mining problem definition and a preliminary plan.

2. Data Understanding. Starts with an initial data collection and proceeds with activities to get familiar with the data, to identify data quality problems, to discover first insights into the data, or to detect interesting subsets to form hypotheses for hidden information.

3. Data Preparation. The data preparation phase covers all activities to construct the final dataset from the initial raw data applying various preprocessing operations like data cleaning, normalization, transformation, reduction, etc.

4. Modeling. Modeling techniques are selected and applied. Since some techniques like neural nets have specific requirements regarding the form of the data, there can be a loop back to the data preparation phase.

5. Evaluation. Once one or more models have been built that appear to have high quality based on whichever evaluation functions have been selected, these need to be tested to ensure they generalize against unseen data and that all key business issues have been sufficiently considered. The result is the selection of the champion model(s).

6. Deployment. Generally, this will mean deploying a code representation of the model into an operating system to score or categorize new unseen data as it arises and to create a mechanism for the use of that new information in the solution of the original business problem. Importantly, the code representation must also include all the data preparation steps leading up to modeling so that the model will treat new raw data in the same manner as during model development.

3.2 Proposed Modeling Methodology

Our methodology is based on the general CRISP-DM methodology described above, but it splits the modeling phase into a cascade of models. It consists of three classification models concerning the characteristics of the problem (see Fig. 3). In the first phase, we construct a data model (*Model Status* in Fig. 3) that classifies users into two classes: *Open* or *Non-Open*. With this model, we can filter out users who do not open e-mail neither with the best conditions. If the *Model Status* classifies a new case (a target customer is considered for sending a particular marketing campaign email, or specific electronic newsletter) into the category *Non-Open*, the automatic system does not send the email in such a case, and no other classification model is considered. If on the other side, the *Model Status* classifies the new case into the *Open* category, the case continuous into the workflow depicted in Fig. 3 into two other classification models (*Model Time* and *Model Nday*) in parallel.

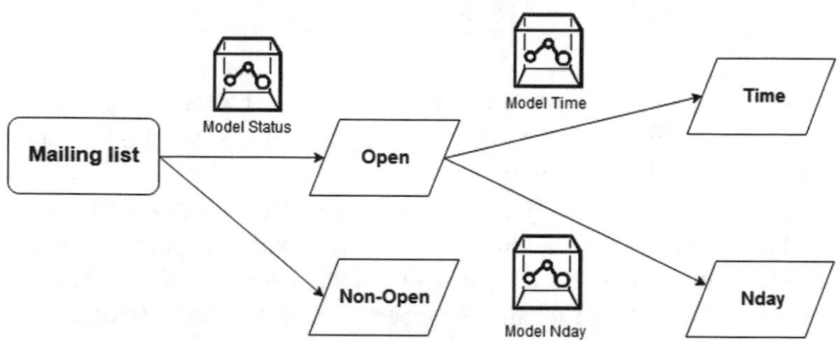

Fig. 3. Proposed modeling methodology diagram.

Model Time classifies users into 24 classes representing hours of the day when they preferably open their marketing emails. Similarly, *Model Nday* classifies users into seven classes representing the day of the week, when they preferably open their marketing emails. These classes reflect the best hour and day to send the marketing e-mail or newsletter. Based on the resulted classes in both of these classification models marketing email or newsletter will be sent in the analyzed new case in the given day of the week and in particular hour of that day.

We named all three classification models by their respective target attributes used for classification:

- *Status* – binary attribute describing the user action (Open or Non-Open e-mail)
- *Time* – a time of day (discretized on 24 values corresponding to 24 h of the day), when the user opens the marketing e-mail
- *Nday* – the day of the week, when the user opens the e-mail

3.3 Experiments

Before the start of the data modeling, we divided the dataset into two parts in proportion 80:20. We used the larger part of given dataset for training and testing classification models. The second (smaller) part, representing 1,253,763 records, was left for the process of validating individual types of data models created.

In the initial phase of generating data models, we decided to create a Baseline model. A baseline is a method that uses heuristics, simple summary statistics, randomness, or very simple machine learning model to create predictions for a dataset. One can use these predictions to measure the baseline's performance (accuracy). The results of this model are based on the simplest prediction and serve as a reference model for comparing the accuracy of later generated data mining models. The baseline model was set to generate predictions based on the majority class in training set's class distribution. The accuracies of the Baseline models for particular target attributes are in Table 1.

Table 1. The Baseline models accuracy for particular target attributes.

Target attribute	Accuracy of the baseline model
Status	90,52%
Time	35,91%
NDay	56,33%

Because of a large number of available attributes (164), we used the Sequential Feature Algorithms (SFA) - the greedy search algorithm [13] to find the most suitable subgroup of attributes for modeling.

SFAs are a family of greedy search algorithms that are used to reduce the d-dimensional initial space to a consequential k-dimensional subspace, where $k < d$. The feature selection algorithms are designed to automatically select a subset of features that are most relevant to the problem. The features selection goal is twofold; we want to improve computational efficiency or reduce the generalization error of the model by removing irrelevant properties or noise. An envelope approach, such as a sequential features selection, is particularly useful when selecting an embedded feature - such as a regularization penalty - is not usable. SFAs remove or add one feature over time based on classifier performance until a subset of the required size k features is reached.

After selecting the appropriate subset of attributes, we started the data modeling process. In the modeling process, we used three types of machine learning algorithms, which returned the best precision in the test phase. The modeling process was the same for all three phases. In the process of designing this work, various options for setting the values of individual algorithms for creating data models were implemented. We used the following machine learning algorithms [11, 12] to classify available data concerning our target attributes:

- Naive Bayes
- Random Forest
- Decision Tree

3.4 Results

To verify the accuracy of the created data models, we used the process of validation of the selected models on real data, not available for the models in the phase of training and testing. In the phase of understanding the goal, we defined the key success factor of the created solution if the accuracy of individual data models in the validation process reaches at least 80% of correctly classified records. The output of the data preparation process was a data set suitable for generating data models using machine learning algorithms.

Table 2. The results of the validation process.

Type of classifier	Target attribute	Accuracy of the best model
Decision tree	*Status*	93,01%
Decision tree	*Time*	80,54%
Decision tree	*Nday*	88,68%

The data set for validation did not affect the process of training and testing the created models in any way. From the set of available data models generated for each target attribute, we have selected models that maximize the accuracy of the classification of each test record. For all three target attribute options, the decision tree algorithm gave the best results in terms of data model accuracy (see Table 2). These models have undergone a validation process on a selected set of data. The following table shows the selected data models for each target attribute along with the accuracy achieved on the data validation set.

4 Summary and Conclusions

We aimed to improve the open rate of e-mails sent to customers in marketing campaigns. We proposed a modeling methodology based on the analysis of the available data set describing e-mail communication between the customer and the seller.

We started with a short state of the art analysis. By analyzing the existing solutions, we have found many studies dealings with determining the optimal time to send an e-mail. Based on the analysis of available solutions, we could see that to determine the optimal time for sending e-mail is varied based on the actual group of target customers. It is necessary to analyze the available data describing customer behavior on the current list of recipients to whom one wants to send marketing messages.

Based on these findings, we proposed a general methodology using a series of data mining models. When analyzing the available dataset, we used the CRISP-DM methodology. Through the processes involved in the steps of this methodology representing the life cycle of knowledge discovery, we were able to create prediction models for classifying target attributes based on individual customer characteristics.

The contribution of this work is a methodology, which enables to create a series of classification models based on which one can classify individual customers and decide if a marketing email should be sent and when is the best time to send a marketing e-mail. By applying this solution, they should be able to increase the open-rate on the site of potential customers.

In the future, we propose the application of created data models to the live environment of the cooperating company. We also propose extending the created data models with the latest data describing communication between customers and vendors. An equally appropriate option to extend this issue is to analyze data from various areas of e-mail marketing.

Acknowledgment. This work was partially supported by the Slovak Grant Agency of the Ministry of Education and Academy of Science of the Slovak Republic under grant no. 1/0493/16 and The Slovak Research and Development Agency under grants no. APVV-16-0213 and APVV-17-0267.

References

1. Bult, R.J., Wansbeek, T.: Optimal selection for direct mail. Mark. Sci. **14**(4), 378–394 (1995)
2. Bitran, G., Mondschein, S.: Mailing decisions in the catalog sales industry. Manag. Sci. **42**(9), 1364–1381 (1996)
3. Blattberg, R.C., Kim, B.D., Neslin, S.A.: Database Marketing: Analyzing and Managing Customers, pp. 763–768. Springer, New York (2008)
4. Bonfrer, A., Drèze, X.: Real-time evaluation of e-mail campaign performance. Mark. Sci. **28**(2), 251–263 (2009)
5. Kirš, D., Herper, M.: E-mail marketing: a weapon for entrepreneurship in the 21st century, Brno (2010). (in Czech)
6. Kotler, P., Keller, K.L.: Marketing management, 14th edn. Prentice Hall, New Jersey (2012)
7. Kotler, P., Wong, V., Saunders, J., Armstrong, G.: Modern marketing. Prague (2007). (in Czech)
8. Madleňák, R.: Analysis of different types of Internet advertisement - e-mail advertisement. Telecommun. Electron. Mark. **3** (2006). (in Slovak)
9. Mars, M.: Perfect Timing: The Very Best Time to Send Email Newsletters (2018)
10. John: Insights from Mailchimp's Send Time Optimization System (2014)
11. Ivančáková, J., Babič, F., Butka, P.: Comparison of different machine learning methods on Wisconsin dataset. In: Proceedings of the SAMI 2018 - IEEE 16th World Symposium on Applied Machine Intelligence and Informatics, pp. 173–178 (2018)
12. Butka, P., Pócs, J., Pócsová, J.: Use of concept lattices for data tables with different types of attributes. J. Inf. Organ. Sci. **36**(1), 1–12 (2012)
13. Aggarwal, ChC: Data Mining: The Textbook. Springer, Heildeberg (2015)
14. Sarnovsky, M., Bednar, P., Smatana, M.: Data integration in scalable data analytics platform for process industries. In: Proceedings of the INES 2017 - IEEE 21st International Conference on Intelligent Engineering Systems, pp. 187–191 (2017)

Different Hierarchical Clustering Methods in Basic-Level Extraction Using Multidendrogram

Mariusz Mulka(✉) and Wojciech Lorkiewicz

Faculty of Computer Science and Management,
Wroclaw University of Science and Technology, Wroclaw, Poland
{mariusz.mulka,wojciech.lorkiewicz}@pwr.edu.pl

Abstract. This paper focuses on the research on how different agglomerative hierarchical clustering methods can be utilised to extract basic-level categories. Assuming three classical basic-levelness measures, namely category attentional slip, category utility and feature possession, and two hybrid measures, namely cue validity with global threshold and feature-possession, a multidendrogram approach is studied. In particular, different proximity measures and linkage criteria are thoroughly investigated against three datasets representing different characteristics of typical data in cyber-physical systems. Performed investigation highlights how different clustering settings affect basic-levelness measures and indicates that additional pruning of features is required.

Keywords: Basic-level categories · Multidendrogram · Cognitive computing

1 Introduction

Object categorization [1] is the ability to organise individual object perceptions, established through the interaction with the external world, into meaningful categories. As such, each category refers to a shared representations of object kinds [6] that share some commonalities or specific purpose, i.e., objects that are alike but still are discriminable. As such, categories allow for a significant reduction of complexity of internal organization of semantic knowledge, as it is no longer required to process all individual objects. Furthermore, categories play an important role in cognitive processes, i.e., perception, learning and communication.

In general, object categories are organised into hierarchical structures, where higher level categories represent more abstract and general object kinds, and lower level categories represent more specific and detailed object kinds. One particular level of categories, the so-called basic-level, is of particular importance to natural system, as it decompose the perception of the world into maximally informative categories [10].

Obviously, basic-level categories are also of high importance to autonomous artificial systems, where they should aid agent's internal organisation of semantic

J. Świątek et al. (Eds.): ISAT 2019, AISC 1051, pp. 197–207, 2020.
https://doi.org/10.1007/978-3-030-30604-5_18

information. Nevertheless, there is a significant research gap in this topic, i.e., how to introduce and maintain basic-level categories within artificial systems.

This paper follows our previous research [8,9] and focuses on how different agglomerative hierarchical clustering methods can be utilised to extract basic-level categories in autonomous systems. In particular, computational clustering methods are utilised to establish provisional clusters, which are further filtered to identify proper basic-level partitioning. Provisional clusters are established using a multidendrogram approach with different distance functions and linkage criteria. Extraction is performed using a dedicated basic-levelness measures.

Presented research focuses on investigation how different basic-levelness measures are affected by different clustering settings. It should be noted that in our previous research [8,9] we focused solely on referential datasets specifically designed for psycholinguistic experiments on basic-level aspects of cognition. However, in this paper we focus on datasets that represent perceptual data gathered by a typical cyber-physical system. Concluding, we study how different settings and measures affect the process of basic-level extraction in artificial systems.

2 Basic-Level Extraction

Due to focus on extraction of basic-level categories we consider only a hierarchical clustering [10] approach, which establishes a series of data partitions in a tree like structure – dendrogram [3]. We focus on agglomerative methods (bottom-up processing), where each object starts in its own cluster (dendrogram's leafs) and such clusters are further successively grouped into larger and larger clusters until a single, all-encompassing cluster, is obtained (dendrogram's root).

There are many implementations of agglomerative approach [11], for instance Fernández and Gómez [4] proposed an agglomerative algorithm for creating hierarchy of clusters as a multidendrogram. This approach solves a common problem of ties, i.e., eliminates the need to introduce additional tie-resolution mechanisms (typically random in nature) into hierarchical clustering algorithms. In particular, in the multidendrogram approach it is possible to merge more than two clusters in a single step, i.e., if the distance between clusters is tied, which is not possible in a typical implementation of agglomerative methods.

Assumed approaches require definition of proximity measures between objects and clusters. In this research we utilise 4 object proximity measures calculated using a similarity measure, namely Tversky Index (Tv) and Cosine similarity (Cos), or a distance function, namely Euclidean distance Euc and Manhattan distance (Man) [3]. Further, proximity between clusters can be calculated using different linkage criteria, i.e., using extreme elements (single linkage (SIN) and complete linkage (COM)), using some averaging technique (unweighted pair group method with arithmetic mean ($UPGMA$), weighted pair group method with arithmetic mean ($WPGMA$)), and using centroids (unweighted pair group method using centroids ($UPGMC$), weighted pair group method using centroids ($WPGMC$)) and joint between-within (JBW). Concluding, we are investigating 19 different implementations of establishing a hierarchy of clusters using the multidendrogram approach ($UPGMC$, $WPGMC$, JBW works only for Euc [4]).

As aforementioned, the basic-level is a specific level in the hierarchy of clusters that requires utilisation of additional measures. In particular, 3 approaches were found in the literature for assessing basic-levelness [8]: (a) Jones [7] proposed a measure called feature-possession (FP) which assigns each attribute to a specific category located in the hierarchy of clusters, (b) Corter & Gluck [2] proposed probabilistic measure called category utility (CU) which captures the following aspects: a category is only useful to the extent that it can improve ability to accurately predict attributes' values for members of that category and efficiently communicate information about attributes exhibited by members of that category, (c) Gosselin & Schyns [5] proposed measure called category attentional slip (CAS) which focuses on cardinality of redundant tests and length of the optimal strategy needed to establish category. We additionally utilise two hybrid measures (see paper [8]): cue validity with global threshold ($CVGT$), which uses a globally predefined threshold in order to determine whether feature should be included in a specific category, and cue validity with feature-possession ($CVFP$), which uses FP measure to determine feature-category mapping.

In order to determine the basic-level of categorization (assuming a given dendrogram), it is necessary to establish the level with highest (except attentional slip, where the value should be minimized) value of the basic-levelness measure. The selected level is further regarded as the basic-level.

3 Simulation Settings

All of the presented simulations were performed using 3 datasets[1]. Each dataset covers a set of individual objects (external world observations) described by a predefined set of binary properties. Datasets range from a small set, covering only several objects and their properties (zoo dataset), to a larger set, covering nearly 10 times more objects and properties (mushroom dataset). There are no missing features and uncertainties in the assumed datasets.

The first dataset was extracted from Whatbird's (see footnote 1) portal and covers descriptions of different bird species. It contains 944 objects (birds) described by 23 features divided into four types (shape, size, pattern, color). It should be stressed that additional binarization tasks were performed on the data. In particular, each object was mapped into a binary vector containing 267 coordinates in total.

The second dataset is an open dataset (Mushroom (see footnote 1)), and covers descriptions of different mushrooms. It contains 8124 objects described by 23 attributes. It contained non-binary features, therefore binarization process was performed. Finally, a set of 117 features (the class attribute was ignored) was utilised.

The third dataset is an open dataset (Zoo (see footnote 1)) and covers descriptions of different animals. It contains 101 objects described by 18 attributes. It contained non-binary features, therefore binarization process was performed,

[1] Whatbird - https://www.whatbird.com/; Mushroom dataset - https://www.openml.org/d/24; Zoo dataset - https://www.openml.org/d/62.

after which each object was described by 21 features (attributes related to class type and animal names were ignored).

It should be noted that in our previous research [8,9] we focused on datasets specifically designed for extraction of basic-level categories, i.e., for the purpose of psycholinguistic experiments. A natural extension of this research is to focus on extraction of basic-level categories in cyber-physical systems capable of autonomous determination of objects and their features, i.e., through individual interaction with the external world. As such, in this paper we focus on datasets that exemplify typical perceptual data gathered by an artificial system.

Using simulation based-approach we study how the process of extraction of basic-level categories behaves in aforementioned settings. Simulations were performed for each of the available dataset, basic-levelness measures and different multidendrogram implementations[2]. Each simulation's iteration, for a fixed dataset, followed a three step procedure. At first, 50 objects are randomly selected forming a set of analysed objects. At second, the hierarchy of clusters is established (for this random set) using all of the 19 different multidendrogram implementations (involving different distance and linkage criteria). Finally, for each resultant hierarchy a basic-level is established – independently for all of the assumed measures of basic-levelness. Further, we focused on two aspect of the selected basic-level, namely, the value of the measure and the number of extracted basic-level categories. In order to compare values among different multidendrogram implementations we calculate relative quality, i.e., for each basic-level measure and iteration (separately) the maximum measure's value is identified and serves as iteration-wise normalization constant (measure values for 19 settings are divided by this maximum)[3]. Note that such relativized value denotes the percentage of basic-levelness as compared to the maximally identified basic-levelness within a specific iteration. Resultantly it provides means to perform comparison of all of the assumed implementations. This procedure was repeated 1000 times and run independently for each dataset.

It should be noted that the relative quality measures allows only for a relative assessment of the behaviour of different multidendrogram clustering settings and should not be used to compare performance of different basic-levelness measures.

4 Simulation Results

As aforementioned, in this study we analyse the effects of different multidendrogram implementations focusing on the behaviour of basic-levelness measure values (Figs. 1, 2, 3) and the number of extracted basic-level categories (Table 1). At first, we study these effects for each measure individually, and finally highlight general implications of the obtained results. All identified characteristics assume support greater than 75%, i.e., is identified in for more than 750 iterations.

[2] For *CAS* measure we assume that attention randomly slips with probability $p = 0.5$ (analogous to [5]), while for ($CVGT$) global threshold is equal to 0.7.

[3] In case of *CAS* measure we utilise it's complement.

Table 1. Average number of extracted basic-level categories (*STD* in brackets).

	CAS			CU			FP			CVGT			CVFP		
	M	W	Z	M	W	Z	M	W	Z	M	W	Z	M	W	Z
Tv_UPGMA	2.4 (1.5)	1.4 (0.7)	3.3 (0.9)	4.2 (1.1)	5.1 (1.9)	5.4 (1)	1.3 (0.7)	1 (0.1)	1.2 (0.8)	3 (1.2)	1 (0)	1.5 (1.3)	1.9 (1)	1.1 (0.4)	2 (1.5)
Tv_WPGMA	2.6 (2)	1.3 (0.6)	3.2 (1.1)	4.2 (1.2)	4.6 (1.7)	5.7 (1.1)	1.2 (0.6)	1 (0.1)	1.4 (1.2)	3 (1.3)	1 (0)	1.4 (1.3)	1.8 (1.1)	1.1 (0.3)	2.4 (1.8)
Tv_COM	5.3 (3.6)	1 (0)	5 (2.7)	4 (1.1)	4.2 (1.3)	3.9 (1.5)	1 (0.1)	1 (0)	1.1 (0.3)	2.7 (1.5)	1 (0)	1.8 (1.2)	1.1 (0.3)	1 (0)	1.5 (0.8)
Tv_SIN	2.5 (1.4)	2.1 (0.6)	3.9 (2.1)	5.9 (2)	16.1 (6.6)	7.6 (2.7)	1.5 (1)	1 (0.1)	1.3 (1.1)	2.6 (1.4)	1 (0)	1.1 (0.4)	2.3 (1.1)	1.1 (0.3)	1.9 (1.7)
Cos_UPGMA	2.5 (1.6)	1.4 (0.6)	3.2 (0.9)	4.2 (1.1)	5 (1.9)	5.1 (1.1)	1.3 (0.7)	1 (0.1)	1.4 (1)	3 (1.2)	1 (0)	1.7 (1.5)	1.9 (1)	1.1 (0.3)	2.3 (1.6)
Cos_WPGMA	2.6 (1.6)	1.2 (0.5)	3.1 (1)	4.2 (1.2)	4.5 (1.6)	5.6 (1.1)	1.2 (0.6)	1 (0.1)	1.5 (1.2)	3 (1.3)	1 (0)	1.4 (1.3)	1.8 (1.1)	1.1 (0.3)	2.5 (1.8)
Cos_COM	5.8 (3.1)	1 (0)	4.2 (1.9)	3.9 (1.2)	4.2 (1.3)	3.8 (1.4)	1 (0.2)	1 (0)	1 (0.2)	2.6 (1.5)	1 (0)	1.9 (1.2)	1.1 (0.4)	1 (0)	1.4 (0.8)
Cos_SIN	2.5 (1.6)	2 (0.4)	3.1 (2.1)	6 (1.9)	16.1 (6.5)	7.4 (2.6)	1.5 (1)	1 (0.1)	1.4 (1.3)	2.6 (1.5)	1 (0)	1.1 (0.5)	2.2 (1.1)	1.1 (0.3)	2.3 (2)
Euc_UPGMA	2.5 (1.5)	2 (0.7)	3 (3.2)	4.2 (1.1)	7.5 (2.5)	2.9 (0.7)	1.3 (0.7)	1 (0.1)	1 (0.2)	3 (1.2)	1 (0.1)	2.5 (0.9)	1.9 (1)	1.1 (0.3)	1.3 (0.5)
Euc_WPGMA	2.6 (2)	1.9 (0.8)	3.1 (3.7)	4.2 (1.2)	5.9 (2.5)	3.2 (0.9)	1.2 (0.6)	1 (0.1)	1.1 (0.3)	3 (1.3)	1 (0.1)	2.4 (1.1)	1.8 (1.1)	1.1 (0.4)	1.5 (0.8)
Euc_UPGMC	2 (0)	2 (0)	2.2 (0.5)	20.3 (5.1)	9.9 (2.5)	9 (2.5)	1.1 (0.4)	1 (0.1)	1.8 (1.6)	1.5 (0.6)	1 (0.1)	1 (0.4)	1.6 (0.6)	1.1 (0.3)	2.5 (2.2)
Euc_WPGMC	2 (0)	2 (0)	2.3 (1.7)	14.1 (3.7)	12.8 (6.7)	8.8 (3)	1.1 (0.5)	1 (0.1)	1.6 (1.5)	1.5 (0.7)	1 (0.1)	1.1 (0.6)	1.7 (0.8)	1.1 (0.3)	2.2 (2.3)
Euc_COM	5.4 (3.3)	1.3 (0.9)	5.4 (4.1)	3.9 (1.1)	4.7 (1.7)	3 (0.9)	1 (0.1)	1 (0)	1.1 (0.3)	2.6 (1.5)	1 (0)	2.2 (1)	1.1 (0.4)	1 (0.2)	1.3 (0.7)
Euc_SIN	2.5 (1.3)	2.3 (0.8)	4.4 (2.9)	6 (1.9)	16.1 (7.4)	4.9 (1.6)	1.4 (0.9)	1 (0.1)	1.1 (0.5)	2.6 (1.5)	1 (0)	1.7 (1.3)	2.2 (1.1)	1 (0.2)	1.7 (1.1)
Euc_JBW	2.3 (0.7)	1 (0)	2.2 (0.4)	3.4 (0.6)	3.7 (0.7)	2.7 (0.5)	1 (0)	1 (0)	1 (0)	2.4 (0.8)	1 (0)	2.2 (0.6)	1.2 (0.5)	1 (0)	1.5 (0.5)
Man_UPGMA	2.5 (1.6)	2 (0.7)	3.1 (3)	4.2 (1.1)	7.3 (2.4)	2.8 (0.6)	1.3 (0.7)	1 (0.1)	1 (0.2)	3 (1.2)	1 (0.1)	2.4 (0.9)	1.9 (1)	1.1 (0.3)	1.3 (0.5)
Man_WPGMA	2.6 (1.6)	1.9 (0.9)	3.2 (3.2)	4.2 (1.2)	5.9 (2.4)	3.1 (0.9)	1.2 (0.6)	1 (0.1)	1.1 (0.3)	3 (1.3)	1 (0.1)	2.3 (1.1)	1.8 (1)	1.1 (0.3)	1.5 (0.7)
Man_COM	5.4 (2.7)	1.3 (1)	5.6 (3.2)	4.1 (1.2)	4.8 (1.8)	3.2 (1.1)	1 (0.1)	1 (0)	1 (0.2)	2.7 (1.6)	1 (0)	2.1 (1.1)	1.1 (0.3)	1 (0.2)	1.2 (0.5)
Man_SIN	3 (1.9)	2.7 (1.5)	8.4 (3.7)	6.2 (2.1)	16.6 (7.6)	5.1 (2.2)	1.4 (0.9)	1 (0.1)	1.1 (0.4)	2.5 (1.5)	1 (0)	1.5 (1.2)	2 (1)	1.1 (0.2)	1.5 (0.8)

CAS measure (Fig. 1). Relative quality values for *CAS* measure are highest for *Man_SIN* setting. In general, centroids and *Euc_SIN* settings seem to result in higher relative quality values (especially visible for Whatbird and Mushroom datasets). The worst relative quality values were obtained for averages and complete linkage (especially with similarity functions) settings. Moreover, in case of Whatbird dataset, the relative quality was higher or equal to 80% when centroids and single linkage (with distance functions) were used, while it was lower or equal to 65% when averages, complete linkage and joint between-within were used. In case of Zoo dataset, the relative quality was higher or equal to 90% when *Man_SIN* setting was used. The worst relative quality values for Zoo dataset were obtained for averages and complete linkage (with similarity functions), i.e., the relative quality was lower than 55%. In case of Mushroom dataset, the relative quality was higher or equal to 85% when centroids and single linkage were used. The worst relative quality values for Mushroom dataset were obtained for joint between-within setting, i.e., the relative quality was lower than 65%.

The number of extracted basic-level categories (Table 1) highlights the fact, that in the case of assumed settings the *CAS* measure has a tendency to select a rather stable number of basic-level categories (between 2–5). Visibly more clusters were selected when complete linkage criteria were utilised for Mushroom and Zoo datasets. However, in case of Whatbird dataset complete linkage resulted in significantly more general categories (typically just a single large category), while single linkage resulted in slightly more fine-grained categories (typically with two categories). Additionally, higher number of categories was observed for *Man_SIN* setting, especially in the case of Zoo and Whatbird datasets.

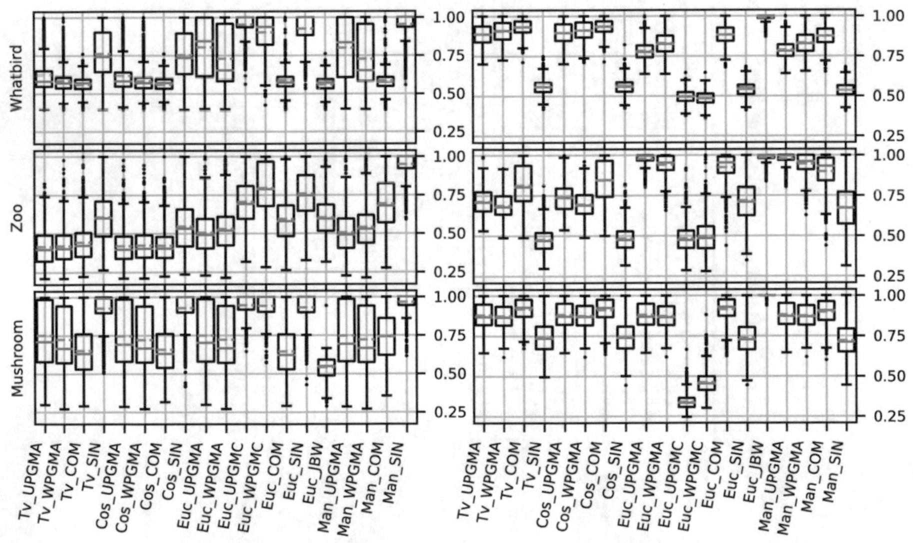

Fig. 1. Relative quality for different hierarchical structures (*CAS* and *CU* measure).

CU measure (Fig. 1). In all datasets the relative quality was higher or equal to 75% when average linkage, complete linkage and joint between-within were used (regardless of used proximity measure). In general, the relative quality is slightly lower when single linkage or Euclidean measure were used. The lowest relative quality values were obtained for centroids settings (the value was below 50%). The best results were obtained for Euclidean distance and joint between-within linkage criteria (the value was around 100%). In general, lower relative quality values were observed in case of Zoo dataset (especially for Tversky index and Cosine similarity).

Regardless of used dataset the *CU* measure extracted reasonable number of basic-level categories (Table 1). Nevertheless, a higher number of basic-level categories was obtained again when single linkage and centroids were used (note that in those cases the obtained relative quality was significantly lower). In case of complete linkage and joint between-within settings the number of basic-level categories was low (3–4 categories).

FP measure (Fig. 2). Relative quality values for *FP* measure are highest for complete linkage (despite internal similarity measure) and joint between-within settings (especially visible for Whatbird and Mushroom datasets). In general, Euclidean and Manhattan distance seem to result in higher relative quality values (especially visible for Zoo dataset). The worst relative quality values were obtained for single linkage (with similarity functions) and centroids settings. Moreover, in case of Mushroom dataset all ways of creating hierarchy of clusters were relatively close to each other (the relative quality was higher than 75%), as compared to other datasets (above 50%). In case of Zoo dataset, the relative quality was higher or equal to 75% when complete linkage (with distance functions), unweighted average (with distance functions) and joint between-within

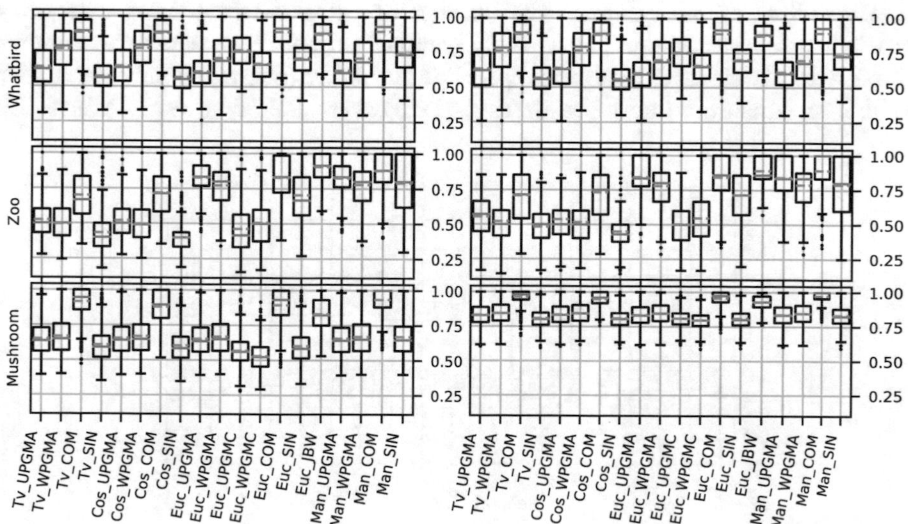

Fig. 2. Relative quality for different hierarchical structures (*CVFP* and *FP* measure).

were used. The worst relative quality values for Zoo dataset were obtained for averages (with similarity functions), single linkage (with similarity functions), centroids were used – the relative quality was lower than 68%.

Regardless of used dataset the number of extracted basic-level categories was in most of cases equal to 1 or 2 (in case of Whatbird it was almost always 1). It resulted from the fact that many features (common to objects) were located at root level, i.e., colours on Whatbird dataset.

CVFP measure (Fig. 2). Relative quality values for *CVFP* measure are similar to results obtained for the *FP* measure, yet tend to have a bit lower values (especially in case of Zoo and Mushroom datasets). In general, highest values are obtained for complete linkage and joint between-within settings, whereas lowest for single linkage. In case of Mushroom dataset, the relative quality was higher or equal to 80% when complete linkage and joint between-within settings were used (regardless of used proximity measure). The quality was smaller when averages, single linkage were used - the lowest for centroids (the relative quality was lower than 65%). In case of Whatbird dataset, the relative quality was higher or equal to 80% when complete linkage and joint between-within were used (regardless of used proximity measure). The lowest quality was obtained when unweighted centroid and single linkage (with similarity function) were used (the relative quality was lower than 68%). In case of Zoo dataset, the relative quality was higher of equal to 70% when unweighted average linkage (with distance function), complete linkage (with distance function) and joint between-within were used. The quality was smaller when complete linkage (with similarity functions), weighted average (with distance functions) and single linkage (with distance function) were used - lowest value was obtained when centroids and single linkage (with similarity functions), averages (with similarity functions) were used (the relative quality was lower than 60%).

Regardless of used dataset the number of extracted basic-level categories was in most of cases equal to 1–3. In comparison to *FP* measure the number of extracted basic-level categories was slightly higher (average number is higher by 0.5 for Mushroom and Zoo datasets). However, assigning more features by

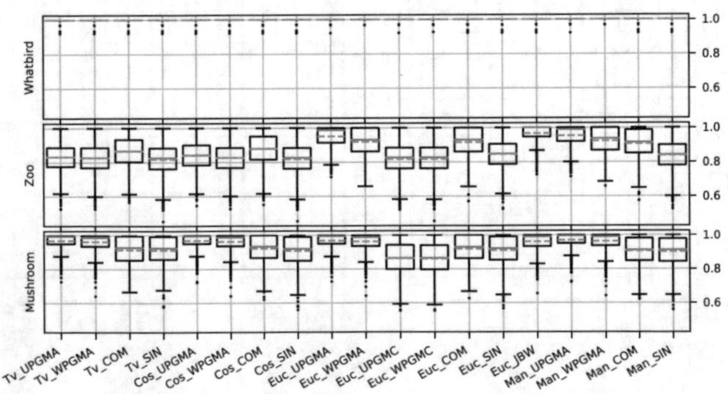

Fig. 3. Relative quality for different hierarchical structures (*CVGT* measure).

FP measure to the root of hierarchy of clusters caused selecting lower number of clusters (especially in case of Whatbird dataset).

CVGT measure (Fig. 3). Relative quality values for *CVGT* measure are generally high. At first, for the Whatbird dataset all values are equal to 100%, what is a direct result from the fact that all settings resulted in *CVGT* measure to select the whole set of objects as a single basic-level category (Table 1). Obviously this is a highly unwanted behaviour that results from fact that many features were common to more than 70% of objects (especially for Whatbird dataset where there were many features common to many objects for instance features related to colour). In case of other datasets (Mushroom and Zoo), the relative quality was higher or equal to 75% for all ways of creating hierarchy of clusters. However, highest values were obtained for average (Euclidean and Manhattan distance functions) and joint between-within settings. Lowest values were obtained for centroid, complete linkage and single linkage settings.

In case of Mushroom dataset the number of extracted basic-level categories was between 2 and 3 (Table 1), while in Zoo dataset in most cases the number of basic-level categories was between 1 and 2 categories, while in Whatbird dataset almost always 1 category was selected because many features were common to at least 70% of selected objects.

Implications. In case of *CU*, *FP*, *CVFP* measures, the basic-levelness was highest when complete linkage (especially with distance function) and joint between-within were used. While, in case of *CAS* measure, the basic-levelness was highest when centroids and single linkage (especially with distance function) were used. The obtained basic-levelness was mostly the lowest when single linkage (especially with similarity function) and centroids were used for *CU*, *FP*, *CVFP* measures. While, in case of *CAS* measure the lowest basic-levelness values were obtained for joint between-within, averages, complete linkage (especially with similarity functions) settings. In all settings the *CVGT* measure resulted in a similar values of basic-levelness, regardless of the used way of establishing the hierarchy of clusters.

Let us now focus on the influence of linkage criteria on basic-levelness measures. High relative quality values for complete linkage setting seem to be a direct consequence of the fact that such linkage criteria tends to create clusters with circle-like shape. As such, it is consistent with prototype theory, where a prototype is in the middle of a category and other objects surround it. In case of joint between-within setting, the high relative quality values seem to be a direct consequence of the fact that this way of merging clusters tends to maximize inner similarity and minimize similarity to other clusters (which are important aspects of basic-level categories). Furthermore, low relative quality values for single linkage setting seem to be a result of the fact that clusters prepared by single linkage were chain-like or snowflake-like. As such, in the former case it might be difficult to find a single prototype describing a whole category, whilst in the latter there were many categories around one big one (so atypical objects form own categories). In case of centroids settings it is hard to clearly identify whether basic-levelness should be high or low. On the one hand the centroid can be compared to the prototype of category which might suggest that a category is built around it. On the other hand the centroid can change radically after merging clusters. Clusters established using average linkages clusters can have miscellaneous shapes, so values of basic-levelness could be high in some cases.

It should be noted that, contrary to other basic-levelness measures, *CAS* measure seems to prefer settings that establish clusters in shapes which are inconsistent with prototype theory. In particular, highest basic-levelness values were obtained for single linkage settings, which has a tendency to create elongated clusters that are hard to cover by a prototype. It is worth noting, that according to our previous research [9], the *CAS* measure obtained low precision and recall for extracting basic-level categories in hierarchies of clusters prepared as dendrogram or multidendrogram.

No consistent relation between proximity measures and the number of extracted basic-level categories was observed. The number of extracted basic-level categories was unreasonable for *FP*, *CVGT*, *CVFP* measures (because those measures mostly selected root of hierarchy of clusters as basic-level). The most problematic was Whatbird dataset because it contained many features irrelevant for determining basic-level categories (such as colours), such features caused selecting lower number of basic-level categories in case of *CAS*, *FP*, *CVGT*, *CVFP* measures. The *CU* measure coped with such features by calculating squared difference between cue validity and probability of feature occurrence. Therefore, a cyber-physical system has to cope with those features in order to use *CAS*, *FP*, *CVGT*, *CVFP* measures for extracting basic-level categories.

5 Conclusions

This paper focused on investigation on how different multidendrogram clustering settings influence selected basic-levelness measures. In particular, we have identified 19 settings (different proximity and linkage methods) and thoroughly investigated them against 3 datasets, which exemplify typical perceptual data that is utilised by a cyber-physical system. This research was carried out independently for 5 basic-levelness measures.

We can note that obtained results indicate that for category utility, feature possession, cue validity with feature-possession basic-levelness measures it is preferable to utilise joint between-within and complete linkage settings. While, for category attentional slip measure it is preferable to utilize centroid-based and single linkage settings. However, it should be stated that this measure seems to result in categories with unwanted properties. Furthermore, no particular preference in linkage criteria was found for cue validity with global threshold measure.

Additional findings indicate that, due to the real world characteristics (typical for cyber-physical systems) of the utilised datasets, analysed measures have a tendency to associate a large number of features with clusters of the highest level. As such, nearly all basic-levelness measures (except category utility) have a tendency to result in overly grown levels, i.e., with small number of clusters (1–2). Resultantly, additional research is needed to develop better solutions.

Acknowledgment. This research was carried out at Wrocław University of Science and Technology (Poland) under Grant 0401/0190/18 titled Models and Methods of Semantic Communication in Cyber-Physical Systems.

References

1. Bornstein, M.H., Arterberry, M.E.: The development of object categorization in young children: hierarchical inclusiveness, age, perceptual attribute, and group versus individual analyses. Dev. Psychol. **46**(2), 350–365 (2010)
2. Corter, J.E., Gluck, M.A.: Explaining basic categories: feature predictability and information. Psychol. Bullet. **111**(2), 291–303 (1992)
3. Everitt, B., Landau, S., Leese, M., Stahl, D.: Cluster Analysis Wiley Series in Probability and Statistics. Wiley, New Jersey (2011)
4. Fernández, A., Gómez, S.: Solving non-uniqueness in agglomerative hierarchical clustering using multidendrograms. J. Classif. **25**(1), 43–65 (2008)
5. Gosselin, F., Schyns, P.: Debunking the Basic Level. In: Cognitive Science Society (US) Conference/Proceedings, pp. 277–282 (1997)
6. Harnad, S.: To Cognize is to Categorize: Cognition is Categorization, Handbook of Categorization in Cognitive Science (Second Edition), Elsevier, 21–54 (2017)
7. Jones, G.V.: Identifying basic categories. Psych. Bullet. **94**(3), 423–428 (1983)
8. Mulka, M., Lorkiewicz, W.: Measures for extracting basic-level categories. In: Proceedings CISP-BMEI, pp. 1–6. IEEE (2018)
9. Mulka, M., Lorkiewicz, W., Katarzyniak, R.P.: Extraction of basic-level categories using dendrogram and multidendrogram. In: Proceedings ICNC-FSKD, Springer (2019). (In press)
10. Rosch, E.: Basic objects in natural categories. Working paper. Language Behavior Research Laboratory, University of California (1975)
11. Saxena, A., Prasad, M., Gupta, A., Bharill, N., Patel, O.P., Tiwari, A., Er, M.J., Ding, W., Lin, C.T.: A review of clustering techniques and developments. Neurocomputing **267**, 664–681 (2017)

Elimination of Redundant Association Rules

Bemarisika Parfait[1,2(✉)] and Totohasina André[1]

[1] Laboratoire de Mathématiques et d'Informatique, ENSET, Université d'Antsiranana, Antsiranana, Madagascar
bemarisikap7@yahoo.fr, andre.totohasina@gmail.com
[2] Laboratoire d'Informatique et de Mathématiques, EA2525, Université de La Réunion, Saint-Denis, France

Abstract. The redundancy concept has been developed of various approaches. However, these approaches concern only the positive association rules, the negative association rules are less studied, and this, with less selective pair, support-confidence. To do remedy these limits, we propose a new approach allowing to generate all non-redundant positive and negative rules, and this, using the new selective pair, support-M_{GK}.

Keywords: Association rules · Positive/Negative rules · Redundant rules

1 Introduction and Motivations

In this paper, we focus on the elimination of non-informative redundant association rules. Given a set $\mathcal{I}$ of items from a database $\mathcal{B}$, an association rule is a logical implication between two disjoint itemsets $X, Y \subseteq \mathcal{I}$, with the form $X \rightarrow Y$, where X and Y are respectively called the premise and the conclusion (or consequent) of rule. The rule r_1 is said to be less informative redundant compared to r_2 if (i) it shares the same information as r_2, (ii) its premise is superset of premise of r_2 and its conclusion is a subset of conclusion of r_2.

In the context of Big Data, the redundancy concept has been developed by several approaches [2,9–11,13–15]. However, these approaches concern only the positive rules, the negative rules[1] are less studied, and this, with less selective pair, support-confidence [1]. This limit does not allow to understand all the needs of the context of Big Data that we meet in the real world, it also needs the negative association rules. To overcome these limits, we propose a new approach for informative base of non-redundant positive and negative association rules, and this, using a selective pair, support-M_{GK} [7]. Based on this model, we also propose a new efficient algorithm to automate the extraction process.

[1] For all $X, Y \subseteq \mathcal{I}$, an association rule is negative if at least one two patterns is negative, of the form $X \rightarrow \overline{Y}$, $\overline{X} \rightarrow Y$ and $\overline{X} \rightarrow \overline{Y}$, where $\overline{\mathcal{A}} = \neg\mathcal{A} = \mathcal{I} \backslash \mathcal{A}$.

J. Świątek et al. (Eds.): ISAT 2019, AISC 1051, pp. 208–218, 2020.
https://doi.org/10.1007/978-3-030-30604-5_19

The rest of this paper is organized as follows. The preliminary concepts are stated in Sect. 2. In Sect. 3, we develop our approach allowing to generate the non-redundant association rules. Sect. 4 represents the NON-REDUNDANT-RULES algorithm. Experimental results are described and discussed in Sect. 5. A conclusion and perspectives are represented in Sect. 6.

2 Preliminaries

In this paper, we place ourselves in a transactional database, which is a triplet $\mathcal{B} = (\mathcal{T}, \mathcal{I}, \mathcal{R})$, where $\mathcal{T}$ (resp. $\mathcal{I}$) represents a finite set of transactions (resp. items) and $\mathcal{R}$ is a binary relation between $\mathcal{T}$ and $\mathcal{I}$ (i.e. $\mathcal{R} \subseteq \mathcal{T} \times \mathcal{I}$). If an item i occurs in a transaction t, we write it as $i\mathcal{R}t$. Table 1 shows an example of $\mathcal{B}$ consisting of 5 items, $\mathcal{I} = \{A, B, C, D, E\}$, and 6 transactions, $\mathcal{T} = \{1, 2, 3, 4, 5, 6\}$. A set $X \subseteq \mathcal{I}$ is called an itemset, and a set $Y \subseteq \mathcal{T}$ is called a tidset. A k-itemset is an itemset of size k, i.e. AB is a 2-itemset.

Table 1. Example of a transactional database $\mathcal{B}$

TID	Items	Global items	Equivalent binary
1	ACD	$\mathrm{A}\overline{B}\mathrm{CD}\overline{E}$	10110
2	BCE	$\overline{A}\mathrm{BC}\overline{D}\mathrm{E}$	01101
3	ABCE	$\mathrm{ABC}\overline{D}\mathrm{E}$	11101
4	DE	$\overline{A}\mathrm{B}\overline{C}\,\overline{D}\mathrm{E}$	01001
5	ABCE	$\mathrm{ABC}\overline{D}\mathrm{E}$	11101
6	BCE	$\overline{A}\mathrm{BC}\overline{D}\mathrm{E}$	01101

Let $(\mathcal{P}(\mathcal{I}), \subseteq)$ be the set of all possible itemsets (called partial orders), and $(\mathcal{P}(\mathcal{T}), \subseteq)$, the set of all possible tidsets, where $\mathcal{P}(O)$ denotes the power set of O. We define two mappings t and i (dual of t) as follows. For an itemset $X \subseteq \mathcal{I}$, $t(X) = \{y \in \mathcal{T} | \forall x \in \mathcal{I}, x\mathcal{R}y\}$ represents the set of all tids that contains X (i.e. extention of X). For a tidset $Y \subseteq \mathcal{T}$, $i(Y) = \{x \in \mathcal{I} | \forall y \in \mathcal{T}, x\mathcal{R}y\}$ is the itemset that is contained in all the transactions in Y. Both mappings t and i form a Galois connexion between $\mathcal{P}(\mathcal{I})$ and $\mathcal{P}(\mathcal{T})$ [8]. Consequently, both compound operators of t and i are closure operators, in particular $\gamma(X) = iot(X) = i(t(X))$ is a closure operator. An itemset X is closed if $X = \gamma(X)$. It is generator if it is minimal (with set inclusion) in its equivalence class, $[X] = \{Y \subseteq \mathcal{I} | \gamma(Y) = \gamma(X)\}$ which is a set sharing the same closure. From Table 1, we obtain $\gamma(AB) = \gamma(ABC) = \gamma(ABE) = \gamma(ABCE)$, then AB is generator and $ABCE$ is closed.

The support (relative) of X is defined by $supp(X) = \frac{|t(X)|}{|\mathcal{T}|}$, where $|\mathcal{A}|$ denotes the cardinality of $\mathcal{A}$. The same support can be written as $supp(X) = P(t(X))^2$, where P is the discrete probability on $(\mathcal{T}, \mathcal{P}(\mathcal{T}))$. The support and confidence of

[2] For the sake of simplification, we will write $P(X)$ instead of $P(t(X))$, $\forall X \subseteq \mathcal{I}$.

$X \rightarrow Y$ are defined as $supp(X \cup Y) = \frac{|t(X \cup Y)|}{|\mathcal{T}|}$ and $conf(X \rightarrow Y) = P(Y|X) = \frac{|t(X \cup Y)|}{|t(X)|}$ respectively. As demonstrated in [4,5], the confidence measure is not selective. For this, we use a new selective measure, M_{GK} [7]. For all $X, Y \subseteq \mathcal{I}$ such as $X \cap Y = \emptyset$, the M_{GK} of an association rule $X \rightarrow Y$ is defined by:

$$M_{GK}(X \rightarrow Y) = \begin{cases} \frac{P(Y|X)-P(Y)}{1-P(Y)} & \text{if } P(Y|X) > P(Y) \\ \frac{P(Y|X)-P(Y)}{P(Y)} & \text{if } P(Y|X) \leq P(Y). \end{cases} \quad (1)$$

Relative to $X \rightarrow Y$, M_{GK} varies between -1 in limit repulsion between X and Y (i.e. $t(X) \cap t(Y) = \emptyset$), pass 0 in independence between X and Y (i.e. $P(Y|X)) = P(Y)$), and 1 in logical implication between X and Y (i.e. $t(X) \subset t(Y)$). If $M_{GK}(X \rightarrow Y) = 1$, $X \rightarrow Y$ is called *exact rule*, else it is called *approximative.*

For all $X, Y \subseteq \mathcal{I}$, we have $P(Y|X) \leq 1 \Rightarrow P(Y|X) - P(Y) \leq 1 - P(Y) \Rightarrow \frac{P(Y|X)-P(Y)}{1-P(Y)} \leq 1, \forall P(Y) \neq 1$. If $P(Y|X) > P(Y)$ (i.e. X favors Y), we have $P(Y|X) - P(Y) > 0 \Rightarrow \frac{P(Y|X)-P(Y)}{1-P(Y)} > 0, \forall P(Y) \neq 1$. Finally, if X favors Y, we have $0 < M_{GK}(X \rightarrow Y) \leq 1$ (dependency positive). Thus, $X \rightarrow Y$ is can be interesting. If $P(Y|X) \leq P(Y)$ (X disfavors Y), we have $P(Y|X) \leq P(Y) \Rightarrow -P(Y) \leq P(Y|X) - P(Y) \leq 0 \Rightarrow -1 \leq \frac{P(Y|X)-P(Y)}{P(Y)} \leq 0, \forall P(Y) \neq 0$. Finally, if X disfavors Y, we have $-1 \leq M_{GK}(X \rightarrow Y) \leq 0$ (dependency negative). Thus, $X \rightarrow Y$ is not interesting. In particular, if $P(Y|X) = P(Y)$ (indenpendency), we have $M_{GK}(X \rightarrow Y) = 0$, then, no rule can be interesting. From these results, we conclude that M_{GK} offers a robust and efficient tool to prune systematically the unintersting association rules (independency, or dependency negative).

3 Pruning Redundant Association Rules

Due to lack of space, the proof of our results are omitted (explained intuitively).

To consider positive and negative rules increases exponentially the number of rules. Given a m-itemset, this number is $\mathcal{O}(5^m - 2(3^m) + 1)$ [5], in the worst case. We restore 8 types such that $X \rightarrow Y$, $Y \rightarrow X$, $\overline{X} \rightarrow \overline{Y}$, $\overline{Y} \rightarrow \overline{X}$, $X \rightarrow \overline{Y}$, $\overline{X} \rightarrow Y$, $\overline{Y} \rightarrow X$ and $Y \rightarrow \overline{X}$ many are uninteresting and redundant. It is necessary to design an efficient method to prune the search space.

As mentioned, the uninteresting association rules are systematically pruned with M_{GK} measure. In the following, we discuss the context of redundant rules.

Definition 1. *$r_1 : X_1 \rightarrow Y_1$ is redundant with respect to $r_2 : X_2 \rightarrow Y_2$ iff: (i) ($supp(r_1) = supp(r_2)$ and $M_{GK}(r_1) = M_{GK}(r_2)$), (ii) ($X_2 \subseteq X_1$ and $Y_1 \subset Y_2$).*

To do this, we can utilize the equivalency rules (see Theorems 1 and 2). We point that these results have demonstrated in [5], but recalled and used for the rest of this paper. The proof for these theorems strongly relies on a key Lemma 1.

Lemma 1. *$\forall X, Y \subseteq \mathcal{I}$, (1) X favors $Y \Leftrightarrow Y$ favors $X \Leftrightarrow \overline{X}$ favors $\overline{Y} \Leftrightarrow \overline{Y}$ favors $\overline{X}$. (2) X disfavors $Y \Leftrightarrow X$ favors $\overline{Y} \Leftrightarrow \overline{Y}$ favors $X \Leftrightarrow Y$ favors $\overline{X} \Leftrightarrow \overline{X}$ favors Y.*

Theorem 1. *For two disjoint itemsets $X, Y \subseteq \mathcal{I}$ such as X favors Y (i.e. $X \subseteq Y$), then $X \rightarrow Y$ (resp. $\overline{X} \rightarrow \overline{Y}$) is equivalent to $\overline{Y} \rightarrow \overline{X}$ (resp. $Y \rightarrow X$).*

According to Theorem 1, as a result that if $X \rightarrow Y$ (resp. $\overline{X} \rightarrow \overline{Y}$) is valid, then $\overline{Y} \rightarrow \overline{X}$ (resp. $Y \rightarrow X$) is also valid. Thus, it's not necessary to consider $\overline{Y} \rightarrow \overline{X}$ and $Y \rightarrow X$ when we have $X \rightarrow Y$ and $\overline{X} \rightarrow \overline{Y}$ and vice-versa.

Theorem 2. *For two disjoint itemsets $X, Y \subseteq \mathcal{I}$ such as X disfavors Y (i.e. $X \subseteq \overline{Y}$), then $X \rightarrow \overline{Y}$ (resp. $\overline{X} \rightarrow Y$) is equivalente to $Y \rightarrow \overline{X}$ (resp. $\overline{Y} \rightarrow X$).*

As a result that if $X \rightarrow \overline{Y}$ and $\overline{X} \rightarrow Y$ are valid, then $Y \rightarrow \overline{X}$ and $\overline{Y} \rightarrow X$ are redundant. In addition, if X favors Y, then $M_{GK}(Y \rightarrow X) = \frac{|t(X)|}{|t(\overline{X})|}\frac{|t(\overline{Y})|}{|t(Y)|}M_{GK}(X \rightarrow Y)$, else $M_{GK}(X \rightarrow \overline{Y}) = \frac{|t(\overline{X})t(\overline{Y})|}{|t(X)t(Y)|}M_{GK}(\overline{X} \rightarrow Y)$ [5]. Thus, $Y \rightarrow X$ and $\overline{X} \rightarrow Y$ can be derived to $Y \rightarrow X$ and $X \rightarrow \overline{Y}$, respectively.

Thanks to these properties, we only retain 4 rules such as $X \rightarrow Y$, $\overline{X} \rightarrow \overline{Y}$, $X \rightarrow \overline{Y}$ and $\overline{X} \rightarrow Y$. For this, we only study 2 rules: $X \rightarrow Y$ and $X \rightarrow \overline{Y}$.

To do this, our first model concerns the base of $X \rightarrow Y$ such that $M_{GK}(X \rightarrow Y) = 1$. The similar approaches to this has been developed in [6,9]. However, these approaches select the premise from to pseudo-closed [9,13] that intuitively returns the maximal itemsets, which are incompatible of the informativity concept. To resolve this problem, we propose a new informative base, called EXACT-POSITIVE-ASSOCIATION-RULES (EPAR) selecting the premise (respectively consequent) from to the generator itemsets (respectively closed itemsets).

Definition 2. *Let $\mathcal{F}_F$ be a set of frequent closed in $\mathcal{B}$ and, for each frequent closed F, let $\mathcal{G}_F$ denotes the set of generators of F, the EPAR is formalized as:*

$$EPAR = \{G \rightarrow F \backslash G | G \in \mathcal{G}_F \wedge F \in \mathcal{F}_F \wedge G \neq F\}. \tag{2}$$

From Table 1, we see that $[AC] = \{A, AC\}$, which gives the candidate $A \rightarrow C$. Indeed, $P(C|A) = \frac{|t(A \cup C)|}{|t(A)|} = \frac{|t(A)|}{|t(A)|} = 1 \Leftrightarrow M_{GK}(A \rightarrow C) = 1$, hence $A \rightarrow C \in EPAR$. The EPAR basis contains only minimal exact positive association rules. All valid exact association rules can be derived from this EPAR basis.

Our second model based on the approximate positive rules, $X \rightarrow Y$, such that $0 < M_{GK}(X \rightarrow Y) < 1$. The naïve approaches [6,9] uses the pseudo-closed [9,13] concept which are icompatible of the informativity concept. To remedy this problem, we propose a new base, called APPROXIMATE-POSITIVE-ASSOCIATION-RULES (APAR), which selects the premise from to generator itemsets, and the conclusions from other closed itemsets containing this frequent closed.

Definition 3. *Let $minmgk \in]0, 1]$ be a minimum M_{GK}. Let $\mathcal{F}_F$ be a closed frequent and, for each frequent closed F, $\mathcal{G}_F$ is a set of generators of F, let $\widetilde{F}$ other frequent closed itemset containing closed F, the APAR is formalized as:*

$$APAR = \{r : G \rightarrow \widetilde{F} \backslash G | F \subset \widetilde{F} \in \mathcal{F}, G \in \mathcal{G}_F, M_{GK}(r) \geq minmgk\} \tag{3}$$

For example, from Table 1, consider $minsup = 0.2$ and $minmgk = 0.4$, we have $[BE] = \{B, E, BE\}$ and $[ABCE] = \{AB, AE, ABC, ABE, ACE, ABCE\}$. Since $BE \subset ABCE$, giving candidates $B \to ACE$ and $E \to ABC$. As a result $P(ACE|B) = \frac{|t(ABCE)|}{|t(B)|} = P(ABC|E) = \frac{|t(ABCE)|}{|t(E)|} = 2/5$ implying $M_{GK}(B \to ACE) = M_{GK}(E \to ABC) = 0.4$, hence $B \to ACE$ and $E \to ABC$ are added to APAR. The APAR basis contains only minimal informative approximate rules. All valid approximate association rules can be deduced from this APAR basis.

Our third model based on the exact negative rules of type $X \to \overline{Y}$ such that $M_{GK}(X \to \overline{Y}) = 1$. A comparable approach has been defined recently in [6]. However, this approach selects the premise from the set of maximal itemsets. This is a contradiction to the informativity concept. To resolve this problem, we propose a new informative base, called Exact-Negative-Association-Rules (ENAR). To do this, we adapt the concept of positive border from [12]. Let $\mathcal{F}$ be a set of frequent itemsets in database $\mathcal{B}$. The positive border of $\mathcal{F}$, denoted by $Bd^+(\mathcal{F})$, is defined as $Bd^+(\mathcal{F}) = \{X \in \mathcal{F} | \nexists Y \supset X, Y \in \mathcal{F}\}$, a set of frequent maximal itemsets in database $\mathcal{B}$. In ENAR basis, the premise (resp. conclusion) itemset is selected from the set of generators for this border positive (resp. the minimal transversal, denoted $\overline{Bd^+(\mathcal{F})}$). For example from Table 1 if $minsup = 0.2$, we have $Bd^+(\mathcal{F}) = \{ABCE\}$, hence $\overline{Bd^+(\mathcal{F})} = \{\overline{ABCE}\} = \{D\}$. Thus, $ABCE$ is frequent maximal, its minimal transversal in $\mathcal{B}$ is D.

Definition 4. *Given $Bd^+(\mathcal{F})$ a positive border and, for each $X \in Bd^+(\mathcal{F})$, let $\mathcal{G}_X$ denotes a set of generators of X. The informative base ENAR is defined as:*

$$ENAR = \{G \to \{\overline{y}\} | X \in Bd^+(\mathcal{F}), G \in \mathcal{G}_X, \forall y \in \overline{Bd^+(\mathcal{F})}\}. \tag{4}$$

Example, we finded $ABCE \in Bd^+(\mathcal{F})$ and $\overline{Bd^+(\mathcal{F})} = \{D\}$. Because AB and AE are generators, we have two candidates (induced rules) $AB \to \overline{D}$ and $AE \to \overline{D}$. Indeed, we obtain $supp(AB\overline{D}) = supp(AB) - supp(ABD) = 2/6 - 0$ implies $P(\overline{D}|AB) = \frac{2/6}{2/6} = 1$ equivalent to $M_{GK}(AB \to \overline{D}) = 1$. Hence, $AB \to \overline{D}$ is added to ENAR. Likewise for rule $AE \to \overline{D}$. The ENAR basis contains only minimal exact negative rules. All valid exact rules can be derived from this.

Our fourth and last model address on the approximate negative association rules of type $X \to \overline{Y}$ such that $M_{GK}(X \to \overline{Y}) < 1$. A similar approach is the one defined in [6]. However, this approach uses the pseudo-closed [9,13] concept. As mentioned, this pseudo-closed concept returns intuitively the maximal elements that are incompatibles of the informativity concepts. To resolve this limit, we propose a new informative base, called Approximate-Negative-Association-Rules (ANAR) selecting the premise and consequent from to generator of incomparable closed. For this, we define the following Property 1.

Property 1. For all $X, Y \subseteq \mathcal{I}$, we have $M_{GK}(X \to \overline{Y}) = -M_{GK}(X \to Y)$.

This Property 1 allows to prune any rule of type $X \to \overline{Y}$ if $M_{GK}(X \to Y)$ is negative. Based on these formalizations, we propose introduce the Definition 5.

Definition 5. *Let $mgk = minmgk \in]0,1]$. Let $\mathcal{F}_F$ be the set of frquent closed itemsets in database $\mathcal{B}$ and, for each frequent closed itemset F, let $\mathcal{G}_F$ denotes the set of generators of F, and $\mathcal{G}_{\widetilde{F}}$ the set of generators of other frequent closed $\widetilde{F}$ such that F and $\widetilde{F}$ are incomparable itemsets. The ANAR basis is defined as:*

$$ANAR = \{r : G \to \overline{\widetilde{G}} | F \not\subseteq \widetilde{F} \in \mathcal{F}_F, (G, \widetilde{G}) \in \mathcal{G}_F \times \mathcal{G}_{\widetilde{F}}, M_{GK}(r) \geq mgk\} \quad (5)$$

For example, from Table 1 if $minsup = 0.2$, we have $[AC] = \{A, AC\}$ and $[BE] = \{B, C, E, BE\}$. As a result AC and BE are two closed incomparables, and that AC disfavors BE (i.e. AC favors $\overline{BE}$). We see that C is comparable of AC, no rule is not studed for this. Contrary to B and E, which gives two induced rules candidates such as $A \to \overline{B}$ and $A \to \overline{E}$. Given $minmgk = 0.2$, we have $M_{GK}(A \to \overline{B}) = M_{GK}(A \to \overline{E}) = 0.2$, hence $\{A \to \overline{B}, A \to \overline{E}\} \in ANAR$. We notice that the ANAR basis contains only minimal approximate negative rules. All valid approximate rules can be derived from this ANAR basis.

4 Non-Redundant-Rules Algorithm

We first introduce our theorical for pruning the search space. To do this, we partition the set of candidates into two disjoint subsets, denoted $\mathcal{L}^+_{XY} = \{X, Y \subseteq \mathcal{I} | \ M_{GK}(X \to Y) > 0\}$ and $\mathcal{L}^+_{X\overline{Y}} = \{X, Y \subseteq \mathcal{I} | \ M_{GK}(X \to \overline{Y}) > 0\}$. The efficiency of this method mainly from the following Theorems 3, 4 and 5.

Theorem 3 (Transitivity). *$\forall X \subseteq Y \subseteq Z \Rightarrow M_{GK}(X \to Y) M_{GK}(Y \to Z) = M_{GK}(X \to Z)$.*

This Theorem 3 infers that if $X \to Z$ is valid association rule, then $X \to Y$ and $Y \to Z$ are less informative redundant rules, they are pruned in our approach.

Theorem 4 (Pruning space). *Anny rule of $\mathcal{L}^+_{XY}$ (resp. $\mathcal{L}^+_{X\overline{Y}}$) is only derivable from to $X \to Y$ (resp. $X \to \overline{Y}$).*

This restriction results from the economic calculation of the M_{GK} of all candidate rules that the Theorems 5 and 6 below show.

Theorem 5 (Pruning space-$\mathcal{L}^+_{XY}$). *$\forall \ X \to Y \in \mathcal{L}^+_{XY}$ and $\forall \ a \subseteq \mathcal{I}$ such that $X \cap \{a\} = \emptyset$, we have $M_{GK}(X \to Y \backslash X) \leq M_{GK}(X \cup \{a\} \to Y \backslash (X \cup \{a\}))$.*

For example, if $ABC \to D$ is invalid, then $A \to BCD$ and $AB \to CD$ are also invalid. In other words, if $A \to BCD$ is valid, then $AB \to \overline{CD}$ and $ABC \to D$ are also valid. Algorithmically, it is not necessary to consider the rules $A \to BCD$ and $AB \to CD$ if $ABC \to D$ is not valid on the $\mathcal{L}^+_{XY}$.

Theorem 6 (Pruning space-$\mathcal{L}^+_{X\overline{Y}}$). *$\forall \ X \to Y \in \mathcal{L}^+_{X\overline{Y}}$ and $\forall \ a \subseteq \mathcal{I}$ such as $X \cap \{a\} = \emptyset$, we have $M_{GK}(X \cup \{a\} \to \overline{Y} \backslash (X \cup \{a\})) \geq M_{GK}(X \to \overline{Y} \backslash X)$.*

Thus, if $ABC \rightarrow \overline{D}$ is valid, then $AB \rightarrow \overline{CD}$ and $A \rightarrow \overline{BCD}$ are also valid. If $A \rightarrow \overline{BCD}$ is invalid, $AB \rightarrow \overline{CD}$ and $ABC \rightarrow \overline{D}$ will be invalid.

These optimizations will be synthesized in Algorithms 1 and 2. Base-Non-Redundant-Rules (Algorithm 1) generates an informative base of positive and negative non-redundant association rules, denoted $\mathcal{B}_{N2R}$.

It proceeds recursively two steps, it first generates all valid rules of $\mathcal{L}^{+}_{GF}$, and secondly generates all valid rules of $\mathcal{L}^{+}_{G\overline{F}}$. NonRedRules function (Algorithm 1 line 35) generates non-redundant rules on $\mathcal{B}_{N2R}$. According to Theorem 5, if the rule $X \rightarrow Y \backslash X$ does not valid, neitheir does $\widetilde{X} \rightarrow Y \backslash \widetilde{X}$ for any $\widetilde{X} \subseteq X$. By rewriting, it follows that for a rule $X \backslash Z \rightarrow Z$ is valid, all rules of the form $X \backslash \widetilde{Z} \rightarrow \widetilde{Z}$ must also valid, where $\widetilde{Z}$ is a non-empty subset of Z ($\widetilde{Z} \subseteq Z$). According to Theorem 6, if $X \rightarrow \overline{Y} \backslash X$ is invalid, $\widetilde{X} \rightarrow \overline{Y} \backslash \widetilde{X}$ will not be either, for all $\widetilde{X} \subseteq X$. In other words, $X \backslash \overline{\widetilde{Z}} \rightarrow \overline{\widetilde{Z}}$ is valid, $X \backslash \overline{Z} \rightarrow \overline{Z}$ is also valid, for all $\widetilde{Z} \subseteq Z$. Derive-Non-Redundant-Rules algorithm (Algorithm 2) derives non-redundant association rules on $\mathcal{B}_{N2R}$.

Algorithm 1. Base-Non-Redundant-Rules

Require: $\mathcal{F}_F$, $\mathcal{G}_F$, $Bd^{+}(\mathcal{F})$, *minsup* and *minmgk*.
Ensure: $\mathcal{B}_{N2R}$, *Informative Base of Non-Redundant Rules.*
1: $\mathcal{B}_{N2R} = \emptyset$;
2: **for all** ($F \in \mathcal{F}_F$) **do**
3: **for all** ($G \in \mathcal{G}_F$) **do**
4: **if** ($G, F \subseteq \mathcal{L}^{+}_{GF}$) **then**
5: **if** ($\gamma(G) = \gamma(F)$) **then**
6: **if** ($G \neq F$ && $supp(G \cup F) \geq minsup$) **then**
7: $\mathcal{B}_{N2R} \leftarrow \mathcal{B}_{N2R} \cup \{G \rightarrow F \backslash G\}$; *[f]Exact Positive Association Rules-EPAR
8: **end if**
9: **else**
10: **for all** ($\widetilde{F} \in \mathcal{F}_F \mid \widetilde{F} \supset F$) **do**
11: **if** ($supp(G \cup \widetilde{F}) \geq minsup$ && $M_{GK}(G \rightarrow \widetilde{F} \backslash G) \geq minmgk$) **then**
12: $\mathcal{B}_{N2R} \leftarrow \mathcal{B}_{N2R} \cup \{G \rightarrow \widetilde{F} \backslash G\}$; *[f]Approximate Positive Rules-APAR
13: **end if**
14: **end for**
15: **end if**
16: **else**
17: **for all** ($X \in Bd^{+}(\mathcal{F})$) **do**
18: $\mathcal{G}_X = generator(X)$;
19: **for all** ($G \in \mathcal{G}_X$) **do**
20: **for all** ($y \in \overline{Bd^{+}(\mathcal{F})}$) **do**
21: **if** ($supp(G \cup \{\overline{y}\}) \geq minsup$) **then**
22: $\mathcal{B}_{N2R} \leftarrow \mathcal{B}_{N2R} \cup \{G \rightarrow \{\overline{y}\} \backslash G\}$; *[f]Exact Negative Rules-ENAR
23: **end if**
24: **end for**
25: **end for**
26: **end for**
27: **for all** ($\widetilde{G} \in \mathcal{G}_{\widetilde{F}} | \widetilde{F} \not\subseteq F \vee F \not\subseteq \widetilde{F}$) **do**
28: **if** ($supp(G \cup \overline{\widetilde{G}}) \geq minsup$ && $M_{GK}(G \rightarrow \overline{\widetilde{G}}) \geq minmgk$) **then**
29: $\mathcal{B}_{N2R} \leftarrow \mathcal{B}_{N2R} \cup \{G \rightarrow \overline{\widetilde{G}} \backslash G\}$; *[f]Approximate Negative Rules-ANAR
30: **end if**
31: **end for**
32: **end if**
33: **end for**
34: **end for**
35: NonRedRules($\mathcal{B}_{N2R}$)
36: $\mathcal{B}_{N2R}$

Algorithm 2. DERIVE-NON-REDUNDANT-RULES

Require: $\mathcal{B}_{N2R}$ *(Base of Non-Redundant Rules), and minsup (support threshold).*
Ensure: $\mathcal{D}_{N2R}$ *(Derive Non-Redundant Rules).*

$\mathcal{D}_{N2R} = \mathcal{B}_{N2R}$;
for all $(X \rightarrow Y \backslash X \in \mathcal{B}_{N2R})$ **do**
  if $(supp(\overline{X} \cup \overline{Y}) \geq minsup)$ **then**
    $\mathcal{D}_{N2R} \leftarrow \mathcal{D}_{N2R} \cup \{\overline{X} \rightarrow \overline{Y} \backslash \overline{X}\}$
  end if
end for
for all $(X \rightarrow \overline{Y} \backslash Y \in \mathcal{B}_{N2R})$ **do**
  if $(supp(\overline{X} \cup Y) \geq minsup)$ **then**
    $\mathcal{D}_{N2R} \leftarrow \mathcal{D}_{N2R} \cup \{\overline{X} \rightarrow Y \backslash \overline{X}\}$
  end if
end for
return $\mathcal{D}_{N2R}$

It first derives the rules of type $X \rightarrow Y \backslash X$. According to Theorem 1, if $X \rightarrow Y \backslash X$ is valid, then $\overline{X} \rightarrow \overline{Y} \backslash \overline{X}$ is also valid, it is added into $\mathcal{D}_{N2R}$. The next step address to derive the rules of type $X \rightarrow \overline{Y} \backslash X$. To do this, because $X \rightarrow \overline{Y} \backslash X$ is valid, then $\overline{X} \rightarrow Y \backslash \overline{X}$ is also valid (see Theorem 2), it is so added into $\mathcal{D}_{N2R}$.

5 Experimental evaluation

The goal is to evaluate the feasibility of our approach and its behavior by looking at the number of rules extracted and the execution time. NON-REDUNDANT-RULES algorithm is implemented in R, carried out on a PC Core i3-4GB of RAM running under Windows system. For the experiments, we have used the following datasets: T20I6D100K[3], T25I10D10K (see Footnote 3), C20D10K[4] and MUSHROOMS[5]. Figure 1 shows the evolution of the number of valid association rules for 4 databases by varying the minimum support *minsup* and keeping the minimum M_{GK} *minmgk* to 0.6.

This observation can be supplemented by the analysis of the figure 2, which reports the evolution of execution time as a function of *minsup* and *minmgk*.

We notice that the number of rules increases as the value of *minsup* decreases. For each databases, EPAR and ENAR basis extract far fewer rules than APAR and ANAR. These performances appear clearly on the epars databases (see graphs 1(a) and 1(b)). This is due to the mode of selection. For EPAR basis, the set of frequent closed is easily confused with that of minimal generators which will then limit the candidates. For ENAR basis, the number of association rules is determined by the number of frequent maximal itemsets that are relatively small, wich will also limit the number of association rules generated.

[3] http://www.almaden.ibm.com/cs/quest/syndata.html
[4] ftp://ftp2.cc.ukans.edu/pub/ippbr/census/pums/pums90ks.zip
[5] ftp://ftp.ics.uci.edu/pub/mach.-lear.-databases/mushroom/agaricus-lepiota.data

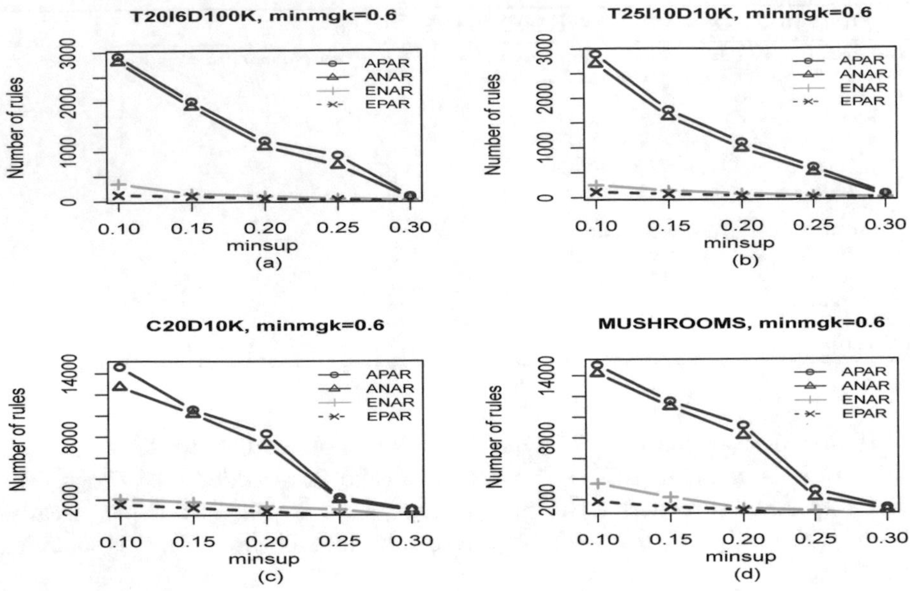

Fig. 1. Number of rules by varying *minsup*, with $minmgk = 0.6$

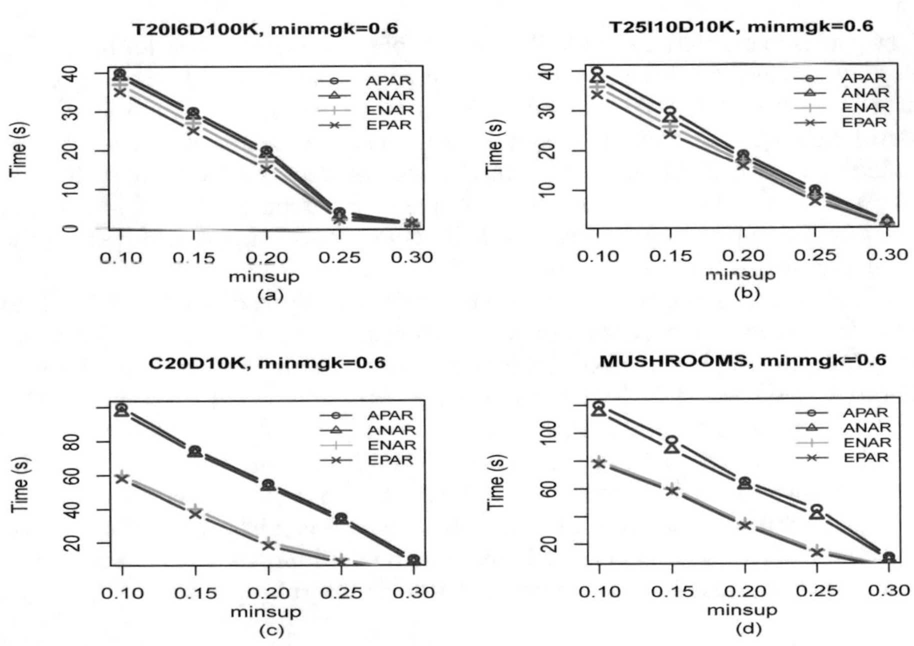

Fig. 2. Reponse times by varying *minsup*, with $minmgk = 0.6$

With the low *minsup* and dense databases (graphs 1(c) and 1(d)), the number of rules generated is very reasonable (not exceed 14000). Therefore, Non-Redundant-Rules is very selective. These performances can be justified as follows. M_{GK} offers an efficient tool to prune the unintersting and redundant association rules, which reduces significantly the size of valid association rules.

We observe that this computation time increases as *minsup* decreases. For epars databases (graphs 2(a) and 2(b)), the running times, for all *minsup*, are similar and very low. We notice that the correlated databases (graphs 2(c) and 2(d)) are very time-consuming, with a slight advantage for EPAR and ENAR basis. More generally, these experiments demonstrate the feasibility for our approach which provides the non-redundant positive and negative rules in increase execution time (not exceed 120 s) even if with very dense databases (graphs 2(c) and 2(d)).

6 Conclusion

We have proposed the new bases of non-redundant positive and negative association rules. For this, various optimizations have been defined. Using these optimizations, we have also proposed a Non-Redundant-Rules algorithm allowing to prune the uninteresting and redundant association rules. Despite its efficience, we could not conduct the comparative study over representative methods. Future work could therefore focuss on this question. Another perspective would be to extend this work to the extraction of association rules generalized.

References

1. Agrawal, R., Srikant, R.: Fast algorithms for mining association rules. In: Proceedings of 20th International Conference VLDB, pp. 487–499 (1994)
2. Bastide, Y., Taouil, R., Pasquier, N., Stumme, G., Lakhal, L.: Mining frequent patterns with counting inference. ACM-SIGKDD Explor. **2**(2), 66–75 (2000)
3. Bemarisika, P.: Extraction de règles d'association selon le couple support-M_{GK}: graphes implicatifs et application en didactique des mathématiques. Université d'Antananarivo (2016)
4. Bemarisika, P., Totohasina, A.: Erapn, an algorithm for extraction positive and negative association rules in Big Data. In: DaWaK, pp. 329–344. Springer (2018)
5. Bemarisika, P., Totohasina, A.: An efficient approach for extraction positive and negative association rules from Big Data. In: CD-MAKE, pp. 79–97. Springer (2018)
6. Feno, D.R., Diatta, J., Totohasina, A.: Galois lattices and based for M_{GK}-valid association rules. In: Proceedings of the 4th International Conference on CLA'06, pp. 127–138 (2006)
7. Feno, D.R.: Mesure de qualité des règles d'association: Normalisation et caractérisation des bases. Ph.D. thesis, Université de La Réunion (2007)
8. Ganter, B., Wille, R.: Formal Concept Analysis: Mathematical Foundations. Springer, Heidelberg (1999)

9. Guigues, J.L., Duquenne, V.: Familles minimales d'implications informatives résultant d'un tableau de donnés binaires. Maths. et Sciences Humaines, pp. 5–18 (1986)
10. Kryszkiewicz, M.: Concise representations of association rules. In: Hand, D.J., Adams, N.M., Bolton, R.J. (eds.) Pattern Detection and Discovery, pp. 92–103 (2002)
11. Latiri, C., Haddad, H., Hamrouni, T.: Towards an effective automatic query expansion process using an association rule mining approach. J. Intell. Inf. Syst. **39**(1), 209–247 (2012)
12. Mannila, H., Toivonen, H.: Levelwise search and borders of theories in knowledge discovery. Data Min. Knowl. Disc. **1**(3), 241–258 (1997)
13. Pasquier, N.: Frequent closed itemsets based condensed representations for association rules. CNRS (UMR6070), France, pp. 248–273 (2009)
14. Séverac, F., et al.: Non-redundant association rules between diseases and medications: an automated method for knowledge base construction. Med. Inform. Decis. Mak. **15**(1), 29 (2015)
15. Zaki, M.J.: Mining non-redundant association rules. In: Proceedings of the Knowledge Discovery and Data Mining (2004)

Artificial Intelligence Methods and Its Applocations

A Kernel Iterative K-Means Algorithm

Bernd-Jürgen Falkowski(✉)

Fachhochschule für Ökonomie und Management FOM,
Arnulfstrasse 30, D80335 München, Germany
bernd.falkowski@fh-stralsund.de

Abstract. In this paper Mercer kernels with certain invariance properties are briefly introduced and an apparently not well-known construction using certain cohomology groups is described. As a consequence some kernels arising from this are given. Hence a kernel version of an iterative k-means algorithm due to Duda et al. is exhibited. It resembles the usual k-means algorithm but relies on a different update procedure and allows an elegant computation of the target function.

Keywords: k-means algorithm · Mercer kernels · Cohomology groups

1 Introduction

In the first part of this paper the structure of Mercer kernels enjoying certain invariance properties is described in terms of first order cocycles. These results have been known for quite some time although they do not seem familiar to the community of researchers dealing with clustering algorithms. Hence they are recalled (essentially from [6–8, 15]) for the benefit of the reader. In particular a connection to an abstract Levy-Khinchine formula of probability theory and Fourier transforms of probability measures (positive definite funcuions) is established. Moreover, apart from a characterization of positive definite kernels invariant under the Euclidean Motion Group as radial functions, some examples of little known kernels are exhibited. This is of interest on the one hand for experimental purposes. On the other hand these kernels might also be relevant for commercial applications, see [16] (it is well known that e.g. Amazon applies various forms of clustering systems). Unfortunately no detailed proofs can be given since some sophisticated tools involving the theory of semi-simple Lie groups would be needed, cf. [6]. These considerations motivate the investigation of a Mercer kernel version of an elegant iterative k-means algorithm due to Duda et al. cf. [5], p. 549, that allows a neat calculation of the target function in the second part of the paper. The particular properties of Mercer kernels guarantee that they can be seen as abstract versions of scalar products. Thus they provide the algebraic versions of *length* and *angle* in a Hilbert space (the reproducing kernel Hilbert space), cf. [19]. As a practical consequence various similarity measures can derived (including in particular the cosine measure). This allows for an easy adaptation of the algorithm to the particular problem

J. Świątek et al. (Eds.): ISAT 2019, AISC 1051, pp. 221–232, 2020.
https://doi.org/10.1007/978-3-030-30604-5_20

considered by choosing a suitable measure. Moreover the sample space is not necessarily required to carry a vector space structure which could be relevant for practical applications, cf. [16].

2 Certain Mercer Kernels and 1-Cohomology

First of all some definitions are required in order to be able to establish the connection between Mercer kernels and 1-Cohomology. In particular the role of logarithms (conditionally positive definite kernels) of positive definite kernels needs to be clarified.

2.1 Conditionally Positive Definite Kernels and Mercer Kernels

Definition 1. *Given a topological space X and a continuous function $K : X \times X \to C$ where C denotes the complex numbers. Then K is called a positive definite (p.d.) kernel if it satisfies*
(a) $K(x, y) = \bar{K}(y, x)$ for all $x, y \in X$
(b) $\sum_{i=1}^{n} \sum_{j=1}^{n} \alpha_i \bar{\alpha}_j K(x_i, x_j) \geq 0$ for all $(\alpha_i, x_i) \in C \times X$ If (a) above holds and (b) above holds under the additional condition
(c) $\sum_{i=1}^{n} \alpha_i = 0$
then K is called conditionally positive definite (c.p.d)

Hence, by remark 3.7 in [2], p.35, a Mercer kernel is just a real valued p.d. kernel.

Example 1. If X is a complex Hilbert space with scalar product denoted by $< ., . >$ then the kernel K defined by $K(x, y) :=< x, y >$ for all $x, y \in X$ is p.d.

Example 2. If K is c.p.d. on $X \times X$ then for any $z \in X$ the kernel L_z defined by $L_z(x, y) := K(x, y) - K(x, z) - K(z, y) + K(z, z)$ is p.d.: Given arbitrary points $x_1, x_2, ..., x_n \in X$ and scalars $\alpha_1, \alpha_2, ..., \alpha_n$ set $x_0 := z$ and $\alpha_0 := -\alpha_1 - \alpha_2 - ... - \alpha_n$ then the claim follows from the definition of p.d.

Example 2 shows how a p.d. kernel may be constructed from a c.p.d. kernel. There is, however, another easy (well-known) way to construct a p.d. kernel from c.p.d. one:

Lemma 1. *Suppose that a p.d. kernel K and a polynomial p with positive coefficcients are given. Then p(K) is also a p.d. kernel.*

Corollary: For every c.p.d. L and every $t > 0$, the kernel defined as $K(x, y) := exp(tL(x, y))$ is p.d.

A proof of **Lemma** 1 can be found in [17], p. 76, whilst the proof of its corollary follows by considering Example 2, for details see [15].

The converse of the corollary is true as well:

Lemma 2. *If the kernel $K_t(x, y) := exp(tL(x, y))$ is p.d. for every $t > 0$ then L is c.p.d.*

This follows by differentiating with respect to t.

After these definitions it is now possible to investigate kernels enjoying certain invariance conditions more closely thus arrivving at some well-known kernels as well as some more exotic ones.

2.2 Group Actions and Invariant Kernels

Unfortunately another definition is needed:

Definition 2. *Let G be a topological group with identity e and let X be a topological space as before. G is said to act continously on X if*

1. *for every fixed* $g \in G$ *the map* $g \to gx$ *is one to one and onto.*
2. $ex = x$ *for all* $x \in X$
3. $g_1(g_2x) = (g_1g_2)x$ *for all* $g_1, g_2 \in G, x \in X$
4. $(g, x) \to gx$ *is continuous.*
5. *for every fixed* $g \in G$ *the map* $x \to gx$ *is a homeomorphism of X.*

A p.d. (c.p.d,) kernel K (L) is said to be invariant under G if $K(gx, gy) = K(x, y)$ *and* $L(gx, gy) = L(x, y)$ *for all* $g \in G$.

The following theorem that is essentially a consquence of the Kolmogorv consistency theorem and that strongly resemmbles the theorem on a Reproducing Kernel Hilbert Space, see below, is of crucial importance, for details and a proof see [15].

Theorem 1. *Let X be a topological space and let G be a group acting continuously on it. Suppose further that K is a p.d. kernel on* $X \times X$ *invariant under G.*

Then there exists a complex Hilbert space H and a weakly continuous unitary respresentation $g \to U_g$ *(i.e.* $U_{g_1g_2} = U_{g_1}U_{g_2}$ *and the map* $g \to <U_g v_1, v_2>$ *is continuous for every* $v_1, v_2 \in H$*) and a continuous map* $v : X \to H$ *such that the vectors v span H and*

1. $K(x, y) = <v(x), v(y)>$
2. $v(gx) = U_g v(x)$

Corollary: If $X = G$ *and G acts on itself by left multiplication, then a G-invariant p.d. kernel K may be viewed as a p.d. function* ϕ *in the usual sense. This is obtained by setting* $\phi(g) := K(g, e)$, $(K(g_1, g_2) = K(g_2^{-1}g_1, e))$. *On setting* $v_e := v_0$ *in the theorem a well-known theorem on p.d. functions due to Gelfand and Raikov (analogous to the GNS-construction for* $C^\star$ *– Algebras, cf. [4]) follows: Every p.d. continuous function* ϕ *may be written as* $\phi(g) = <U_g v_0, v_0>$.

2.3 1-Cohomology and Invariant c.p.d. Kernels

Definition 3. *Let G be a topological group acting continuously on the topological space X and let* $g \to U_g$ *be a weakly continuous unitary representation of G in a Hilbert space H. A map* $\delta : X \to H$ *is called a first order cocycle with origin* x_0 *if* $U_g\delta(x) = \delta(gx) - \delta(gx_0)$.

If $G = X$*, G acts on itself by left multiplication, and* $x_0 = e$ *then* δ *is just called a (first order) cocycle associated with U.*

Example 3. Suppose that v is a fixed vector in H. Then a trivial cocycle (coboundary) may be defined by setting $\delta(g) := U_g v - v$.

Theorem 2. *Let G be a topological group acting continuously on the topological space X and let L be a c.p.d. kernel on $X \times X$ invariant under G. Then for any fixed $x_0 \in X$ there exists a weakly continuous unitary representation $g \to U_g$ of G in H and a continuous cocycle $\delta : X \to H$ with origin x_0 such that*

$$<\delta(x), \delta(y)> = L(x, y) - L(x, x_0 - L(x_0, y) + L(x_0, x_0)$$

for all $x, y \in X$.

Conversely, if $g \to U_g$ is a weakly continuous unitary representation of G in H and δ is a continuous cocycle with origin x_0 and values in H and if L is kernel such that the above equation holds, then L is a $G-$invariant continuous c.p.d. kernel.

Proof: See Corollary 1.4, theorem 3.4, and remark 3.5 in [15].

Corollary: Taking again $X = G$ and G acting on itself by left multiplication, then a G-invariant c.p.d. kernel L may be viewed as a c.p.d. function $\psi(g) := L(g, e)$ in the usual sense and all c.p.d. arise in this manner. If $\psi(e) = 0$ (i.e. ψ is normalized) then it satisfies

$$<\delta(g), \delta(h)> = \psi(h^{-1}g) - \psi(g) - \psi(h^{-1})$$

for all $g, h \in G$.

From this it immediately follows that

$$Re(\psi(g)) = -1/2 < \delta(g), \delta(g) >$$

The above theorem shows that G-invariant continuous c.p.d. kernels are completely described by the 1-cohomology of G. Thus using the corollary to Lemma 2 and Example 2 a large class of Mercer kernels is completely described in this manner as well.

Of course, the 1-cohomology of groups is by no means easy to compute in general. Nevertheless some interesting results exist.

2.4 Special Mercer Kernels

For R^n, a case of particular interest, the 1-cohomology is completely known. In fact R^n is abelian with respect to addition and hence a type I group thus allowing a direct integral decomposition of weakly continuous unitary reprentations over irreducibles, cf. [15]. A similar statement holds for the associated cocycles. In fact the following theorem holds.

Theorem 3. *Let U be an arbitrary weakly continuous unitary representation of R^n in a Hilbert space H. Then there exists a continuous homomorphism $\eta : G \to H$, a measure space (Ω, μ) and a measurable map $\chi : \Omega \to \hat{G}$ (the character group of R^n that is of course isomorphic to R^n itself) such that a first order cocycle associated with U may be written as*

$$\delta(\mathbf{x}) = \eta(\mathbf{x}) + \int \mathrm{c}(\omega)[<\chi(\omega), \mathbf{x}> -1]\mathrm{d}(\mu)$$

for $\mathbf{x} \in \mathrm{R}^{\mathrm{n}}$. *Here* $<\chi(\omega), \mathbf{x}>$ *denotes the value of the character* $\chi(\omega)$ *at the point* $\mathbf{x}$ *and the meaning of the other symbols has been explained.*

Corollary: Every real valued c.p.d. function ψ *on* R^n *may be written as*

$$\psi(\mathbf{x}) = -1/2 <\mathbf{Ax}, \mathbf{x}> + \int |\mathrm{c}(\omega)|^2 [\mathrm{Re} <\chi(\omega), \mathbf{x}> -1] \mathrm{d}(\mu)$$

where **A** *is a positive definite matrix.*

Remark: The above corollary contains an abstract version of the well-known Levy-Khinchine formula, that describes the logarithms of characteristic functions of infinitely divisible probability measures.

The corollary to the above theorem now allows the explicit description of some Mercer kernels in terms of their logarithms.

1. The well-known Inner Product Kernel (taking the second sumand in the c.p.d. function equal to zero and the Matrix $\mathbf{A} = \mathbf{I}$ gives as c.p.d. function $\psi(\mathbf{x}) = -1/2 <\mathbf{x}, \mathbf{x}>$. Applying Example 2 gives (with $\mathrm{K}(\mathbf{x}, \mathbf{y}) = \psi(\mathbf{x}\text{-}\mathbf{y})$, $\mathbf{z} = \mathbf{0}$) $L_0(\mathbf{x}, \mathbf{y}) := <\mathbf{x}, \mathbf{y}>$ is p.d.)
2. The same choice of parameters and consequently the same c.p.d. function gives when combined with the corollary to Lemma 1 the Gaussian kernel $G(\mathbf{x}, \mathbf{y}) := exp - t < \mathbf{x} - \mathbf{y}, \mathbf{x} - \mathbf{y} >$ as positive definite and hence a Mercer kernel for any $t > 0$. Again this is well-known.
3. Kernels derived from coboundaries, for details see [7]
4. Subordinate kernels by considering the c.p.d. function $\psi(\mathbf{x}) = -|<\mathbf{x}, \mathbf{x}>|^{\alpha}$ for $0 < \alpha < 2$, see [7], and again appealing to the corollary of Lemma 1.
5. A further interesting example using an explicit version of the c.p.d function ψ may be found in [11]

It is interesting to note that explicit knowledge of the 1-cohomology may not be needed in general. Indeed, considering the Euclidean group of all proper rigid motions in R^n leads to the following theorem, cf. [8].

Theorem 4. *Every p.d. kernel* K *on* R^n *that is invariant under the Euclidean motion group* G *is given by a p.d radial function on* R^n.

Moreover, exploiting the Iwasawa decomposition of certain semi-simple Lie groups with finite centre (the real and complex Lorentz groups) additional explicit versions of Mercer kernels on the real line can be exhibited, cf. [8]. Indeed, the following c.p.d. functions (again characterizing the logarithms of Mercer kernels on the real line are obtained)

$$\psi_1 := -(tcotht - 1)$$

$$\psi_2 := -log(cosht)$$

From ψ_1 the following Mercer kernels are deduced:

$$K_{11}(t_1, t_2) := exp(1 - (t_1 - t_2) * coth(t_1 - t_2))$$

respectively

$$K_{12}(t_1, t_2) := -(t_1 - t_2) * coth(t_1 - t_2) + t_2 * coth t_2 + t_1 * coth t_1 + 1$$

From ψ_2 the following Mercer kernels result:

$$K_{21}(t_1, t_2) := 1/cosh(t_1 - t_2)$$

respectively

$$K_{22}(t_1, t_2) := log((cosh t_1 * cosh t_2)/cosh(t_1 - t_2))$$

It appeared useful to elucidate the structure of the above kernels by giving the above abstract descriptions since en passant some kernels appeared that do not seem to be well-known since they exceed the examples given in [17,20,21].

In addition it seems worth mentioning as a certain curiosity that in [10] Fürstenberg quite some time ago established a relation to non-commuting random products on certain Lie groups.

3 Mercer Kernels and an Interative k-Means Algorithm

In oder to avoid having to deal with unacceptably high CPU times in the past the so-called *kernel trick* has been applied for Neural Networks and in particular support vector machinse, see e.g. [12]. However, this lead to large RAM requirements concerning the efficient treatment of big matrices. In view of recent developments in the hardware sector (graphic cards) and improvements of the software (concurrent programming) these problems are likely to be alleviated in the non too distant future though. Nowadays, however, this trick allows neat generalizations of clustering algorithms amongst others.

The kernel trick provides an embedding of a set (that in general is not required to carry a vector space structure) in a Hilbert space. It seems useful at this stage to sketch the construction since it clarifies how the algebraic versions of length and angle appear via kernels within this Hilbert space sttting.

3.1 The Reproducing Kernel Hilber Space

Definition 4. *Given a Mercer kernel on a set* $S := \{x_1, x_2, ..., x_n\}$*, an embedding* η *of* S *in a vector space* $H = \Re^S$ *(the space of functions from* S *to* $\Re$*) may always be constructed by setting* $\eta(x) := K(., x)$*, considering functions* $f = \sum_{i=1}^{n} \alpha_i K(., x_i)$*, and defining addition of such functions and multiplication of such a function by a scalar pointwise. If the inner product is defined by* $< \eta(x), \eta(y) >_H := K(x, y)$ *and extended by linearity, then a Hilbert space* H *(the Reproducing Kernel Hilbert space,RKHS, sometimes also called feature space) is obtained by completion as usual, see e.g. [19].*

From this definition clearly a Mercer kernel is the abstract version of a scalar product. Thus, if elements in the RKHS are denoted by (a possibly indexed) z, then the *generalized concept of length* in the RKHS is given by

$$\|z\| = K(z,z)^{1/2}$$

Hence, using the Schwartz inequality, the *generalized concept of angle* can be defined via

$$cos(z_1, z_2) := \frac{< z_1, z_2 >}{\|z_1\|\|z_2\|}$$

For use below the *generalized concept of mean* is needed as well. In fact it is given as m(S) by

$$m(S) := 1/n \sum_{i=1}^{n} \eta(x_i)$$

This definition may cause difficulties since it is not always desirable to use an explicit version of η if only the Mercer kernel is given. However, if one just wishes to consider distances from the mean, then this difficulty can easily be overcome. Note that

$$\|\eta(x) - m\|^2 = K(x,x) + 1/n^2 \sum_{i=1}^{n}\sum_{j=1}^{n} K(x_i, x_j) - 2/n \sum_{j=1}^{n} K(x, x_j)$$

3.2 Duda's Original k-Means Iterative Algorithm

The problem that is being considered here may be formulated as follows:

Given a set of n samples $S := \{\mathbf{x_1}, \mathbf{x_2}, ..., \mathbf{x_n}\}$, then these samples are to be partitioned into exactly k sets $S_1, S_2, ..., S_k$. Each cluster is to contain samples more similar to each other than they are to samples in other clusters. To this end one defines a target function that measures the clustering quality of any partition of the data. The problem then is to find a partition of the samples that optimizes the target function.

Duda defines his target function as the sum of squared errors. More precisely, let n_i be the number of samples in S_i and let $\mathbf{m_i} = 1/\mathrm{n_i} \sum_{\mathbf{x} \in \mathbf{S_i}} \mathbf{x}$ be their mean, then the sum of squared errors is defined by

$$E_k := \sum_{i=1}^{k} \sum_{\mathbf{x} \in \mathbf{S_i}} \|\mathbf{x} - \mathbf{m_i}\|^2$$

Thus for a given cluster S_i the mean vector $\mathbf{m_i}$ is the best representative of the samples in S_i in the sense that it minimizes the squared lengths of the *error* vectors $\mathbf{x} - \mathbf{m_i}$ in S_i.

The target function can now be optimized by iterative improvement by setting

$$E_k := \sum_{i=1}^{k} E_i$$

where the squared error per cluster is defined by

$$E_i := \sum_{\mathbf{x} \in \mathrm{S_i}} \|\mathbf{x} - \mathbf{m_i}\|^2$$

Suppose that sample $\hat{x}$ in S_i is tentatively moved to S_j then $\mathbf{m_j}$ changes to

$$\mathbf{m_j^\star} := \mathbf{m_j} + 1/(\mathrm{n_j} + 1)(\hat{\mathbf{x}} - \mathbf{m_j})$$

and E_j increases to

$$E_j^\star = E_j + n_j/(n_j + 1)\|\hat{\mathbf{x}} - \mathbf{m_j}\|^2$$

For details see [5], p. 549.

Similarly, under the assumption that $n_i \neq 1$, (singleton clusters should not be removed) $\mathbf{m_i}$ changes to

$$\mathbf{m_i^\star} := \mathbf{m_i} - 1/(\mathrm{n_i} - 1)(\hat{\mathbf{x}} - \mathbf{m_i})$$

and E_i decreases to

$$E_i^\star = E_i - n_i/(n_i - 1)\|\hat{\mathbf{x}} - \mathbf{m_i}\|^2$$

These formulae simplify the computation of the change in the target functuin considerably. Thus it becomes obvious that a transfer of $\hat{x}$ from S_i to S_j is advantageous if the decrease in E_i is greater than the increase in E_j. This is the case if

$$n_i/(n_i - 1)\|\hat{\mathbf{x}} - \mathbf{m_i}\|^2 > \mathrm{n_j}/(\mathrm{n_j} + 1)\|\hat{\mathbf{x}} - \mathbf{m_j}\|^2$$

If reassignment is advantageous then the greatest decrease in the target function is obtained by selecting the cluster for which

$$\mathrm{n_j}/(\mathrm{n_j} + 1)\|\hat{\mathbf{x}} - \mathbf{m_j}\|^2$$

is minimal. Hence a neat algorithm is obtained. Of course, it must be admitted that such a hill climbing algorithm can get stuck in local minima and that its success depends on choosing suitable initial conditions. Note also that Duda's version depends on a vector space structure for the samples. Moreover the target function and thus the concept of similarity is fixed. In order to remedy at least some of these shortcomings the kernel version of the algorithm is presented in the next section.

3.3 The Kernel k-Means Iterative Algorithm

Implicit in the notation of Duda's version of the k-means algorithm was the assumption of a vector space structure of the samples. This will now be discarded and hence one starts with an arbitrary set of n samples $S := \{x_1, x_2, ..., x_n\}$ that are to be partitioned into exactly k clusters $S_1, S_2, ..., S_k$. The algorithm is however to be realized in feature space and this necessitates applying the feature map η.

The means in feature space will be denoted by $c_i := 1/n_i \sum_{x \in S_i} \eta(x)$ and an easy calculation then shows that if tentatively $\hat{x}$ is moved from S_i to S_j then

$$c_j^\star := c_j + 1/(n_j + 1)(\eta(\hat{x}) - c_j)$$

Here and below the same notation as in Duda's original algorithm has been employed.

At this stage it seem desirable to verify the increase in E_j in detail since the decrease in E_i can be computed analogously. Indeed, one has

$$E_j^\star = \sum_{x \in S_j} \|\eta(x) - c_j^\star\|^2 + \|\eta(\hat{x}) - c_j^\star\|^2$$

which gives

$$= \sum_{x \in S_j} \|\eta(x) - c_j - 1/(n_j + 1)(\eta(\hat{x}) - c_j)\|^2 + \|\eta(\hat{x}) - c_j - 1/(n_j + 1)(\eta(\hat{x}) - c_j)\|^2$$

This then reduces to

$$E_j + n_j/(n_j + 1)^2 \|\eta(\hat{x}) - c_j\|^2 + n_j^2/(n_j + 1)^2 \|\eta(\hat{x}) - c_j\|^2$$

and further to

$$E_j + n_j/(n_j + 1)\|\eta(\hat{x}) - c_j\|^2$$

as was to be expected. Without explicitly using the kernel feature space function the increase in E_j can now be expressed as

$$n_j/(n_j + 1)[K(\hat{x}, \hat{x}) + 1/n^2 \sum_{x \in S_j} \sum_{y \in S_j} K(x, y) - 2/n \sum_{y \in S_j} K(\hat{x}, y)]$$

The decrease in E_i can be obtained in a completely analogous fashion, as mentioned before. Thus as above, if reassignment is possible, then the cluster that minimizes the above expression should be selected.

It remains to show how the means for the updated clusters can be computed without using the explicit feature space function. Whilst the above calculations were fairly straightworward for this some extra work is needed and a slightly modified method essentially contained in [17] will be applied.

3.4 Mean Updates in Terms of Kernels

It is useful to define an $n \times k$ indicator matrix $\mathbf{S}$ as follows:

$$\begin{pmatrix} s_{11} & s_{12} & \cdots & s_{1k} \\ s_{21} & s_{22} & \cdots & s_{2k} \\ \cdots & \cdots & \cdots & \cdots \\ s_{n1} & s_{n2} & \cdots & s_{nk} \end{pmatrix}$$

Here $s_{ij} = 1$ if $x_i \in S_j$ and $s_{ij} = 0$ otherwise. Clearly the matrix $\mathbf{S}$ has precisely one 1 in every row whilst the column sums describe the number of samples in every cluster,

Moreover a $k \times k$ diagonal matrix $\mathbf{D}$ is needed:

$$\begin{pmatrix} 1/n_1 & 0 & \cdots & 0 \\ 0 & 1/n_2 & \cdots & 0 \\ \cdots & \cdots & \cdots & \cdots \\ 0 & 0 & \cdots & 1/n_k \end{pmatrix}$$

The entries on the diagonal of $\mathbf{D}$ are just the inverses of the number of elements in each cluster (notation as above).

In addition a vector $\mathbf{X}$ containing the feature version of the training examples will be helpful:

$$\begin{pmatrix} \eta(x_1) \\ \eta(x_2) \\ \cdots\cdots \\ \eta(x_n) \end{pmatrix}$$

From this one obtains

$$\mathbf{X}^{\mathrm{T}}\mathbf{S}\mathbf{D} = (\sum_{x \in S_1} \eta(x), \sum_{x \in S_2} \eta(x), \ldots, \sum_{x \in S_k} \eta(x))\mathbf{D}$$

which gives

$$(1/n_1 \sum_{x \in S_1} \eta(x), 1/n_2 \sum_{x \in S_2} \eta(x), \ldots, 1/n_k \sum_{x \in S_k} \eta(x))$$

Finally a vector $\mathbf{k}$ of scalar products between $\eta(\hat{x})$ and the feature version of the samples is going to be defined in terms of the kernel K:

$$\begin{pmatrix} K(\hat{x}, x_1) \\ K(\hat{x}, x_2) \\ \cdots\cdots \\ K(\hat{x}, x_n) \end{pmatrix}$$

Hence $\mathbf{k}^{\mathrm{T}}\mathbf{S}\mathbf{D}$ is given by

$$(< \eta(\hat{x}), 1/n_1 \sum_{x \in S_1} \eta(x) >, < \eta(\hat{x}), 1/n_2 \sum_{x \in S_2} \eta(x) >, \ldots, < \eta(\hat{x}), 1/n_k \sum_{x \in S_k} \eta(x) >)$$

It is now possible to compute

$$\begin{aligned} & n_j/(n_j+1)\|\eta(\hat{x}) - c_j\|^2 \\ &= n_j/(n_j+1)(\|\eta(\hat{x})\|^2 - 2 < \eta(\hat{x}), c_j > + \|c_j\|^2) \end{aligned}$$

without involving the explicit use of the feature map whilst also including the indicator matrix:

$$n_j/(n_j+1)(K(\hat{x}, \hat{x}) - 2(\mathbf{k}^{\mathrm{T}}\mathbf{S}\mathbf{D})_{\mathrm{j}} + (\mathbf{D}\mathbf{S}^{\mathrm{T}}\mathbf{X}\mathbf{X}^{\mathrm{T}}\mathbf{S}\mathbf{D})_{\mathrm{jj}})$$

Note here that for brevity the j-th vector (jj-th matrix) elements have been indicated by underscores.

Note also that $\mathbf{X}\mathbf{X}^{\mathrm{T}}$ just describes the kernel matrix.

3.5 The Kernel Algorithm

Collecting together the above results the following kernel k-means algoritm is obtained:

begin initialize $\mathrm{n}, \mathrm{k}, \mathrm{c}_1, \mathrm{c}_2, \ldots, \mathrm{c}_{\mathrm{k}}$
  do randomly select a sample $\hat{x}$
    $i \leftarrow \mathbf{argmin}_{1 \leq l \leq k} \, (K(\hat{x}, \hat{x}) - 2(\mathbf{k}^{\mathrm{T}}\mathbf{SD})_{\mathrm{l}} + (\mathbf{DS}^{\mathrm{T}}\mathbf{XX}^{\mathrm{T}}\mathbf{SD})_{\mathrm{ll}}))$ (classify $\hat{x}$)
    if $n_i \neq 1$ **then** compute
    $$\rho_j = \begin{cases} n_j/(n_j+1)(K(\hat{x},\hat{x}) - 2(\mathbf{k}^{\mathrm{T}}\mathbf{SD})_{\mathrm{j}} + (\mathbf{DS}^{\mathrm{T}}\mathbf{XX}^{\mathrm{T}}\mathbf{SD})_{\mathrm{jj}}) & \mathrm{j} \neq \mathrm{i} \\ n_i/(n_i-1)(K(\hat{x},\hat{x}) - 2(\mathbf{k}^{\mathrm{T}}\mathbf{SD})_{\mathrm{i}} + (\mathbf{DS}^{\mathrm{T}}\mathbf{XX}^{\mathrm{T}}\mathbf{SD})_{\mathrm{ii}}) & \mathrm{j} = \mathrm{i} \end{cases}$$
    if $\rho_m \leq \rho_j$ for all j **then** transfer $\hat{x}$ to S_m
    recompute E_k and update the n_i in $\mathbf{D}$ as well as the entries of $\mathbf{S}$
  until no change in E_k in n attempts
  return S
end

It should be obvious from the algorithm that together with the initial classification of the $\hat{x}$, the ρ_j, the ρ_m and the change in the E_k can easily be computed.

In addition, due to the various similarity concepts provided by the particular choice of Mercer kernel, the algorithm posseses a remarkable degree of flexibility.

Note that the above algorithm only differs from the usual k-means algorithm with respect to the update procedure, see e.g. [5], p. 527. Indeed in the original algorithm all samples are reclassified before updates are effected whilst in Duda's original k-means iterative algorithm (and in its kernel version given above) updates are effected after reclassification of each sample.

4 Conclusion

In the first part of the paper the structure of certain Mercer kernels has beeen described in terms of 1-cohomology since these results do not seem to be well-known. Although in practical applications for supervised learning standard kernels (e.g. the low order polynomial kernels) appear most successful, cf. [9], since even the higher order poynomial kernels and the RBF kernel turn out to be susceptible to overfitting, cf. [1,3,18], it seems useful to experiment with other kernels, like the ones described above, as well. This is all the more the case since for some problems invariant kernels may be useful, although Minsky in [14] mentions certain restrictions. It can be hoped, as mentioned before, that with further improvements in the hardware and software region even large numbers of samples can be treated succecssfully. After all, many of the algorithms in use nowadays could hardly have been implemented 20–30 years ago.

It should also be emphasized that kernels do not require a vector space structure of the samples, for examples see [17], which might make them useful for commercial applications, It seems obvious that a lot of experimental work is needed since it is by no means clear which kernel is best suited for any given particular problem.Thus it remains a challenging task to make progress in this direction.

References

1. Bishop, C.M.: Pattern Recognition and Machine Learning. Springer, New York (2006)
2. Cristianini, N., Shawe-Taylor, J.: An Introduction to Support Vector Machines and other Kernel-Based Learning Methods. Cambridge University Press, Cambridge (2000)
3. Cover, T.M.: Geometrical and statistical properties of systems of linear inequalities with applications in pattern recognition. IEEE Trans. Electron. Comput. **14**, 326–334 (1965)
4. Dixmiers, J.: Les $C^{\star}$-algebres et leurs reprsentations. Gauthier-Villars (1969)
5. Duda, R.O., Hart, P.F., Stork, D.G.: Pattern Classification. Wiley, New York (2017). Reprinted
6. Falkowski, B.-J,: Mercer Kernels and 1-Cohomology. In: Proceedings of KES (2001)
7. Falkowski, B.-J,: Mercer Kernels and 1-cohomology of certain of certain semi-simple lie groups. In: Proceedings of KES (2003)
8. Falkowski, B.-J,: On certain group invariant mercer kernels. In: Proceedings of the International Joint Conference on Computational Intelligence, IJCCI (2009)
9. Falkowski, B.-J.: A perceptron classifier, its correctness proof and a probabilistic interpretation. In: Proceedings of HAIS 2017, LNAI, vol. 10334. Springer (2017)
10. Fürstenberg, H.: Noncommuting Random Products. Trans. of the AMS **108**, (1962)
11. Gangolli, R.: Positive Definite Kernels on Homogeneous Spaces. In: Ann. Inst. H. Poincare B, Vol. 3, (1967)
12. Haykin, S.: Neural Networks, a Comprehensive Foundation. Prentice-Hall (1999)
13. Helgason, S.: Differential Geometry and Symmetric Spaces. Academic Press, New York (1963)
14. Minsky, M.L., Papert, S.: Perceptrons, Expanded edn. MIT Press (1990)
15. Parthasarathy, K.R., Schmidt, K.: Positive Definite Kernels, Continuous Tensor Products, and Central Limit Theorems of Probability Theory. Springer Lecture Notes in Mathematics **272**, (1972)
16. Ricci, F., Rokach, L., Shapira, B.: Introduction to Recommender Systems Handbook. Springer (2011)
17. Shawe-Taylor, J., Cristianini, N.: Kernel Methods for Pattern Analysis. Cambridge University Press (2004)
18. Vapnik, V.N.: Statistical Learning Theory. John Wiley and Sons (1998)
19. Wahba, G.: Support Vector Machines, Reproducing Kernel Hilbert Spaces, and Randomized GACV. In: Advances in Kernel Methods, Support Vector Learning. MIT Press, (1999)
20. http://www.cse.msu.edu> cse902 > ppt > Ker... Chitta, Radha: Kernel Clustering. Last visited 20.01.2019
21. https://www.ics.uci.edu> teaching > Spec ... Welling, Max: Kernel K-Means and Spectral Clustering. Last visited 19.06.2018

Research on the Development of Expert Systems Using Artificial Intelligence

Rustam A. Burnashev(✉), Ruslan G. Gabdrahmanov, Ismail F. Amer, Galim Z. Vakhitov, and Arslan I. Enikeev

Institute of Computational Mathematics and Information Technology, Kazan Federal University, Kazan, Russia
r.burnashev@inbox.ru

Abstract. This paper presents a study related to the development of logical expert systems using artificial intelligence on the example of diagnosing leukemia. For the development of the system, various approaches to the design of artificial intelligence systems and medical data used in the diagnosis of leukemia were applied. A distinctive feature of the expert system is that a specialist working with the knowledge base can not only get the answer he needs, but also get access to all the knowledge from the knowledge base by asking the necessary questions to the expert system. The methods are based on obtaining information about the characteristics of the software using specialized software that provides automation. Data sets and logical rules that were used for the initial diagnosis of the disease were identified. In developing and analyzing software requirements, a prototype system was developed. This system includes a perceptron and consists of 3 hidden layers. For software development was used programming languages Python, JavaScript and Prolog.

Keywords: Expert system · Python 3 · JavaScript · Prolog · Numpy · Psycopg2 · PostgreSQL

1 Introduction

An artificial intelligence system is a system capable of performing creative or ambiguous functions that were previously considered exclusively human prerogatives, while AI researchers are free to use methods that are not applied by people if it is necessary to solve specific problems [1].

In the scientific environment, the most common of the programming languages of artificial intelligence aimed at the development of expert systems are Lisp and Prolog systems. With these programming languages, the system developer has more convenience in choosing the methods of knowledge representation and management strategy.

This means that after training, the neural network is able to return the correct result based on data that was missing from the data used for training, as well as noisy, incomplete, distorted data, but with less accuracy. Such a system is already capable of solving rather complex tasks [2].

The medical knowledge base contains inference rules and information about human experience and knowledge in a certain subject area [4]. In self-learning systems, the

J. Świątek et al. (Eds.): ISAT 2019, AISC 1051, pp. 233–242, 2020.
https://doi.org/10.1007/978-3-030-30604-5_21

knowledge base also contains information resulting from the solution of previous problems [3].

Leukemia is a term that unites numerous tumors of the hematopoietic system, arising from hematopoietic cells and affecting the bone marrow. The division of leukemia into two main groups: acute and chronic is determined by the structure of tumor cells: acute are leukemia's, the cell substrate of which is represented by blasts, and chronic - leukemia's, in which the bulk of tumor cells is differentiated and consists mainly of mature elements [9].

In the course of the study, existing methods of diagnosing leukemia, known information about the nature of the course of the disease and about the changes that occur in the body of a sick person were studied.

2 Research Methods

The formulation (specification) of the problem, the development progress is specified the prototype of the expert system is planned to be determined:

- Necessary resources (time, people, computers, money, etc.);
  - Sources of knowledge (books, additional experts, methods) [6, 8];
  - Available, a similar expert system;
- Goals (dissemination of experience, automation of routine actions, etc.);
  - Classes of solved problems, etc.

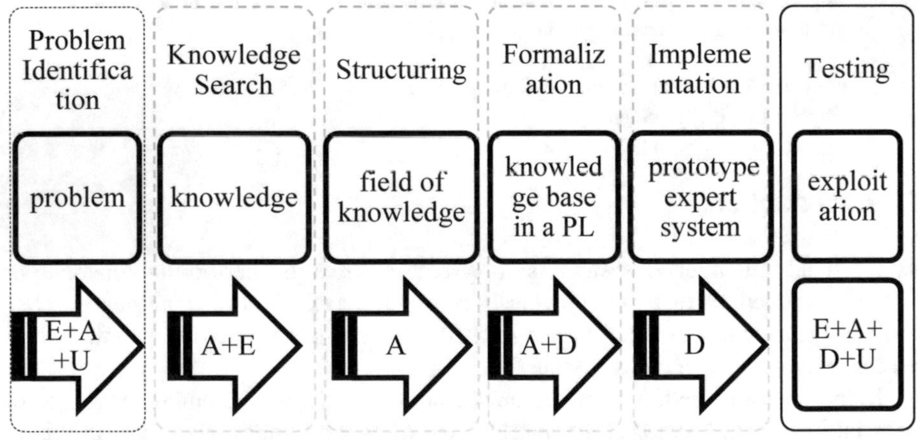

Fig. 1. Stages to develop an expert system

The Fig. 1 involved the following participants:

- E - Expert;
- A - Analyst (knowledge engineer);
- D - Developer (programmer);
- U - Users of the system.

To solve the tasks, the following sequence of actions was performed:

1. Analysis and formation of requirements for the expert system;
2. Apply a knowledge base to collect and analyze software requirements;
3. Storage of facts and rules in the knowledge base;
4. Structuring and analyzing data using Python and Prolog libraries;
5. The choice of parameters used for the diagnosis, the type and architecture of the future system based on the data obtained;
6. Creating a knowledge base for training and testing the developed expert system;
7. Design, training, testing a prototype of an expert system based on dynamically updated knowledge bases.

To form requirements for the software being developed, an analysis was made of the area in which the expert systems will be used (Figs. 2 and 3).

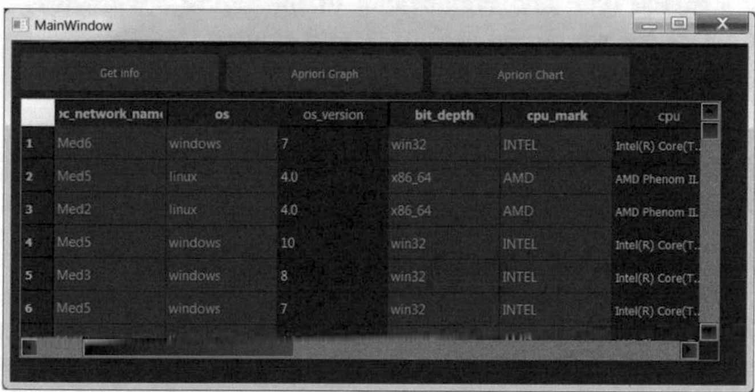

Fig. 2. Apply a knowledge base to collect and analyze software requirements

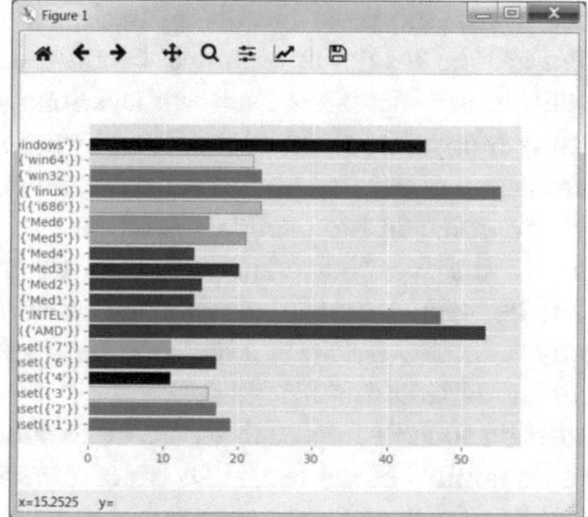

Fig. 3. Analyze of the software requirements

In the process of studying the existing methods of diagnosing leukemia, the literature on this topic and some scientific articles were studied.

The collection of information about changes that occur with the patient's body was carried out through the study of literature, articles, cases from practice.

Scientific works in the field of diagnosis of leukemia using artificial intelligence systems were considered.

The choice of the used parameters for the diagnosis was carried out on the basis of them:

- Reliability;
- Accuracy;
- Availability (data should be collected during standard medical prophylaxis - "physical examination");
- Manufacturability (processing and interpretation complexity).

The choice of the type and architecture of the future system was based on the selected parameters for which the system should achieve:

- Theoretical and practical feasibility;
- The greatest expected accuracy of diagnosis;
- Acceptable from the point of view of practical operation speed and ease of use;
- Available modern computer system requirements.

The Python programming language was chosen for the development, training and testing of the system since it has an extensive set of libraries in the field of AI and data processing, as well as a fairly high computing speed, which will significantly speed up development. PostgreSQL DBMS is used for the logical facts and rules storage.

3 Research

Based on the above mentioned, the most promising basic set of parameters for processing by the AI system are the results of a clinical blood test.

As already noted, the disease manifests itself with various symptoms, changes in the cellular and chemical composition of the blood, the physiological properties of the patient. Separately, it should be noted that almost all indicators of the clinical analysis of blood change, but all are noted in the medical literature [9].

This is probably due to the fact that other indicators do not change sufficiently enough so that they can be used to make reliable and accurate conclusions when examining the analysis by a doctor. Below is a table (Table 1) with an example of the clinical characteristics of blood tests for a patient.

In this regard, the most reasonable approach would be to submit as input data of all parameters obtained from a clinical blood test. Also, since the specified parameters are normally different for people of different age and gender, it is obligatory to supplement the data of a clinical analysis of blood with data on age and sex. Additionally, it is possible to indicate some symptoms, such as sweating and anemia, to improve the accuracy of the diagnosis.

Table 1. Clinical characteristics of the blood test

Analysis rate (adult)	Norm	Sick (acute leukemia)
White blood cell	In the range 4,0–9,0 × 10^9/l	In the range 100 × 10^9/l
Platelet count	In the range 180–320·10^9/l	In the range 150·10^9/l
Color indicator (CI)	0.85–1.15	0.85–1.15
Basophils	0–1%	~0%
Eosinophils	1–2%	~0%
Band	3–6%	1%
Monocytes	4–8%	6%

Existing work in the field of application of AI in the diagnosis of leukemia is extremely small. There is a concept of a system designed to diagnose acute leukemia based on computer microscopy data [2].

Based on all the collected data and the conclusions made, it was decided to choose the following input parameters:

- Age;
- Gender;
- Hemoglobin;
- Red blood cells;
- White blood cells;
- Hematocrit;
- Reticulocytes;
- Platelets and etc.

Based on the above mentioned, age and gender here are corrective parameters. They do not indicate the presence or absence of the disease directly, but adjust other indicators, thereby increasing the accuracy of the diagnosis.

The last 5 indicators refer to the leukocyte formula and are measured in percent (from 0 to 100%).

Many of these factors are interrelated and interact with each other in both well-known and not fully studied images.

The presented list is not complete and final and may be supplemented by new parameters in the process of future research to improve the accuracy of the diagnosis.

These factors were roughly structured as follows:

- Basic blood parameters;
- Clarifying factors;
- Leukocyte formula;
- Relationships between individual parameters.

Since blood is a complex structure with many parameters and interconnections, it was decided to resort to the learning systems, namely neural networks, since they are able to provide the most complete account of all deviations and interconnections.

Since all parameters are initially numeric, the use of convolutional neural networks does not make sense. To create a prototype of the system, a perceptron was chosen, due to its simplicity and extensive experience in creating predictive systems based on it. The use of adaptive resonance networks or others not considered in this article is also allowed.

It is assumed that the neural network itself will learn to identify and correctly interpret these relationships by adjusting the weights with each measurement, in order to achieve a more accurate diagnosis. The created scales, in this case, are the knowledge base of the system.

Since, in this study, the knowledge base and the conclusion from it are of interest, the structure of the expert system was simplified and included the following components (Table 2):

- Knowledge base;
- Working memory, also called a database (PostgreSQL);
- Interpreter (Python 3);
- Knowledge acquisition components;
- User interface.

Table 2. Analysis requirements and testing of the prototype of the expert system

Stages of LC	Name of works	Performance (yes/no)	%
Concept definition (10%)	Cooperation with the customer Target definition Formulation of the problem	Yes	10%
Requirements collection (10%)	Collection of information about the problem Definition of GUI Requirements	Yes	10%
Requirements analysis (10%)	Creating UML diagrams GUI concept	Yes	10%
Development (30%)	GUI development Creating a neural network prototype Neural network training Organization of data processing	Yes	30%
Testing (30%)	Manual testing	Yes	30%
Exploitation (10%)	Documentation and exploitation	Yes	10%
Total		y or n	100%

4 Results

The PostgreSQL DBMS is used to store the generated knowledge for learning, and the psycopg2 library is used to interact with it.

During the design process, several prototypes of the neural network were created, which were integrated into the graphical interface and attached to the database. Thus, a prototype of an expert system was created based on the use of AI technologies that can assist specialists in diagnosing acute and chronic forms of leukemia [7].

The architecture (Fig. 4) of the most successful designed perceptron is as follows:

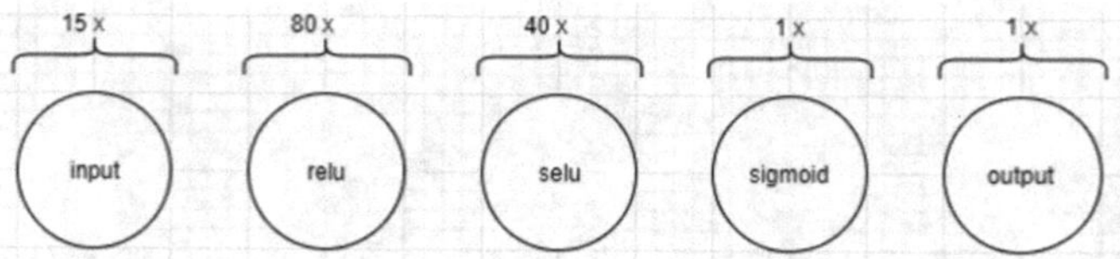

Fig. 4. Neural network architecture diagram

During the development of the system, the data was structured [5] in the knowledge base with the specified architecture, which shows the following accuracy change in the learning process (Fig. 5):

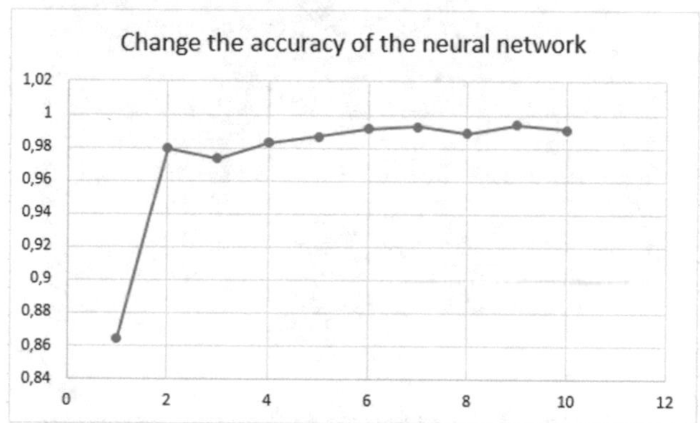

Fig. 5. Change the accuracy of the neural network

The neural network demonstrates relatively high accuracy (98–99%) with a fairly small drop in accuracy due to retraining of the generated data on close to reality, which suggests a theoretical possibility of successfully using similar systems in practice.

Based on the data obtained, a user interface was developed in compliance with modern GUI requirements. The PyQt 5 library was used to develop the GUI shell (Fig. 6).

In the prototype of the expert systems a knowledge base was created, which can be updated as a result of further processing of data specially collected for training the neural network.

In parallel with the theoretical studies of dynamic production systems, a dynamic version of the expert system was created, in which users had the opportunity to dynamically simulate the graphical shell of the template, setting the necessary requirements for the system (Fig. 7).

Fig. 6. Prediction interface of the prototype system

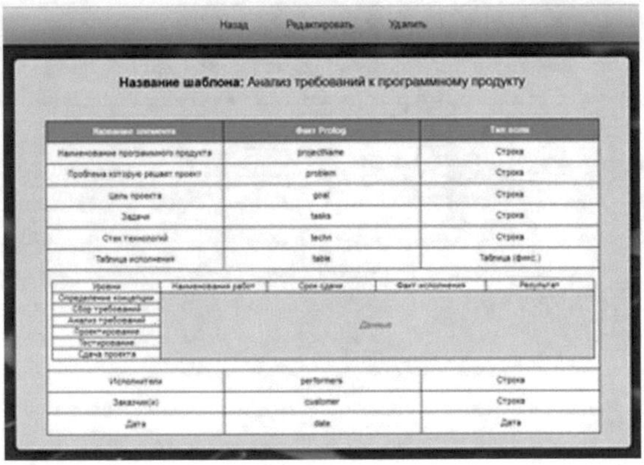

Fig. 7. Dynamic modeling of the expert system

The software tool was used to create a number of specific systems. This is achieved by having a library of attached procedures in the system, which the user can include in the knowledge base at his discretion. A deeper level of reprogramming is associated with the development of new procedures and their inclusion in the knowledge base. This work should be performed by the developers of the system, because it requires familiarity with the internal organization of the system and with the logical programming language Prolog and JavaScript, in which it is written (Fig. 8).

Thus, every fact in the software system is that this attribute has this value. Usually the task of an expert system is to find out the value of some attribute called the purpose of the system. The purpose of the system development and analysis of requirements are the attributes of "goal setting", "technology stack", "testing", "evaluation", etc. In the developed system, the goal can be selected by the user, which makes our system capable of dynamically processing data.

We will use the equal sign to record the fact. The fact, therefore, is a record of the form <attribute-value>.

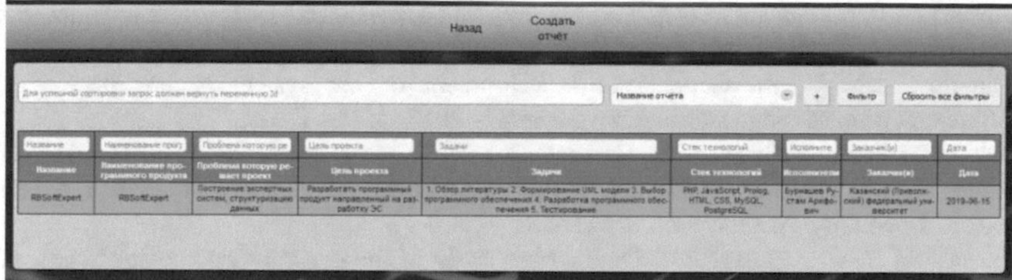

Fig. 8. Editing the knowledge base

In the process of output to the database, the facts established by the system are recorded. In the normal operation of the system, from the beginning to receive a response (this period will be called a session) facts can only be added to the database. Once a fact is established in the output process, it can neither be removed from the database nor changed. We note immediately that both the operation and the removing, and modifying the previously obtained fact - it is possible to realize by applying the attached control procedures. The deletion and modification times of established facts is also in the dynamic expert systems.

5 Conclusion

As the result of the research, a prototype of the medical expert system was developed, designed to help oncologists in predicting the acute and chronic forms of leukemia. In contrast to the database, the knowledge base of the system is what it knows about the tasks for which it is created, and not about each specific task. Knowledge base in all sessions is the same and is in the process of withdrawal it is not changed. Here, however, again it should be noted: the attached procedures can change the knowledge base. A survey study of the forms of diagnosis of leukemia, which are aimed at identifying a malignant disease through a prototype expert system, was conducted. At the stage of knowledge gathering, experimental data were structured, and a generator was developed in the Python programming environment, which allowed processing data for training and teaching an artificial neural network. The client application was implemented in the integrated development environment in the JavaScript and Prolog programming language.

Acknowledgment. This work was supported by the research grant of Kazan Federal University.

References

1. Copeland, J.: What is Artificial Intelligence? (2000)
2. Requena, G., Sánchez, C., Corzo-Higueras, J.L., Reyes-Alvarado, S., Rivas-Ruiz, F.: Melomics music medicine (M3) to lessen pain perception during the pediatric prick test procedure (Eng.) Pediatr. Allergy Immunol. **25**(7), 721–724 (2014). https://doi.org/10.1111/pai.12263. ISSN 1399-3038

3. Burnashev, R.A., Gubajdullin, A.V., Enikeev, A.I.: Specialized case tools for advanced expert systems. Adv. Intell. Syst. Comput. **745**, 599–605 (2018)
4. Kamalov, A.M., Burnashev, R.A.: Development of the system expert medexpert for differential disease diagnostics. Astra Salvensis **2017**, 55–64 (2017)
5. Burnashev, R.A., Yalkaev, N.S., Enikeev, A.I.: J. Fundam. Appl. Sci. **9**, 1403–1416 (2017)
6. Poole, D., Mackworth, A., Goebel, R.: Computational Intelligence: A Logical Approach. Oxford University Press, New York (1998)
7. Burnashev, R.A., Enikeev, A.I.: On the case study of life cycle **10**(10 Special Issue), 1729–1734 (2018)
8. Lorier, J.-L.: Artificial Intelligence Systems. M.: Mir, 568 p. (1991). 20 000 copies. ISBN 5-03-001408-X
9. Petrovsky, B.V. (ed.) Big Medical Encyclopedia (BME), 3rd edn.

Machine-Learning and R in Plastic Surgery – Evaluation of Facial Attractiveness and Classification of Facial Emotions

Lubomír Štěpánek[1(✉)], Pavel Kasal[1], and Jan Měšťák[2]

[1] Faculty of Biomedical Engineering, Czech Technical University in Prague, Kladno, Czech Republic
lubomir.stepanek@fbmi.cvut.cz

[2] First Faculty of Medicine, Charles University and Na Bulovce Hospital, Prague, Czech Republic

Abstract. Although facial attractiveness is data-driven and nondependent on a perceiver, traditional statistical methods cannot properly identify relationships between facial geometry and its visual impression. Similarly, classification of facial images into facial emotions is also challenging, since the classification should consider the fact that overall facial impression is always dependent on currently present facial emotion.

To address the problems, both profile and portrait facial images of the patients ($n = 42$) were preprocessed, landmarked, and analyzed via R language. Multivariate regression was carried out to detect indicators increasing facial attractiveness after going through rhinoplasty.

Bayesian naive classifiers, decision trees (CART) and neural networks, respectively, were built to classify a new facial image into one of the facial emotions, defined using Ekman-Friesen FACS scale.

Nasolabial and nasofrontal angles' enlargement within rhinoplasty increases facial attractiveness ($p < 0.05$). Decision trees proved the geometry of a mouth, then eyebrows and finally eyes affect in this descending order an impact on classified emotion. Neural networks returned the highest accuracy of the classification.

Performed machine-learning analyses pointed out which facial features affect facial attractiveness the most and should be therefore treated by plastics surgery procedures. The classification of facial images into emotions show possible associations between facial geometry and facial emotions.

Keywords: Machine-learning · Facial attractiveness · Facial emotions · Plastic surgery · R language

1 Introduction

While origins of plastic facial surgery are related to the First World War (1914–1918), human facial attractiveness received its attention already from ancient philosophers, namely from Polykleitos and Aristotle (4–3 century BC) [1]. They defined classical rules but only subjectively; the rules were applied only on facial appearance of the

J. Świątek et al. (Eds.): ISAT 2019, AISC 1051, pp. 243–252, 2020.
https://doi.org/10.1007/978-3-030-30604-5_22

Caucasian race and limited on viewing of beauty by a naked eye [1]. In Renaissance, Leonardo Da Vinci turned the ancient ideas into so-called Neoclassical facial canons, serving mainly for contemporary artists and consisting of nine simple mathematical rules such as subtractions or proportions of any two linear facial distances should be equaled to some fixed constants.

According to the rules, some of facial distances should follow the golden ratio $\left(\frac{\sqrt{5}-1}{2}\right)$ [2]; more attractive faces are axially-symmetric [3] or similar to the average (*morphed*) face of a population [4], furthermore, the usually include signs of neoteny, *juvenilization*, i.e. relatively large eyes or low mouth [5]. Finally, sexual dimorphism seems to increase human facial attractiveness – both in male faces with dominating masculine facial geometry and female faces with dominating feminine facial geometry, respectively [6].

All the rules mentioned above within the Neoclassical Canon inclusively are still widely applied in nowadays plastic facial surgery procedures, including rhinoplasty (a correction of nasal size or shape), and – in addition – they are the crucial (or even the only one available) ways of how to plan an operational strategy before it is realized. Any data-driven approaches to procedures covered by plastic facial surgery are the one whose time have to come though [7].

Patients undergoing plastic facial surgeries wish also not only improvements of their "static" facial features such as corrections of nasal size or shape (rhinoplasty), but also changes of the "dynamic" facial expression, e.g. surgical improvements of mouth in order to make a smile more facially attractive and to increase the level of facial attractiveness whenever the patient's face is just smiling [7].

Due to this, the field of plastic facial surgery should be take into account the fact that movements of facial muscles during emotion expression are connected to the facial impressions. The idea that total human face impression is always dependent on presently expressed facial emotion was first taken into consideration by Charles Darwin; he supposed that there is a limited and universal set of facial emotions which could be expressed by all higher mammals [8]. Later, American psychologists, Silvan Solomon Tomkins, improved the idea by claiming that specific facial expressions are uniquely linked to individual emotions and, not only but also, assumed that emotions are easily comprehensible (and comparable, able to be analyzed) across races, ethnic groups, and cultures [9].

A classification of human facial impressions based on six ("clusters" of) emotions – happiness, sadness, surprise, anger, fear, disgust [10] was established in 1971 by two psychologists Paul Ekman and Wallace Friesen and improved into a well-known system called Facial Action Coding System (FACS) in 90's. The system reflects the fact that each movement of facial muscles uniquely determines a consequent emotion such that an observer is able to perceive an impression resulting from the emotion [11].

Recognition techniques applicable on human facial emotions is a part of more general human face image recognition techniques and includes three consequent phases [12]:

1. face detection and localization;
2. extraction of appropriate face features;
3. classification of a facial expression into a facial emotion.

Talking about the first phase (1.), *face detection and localization*, an expert method (e.g. left and right eye are both symmetric and of similar size, etc.) could be applied [13], then a *feature invariant method* (e.g. eyes, nose, and mouth is detected by human perceiver regardless of an angle of view or intensity of current lighting) is possible, and finally an *appearance-based method* (when face image is compared to face templates generated by a machine-learning algorithm) is a choice [14, 15].

Then, the second phase (2.), covers an *extraction of appropriate face features*, which can be done via *Gabor wavelets method* [16], additionally an *image intensity analysis*, a *principal component analysis* (PCA), an *active appearance model* [17], or *graph models* [17], respectively, including also the well-known *Marquardt mask* could be considered.

Lastly but not leastly, the third phase (3.), *classification of facial expression into facial emotion*, is a subset of a classification problem and belongs to the families of machine-learning and deep-learning algorithms. It can be performed using *rule-based classifiers* [18], *model comparing classifiers* [18] or *machine-learning classifiers* [18].

To sum this up, aims of the study are therefore

- firstly, to identify which facial geometric features and their surgical improvements are connected with increased facial attractiveness level in patients who think about undergoing of rhinoplasty or are indicated to undergo rhinoplasty;
- secondary, to adopt a system of facial expression based on FACS and both test it and improve it in order to eventually increase a number of facial emotions, so that it could be used for appropriate classification of facial images into facial emotions – this can be seen as a promising insight useful for analysis of relationships between facial expressions based on facial muscles geometry and muscles movements on the one hand, and facial emotions, respectively, on the other hand.

The second point above could be considered to be crucial for planning of facial surgical interventions – while real structures such as facial muscles are already objects of surgical procedures, changes for the better in facial expressions should be in fact the expected results of the surgical strategies. However, the relations are too complex and not obvious, therefore a machine-learning classification of facial images into facial emotions could be a kickoff of the process of their explanation.

2 Research Material and Methodology

All appropriate patients who attended the Department of Plastic Surgery, First Faculty of Medicine, Charles University, Prague and Na Bulovce Hospital were asked to join the study and were informed enough about all details of the study before their participation.

There were in total precisely 30 patients who underwent the rhinoplasty surgery and were eligible to take part in the study. Both a profile and portrait picture of each of them was taken and saved using a secured database.

Other 12 patients constituted another sample (all of them were students at the Faculty of Biomedical Engineering, Czech Technical University in Prague) whose portrait and profile images were shot just at the moment they expressed a facial expression based on the given incentive. For instance, they were told a joke in order to shew a *fun* emotion. An overview of the facial expressions is in Table 1. The grand total number of their pictures is therefore equal to $12 \times 14 = 168$.

Table 1. Facial emotions and relations to their quality.

Facial emotion	Quality	Facial emotion	Quality
Contact	Positive	Reaction	Neutral
Helpfulness	Positive	Decision	Neutral
Evocation	Positive	Rejection	Negative
Well-being	Positive	Depression	Negative
Fun	Positive	Fear	Negative
Deliberation	Positive	Defence	Negative
Expectation	Positive	Aggression	Negative

2.1 Data of Interests

Facial image data were described one paragraph above, and besides them a seven-level Likert scale following the values of (−3, −2, −1, 0, +1, +2, +3) (the higher score, the more attractive an observed face is considered to be) was used to assess each photography of each patient before and after undergoing the rhinoplasty. A board of 14 independent evaluators does the evaluation was as depicted above.

The facial emotions we used in the study come from the Facial Action Coding System (FACS), but has been improved a bit. We defined 14 facial emotions in total – *contact*, *helpfulness*, *evocation*, *defence*, *aggression*, *reaction*, *decision*, *well-being*, *fun*, *rejection*, *depression*, *fear*, *deliberation,* and *expectation*, respectively [19, 20].

What is more, we expertly defined on our own a quality of facial emotions such that each one of the emotions is either positive, negative or neutral, respectively, according to a mean effect on an observer (and stated/confirmed by an expert). Relations between the facial emotions and the quality of the facial emotions including the way how they were used in the study, are shown in Table 1.

2.2 Landmarking

Landmarks can be described as morphometrically essential points on a plane of a facial image. Overview of the landmarks we are interested in is in the Fig. 1.

Landmarks were plotted manually using proprietary executable program written in C# language, by which the coordinates of all of landmark points were collected. The landmarks were also gathered as well using an experimental application written in R language [21] which is able to bridge a well-known C++ library called `dlib` [22] for

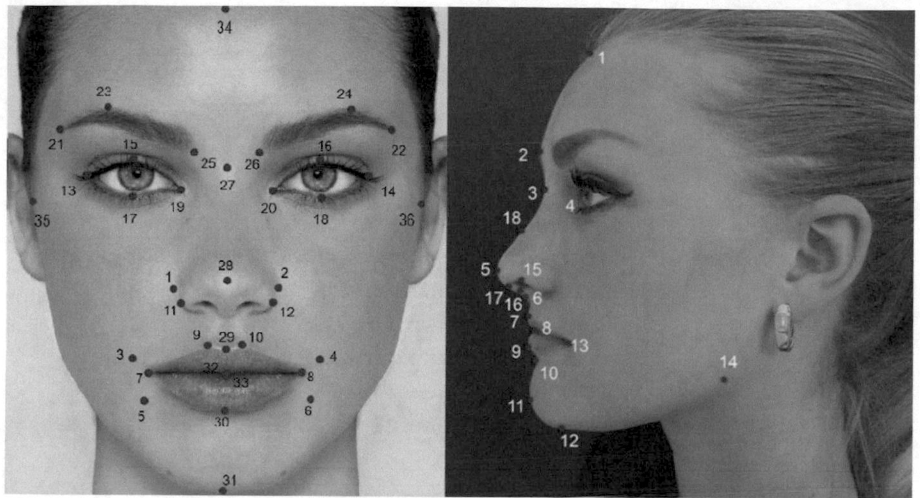

Fig. 1. Landmarks of a face portrait (on the left) and a face profile (on the right).

the R application; the `dlib` library enables besides other to use automatic facial landmarking approaches. After the collection of all n landmarks' coordinates, for i-th landmark, where $\mathrm{i} \in \{1, 2, 3, \ldots, \mathrm{n}\}$, with original coordinates $[\mathrm{x_i}, \mathrm{y_i}]$, new standardized coordinates $[\mathrm{x'_i}, \mathrm{y'_i}]$ were calculated following the formulas (1–2) below

$$x'_i = \left. \frac{x_i - \min_j\{x_j\}}{\max_j\{x_j\} - \min_j\{x_j\}} \right|_{j \in \{1,2,3,\ldots,n\}} \tag{1}$$

$$y'_i = \left. \frac{y_i - \min_j\{y_j\}}{\max_j\{y_j\} - \min_j\{y_j\}} \right|_{j \in \{1,2,3,\ldots,n\}} \tag{2}$$

ensuring that all faces taken in the images are of equal size. The described transformation of coordinates (standardization) allows us to compare feasibly enough any two face profiles one to each other (their transformed coordinates $[x'_i, y'_i]$), and any two face portraits one to each other (their transformed coordinates $[x'_i, y'_i]$), respectively.

The derived metrics and angles computed using the transformed coordinates belonging to the landmarks are in Table 2 (point definitions of the landmarks are shown in Fig. 1).

Table 2. The derived metrics and angles calculated using the transformed coordinates od the landmarks.

Metric/angle	Definition
Nasofrontal angle	Angle between landmarks 2, 3, 18 (profile)
Nasolabial angle	Angle between landmarks 7, 6, 17 (profile)
Nasal tip	Horizontal Euclidean distance between landmarks 6, 5 (profile)
Nostril prominence	Euclidean distance between landmarks 15, 16 (profile)
Cornea-nasion distance	Horizontal Euclidean distance between landmarks 3, 4 (profile)
Outer eyebrow	Euclidean distance between landmarks 21, 22 (portrait)
Inner eyebrow	Euclidean distance between landmarks 25, 26 (portrait)
Lower lip	Euclidean distance between landmarks 30, 33 (portrait)
Mouth height	Euclidean distance between landmarks 6, 8 (profile)
Angular height	Euclidean distance between landmarks 7 (or 8) and 33 (portrait)

2.3 Statistical Analysis

All statistical analyses were performed using R language for statistical computing and graphics [21]. Outputs with p-values below (or very "close" to) 0.05 were assumed to be statistically significant.

In order to detect which predictors, i.e. derived metrics or angles (see Table 2 for more details) statistically significantly influence an average difference of the attractiveness' Likert scores after and before the rhinoplasty undergoing, a multivariate linear regression analysis was carried out [23].

Then, Bayesian naive classifiers [24], classification trees (CART) [25] and neural networks using backpropagation with sigmoidal activating function [26], respectively, were learned and applied to classify an image of a human face (portrait) into one of the facial emotions, and as well into one of the value of the quality of facial emotions (and even into some more parameters of emotions not discussed in this article).

Performance measures of predictive accuracy of the previously mentioned three machine-learning methods are reported as confusion matrices and as 95% confidence intervals for an average predictive accuracy estimate. Grant total sum of each of the confusion matrices is equal to $12 \times 14 = 168$, i.e. a number of all individuals multiplied by a number of pictures taken per individual.

3 Results and Discussion

A table with summary of the multivariate linear regression is shown in Table 3. As we can see from the Table 3, the mean increase of facial attractiveness level after undergoing the rhinoplasty is about 3.8 Likert point, $p = 0.043$. Moreover, per each radian of nasofrontal angle enlargement, there is an expectation of mean increase about 0.353 Likert point in facial attractiveness after undergoing the rhinoplasty (when a patient went through this kind of correction, of course), $p = 0.050$. Similarly, per each radian of nasolabial angle enlargement, there is an expectation of mean increase about 0.439

Likert point in facial attractiveness after undergoing the rhinoplasty (again, this is true if and only if this correction is even applied to a patient), $p = 0.057$.

Table 3. Summary table of the multivariate linear regression.

Predictor	Point estimate	t-value	p-value
$\text{Intercept}_{\text{after-before}}$	3.832	1.696	0.043
$\text{Nasofrontal angle}_{\text{after-before}}$	0.353	0.174	0.050
$\text{Nasolabial angle}_{\text{after-before}}$	0.439	1.624	0.057
$\text{Nasal tip}_{\text{after-before}}$	−3.178	0.234	0.068
$\text{Nostril prominence}_{\text{after-before}}$	−0.145	0.128	0.266
$\text{Cornea-nasion distance}_{\text{after-before}}$	−0.014	0.035	0.694

As we expected, the larger both nasofrontal and nasolabial angles corrections are, the higher score of attractiveness level such a face obtains. The two mentioned angles are the usual corrections routinely done within a rhinoplasty procedure. Of course, these results are limited. For instance, if both angles, nasofrontal and nasolabial one would become straight angles, a nose would "disappear" under these conditions, instead of expected facial attractiveness level increasing, as stated above.

The confusion matrices of the prediction of the emotional quality based both on Bayesian naive classifier and neural network follow (Table 4). Confusion matrices of the prediction of the facial emotions are not reported due to the fact they oversize the page format (there are 14 of both true and predicted classes).

Table 4. Confusion matrices of a predictions of the emotional quality based on Bayesian naïve classifier and on a neural network.

		Predicted class based on **Bayesian naïve classifier**			Predicted class based on **neural network**		
		Negative	Neutral	Positive	Negative	Neutral	Positive
True class	Negative	34	16	16	36	6	6
	Neutral	11	39	8	12	54	16
	Positive	4	10	30	2	4	32

Point estimate and 95% confidence interval of mean prediction accuracy of the facial emotions based on Bayesian naive classifier is 0.325 (0.321, 0.329). Point estimate and 95% confidence interval of mean prediction accuracy of the emotional quality based on Bayesian naive classifier is 0.413 (0.409, 0.417). Since $0.325 > \frac{1}{|\text{facial emotions}|} = \frac{1}{14}$ and $0.413 > \frac{1}{|\text{emotional qualities}|} = \frac{1}{3}$, both classifiers predict more precise then random process. Since the target variables (*facial emotions* and *quality of facial emotions*, respectively) contain multiple classes, the classification task here is so-called "multiclass" and even only moderate prediction accuracy is acceptable under this conditions [27, 28].

Point estimate and 95% confidence interval of mean prediction accuracy of the facial emotions based on decision trees is 0.488 (0.484, 0.492). Point estimate and 95% confidence interval of mean prediction accuracy of the emotional quality based on decision trees is 0.525 (0.521, 0.529). Similarly, in both cases, the classifier predicts more precise than a random process.

Finally, point estimate and 95% confidence interval of mean prediction accuracy of the facial emotions based on neural networks is 0.507 (0.503, 0.511). Point estimate and 95% confidence interval of mean prediction accuracy of the emotional quality based on neural network is 0.726 (0.722, 0.730). Again, in both cases, the classifier predicts far more precise than a random process (and even substantially better than the previous two classifiers, though).

There are examples of decision trees learned in order to predict one of the facial emotions or one of the emotional quality using facial geometry of the photographed facial expression in Fig. 2.

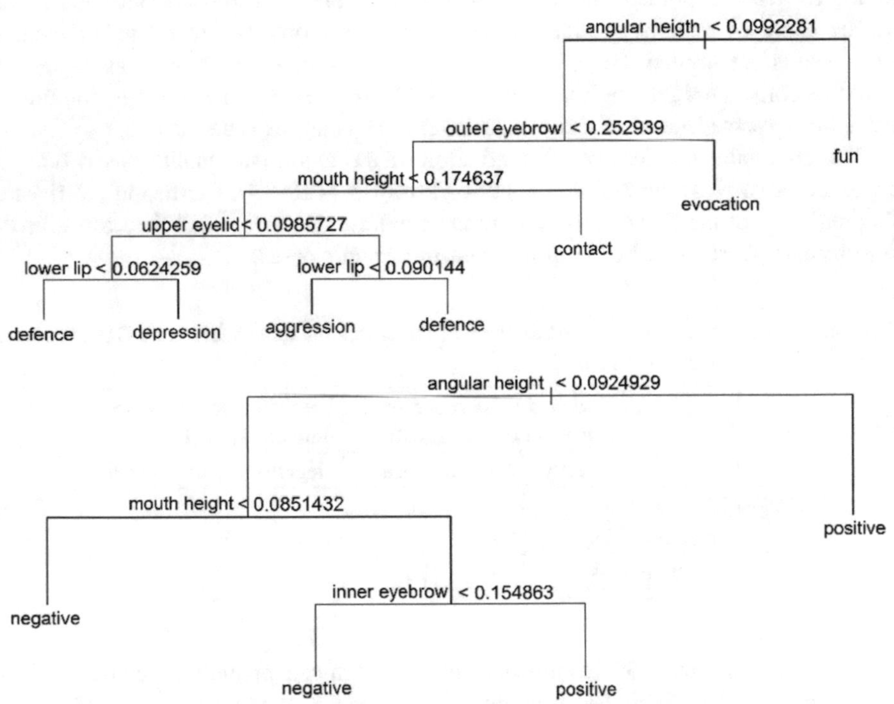

Fig. 2. A decision tree for prediction of the facial emotions (upper plot) and the quality of facial emotions (bottom plot) (statements in nodes are true for left child nodes).

The closer to the root node the derived geometrical metric or angle in the plot is, the more important seems to be in order to explain a "direction" of the classification into the final class of interest. As we can see, the facial expressions are determined by geometry of the mouth, then by geometry of the eyes, respectively.

Furthermore, going deeper into results of Fig. 2 (upper plot), if the angular height – that is a vertical distance between mouth angles and a horizontal line between the lips – is large enough (more precisely, it is larger than 0.0992), and it means that such a face in the image is smiling, then an emotion of that image is classified as a *fun*, as we can expect.

4 Conclusion

The machine-learning analyses we have performed illustrate which geometric facial features, based on significant data evidence, affect facial attractiveness the most – either as predictors increasing facial attractiveness level after undergoing rhinoplasty or as geometric features influencing the classification of facial images into facial emotions –, and therefore should preferentially be treated within rhinoplasty procedures.

Moreover, the learned classification methods confirmed that they are, despite the suggested improvement of FACS scale in terms of increasing the number of facial emotions, able to classify facial images into the defined facial emotions accurately enough.

References

1. Farkas, L.G., Hreczko, T.A., Kolar, J.C., Munro, I.R.: Vertical and horizontal proportions of the face in young adult North American Caucasians. Plast. Reconstr. Surg. **75**(3), 328–338 (1985)
2. Schmid, K., Marx, D., Samal, A.: Computation of a face attractiveness index based on neoclassical canons, symmetry, and golden ratios. Pattern Recogn. **41**, 2710–2717 (2008)
3. Bashour, M.: Is an Objective Measuring System for Facial Attractiveness Possible, 1st edn., Boca Raton, Florida (2007). ISBN 978-158-1123-654
4. Thornhill, R., Gangestad, S.W.: Human facial beauty. Hum. Nature **4**, 237–269 (1993)
5. Little, A.C., Jones, B.C., DeBruine, L.M.: Facial attractiveness: evolutionary based research. Philos. Trans. R. Soc. B Biol. Sci. **366**, 1638–1659 (2011)
6. Perrett, D.I., Lee, J.K., Penton-Voak, I., Rowland, D., Yoshikawa, S., Burt, D.M., Henzi, S. P., Castles, D.L., Akamatsu, S.: Effects of sexual dimorphism on facial attractiveness. Nature **394**(6696), 884–887 (1998)
7. Naini, F.: Facial Aesthetics: Concepts & Clinical Diagnosis, 1st edn. Wiley-Blackwell, Chichester (2011). ISBN 978-1-405-18192-1
8. Darwin, C., Ekman, P., Prodger, P.: The Expression of the Emotions in Man and Animals, 1st edn. Oxford University Press (1998). ISBN 9780195158069
9. Tomkins, S.: Affect Imagery Consciousness, 1st edn. Springer, New York (1963). ISBN 0826144047
10. Ekman, P., Friesen, W.V.: Constants across cultures in the face and emotion. J. Pers. Soc. Psychol. **17**, 124–129 (1971)
11. Ekman, P.: Unmasking the Face: A Guide to Recognizing Emotions from Facial Clues. Malor Books, Cambridge (2003). ISBN 1883536367
12. Fasel, B., Luettin, J.: Automatic facial expression analysis: a survey. Pattern Recogn. **36**(1), 259–278 (2003)

13. Yang, M.-H., Kriegman, D.J., Ahuja, N.: Detecting faces in images: a survey. IEEE Trans. Pattern Anal. Mach. Intell. **24**, 34–58 (2002). ISSN 0162-8828
14. Lanitis, A., Taylor, C., Cootes, T.: Automatic face identification system using flexible appearance models. Image Vis. Comput. **13**, 393–401 (1995)
15. Rowley, H.A., Baluja, S., Kanade, T.: Neural network-based face detection. IEEE Trans. Pattern Anal. Mach. Intell. **20**, 23–38 (1998). ISSN 0162-8828
16. Zhao, X., Zhang, S.: A review on facial expression recognition: feature extraction and classification. IETE Tech. Rev. **33**, 505–517 (2016)
17. Cootes, T., Taylor, C., Cooper, D., Graham, J.: Active shape models-their training and application. Comput. Vis. Image Underst. **61**, 38–59 (1995)
18. Alpaydin, E.: Introduction to Machine Learning, 2nd edn. MIT Press, Cambridge (2010). ISBN 9780262012430
19. Kasal, P., Fiala, P., Stepanek, L., Mestak, J.: Application of image analysis for clinical evaluation of facial structures. Medsoft **2015**, 64–70 (2015)
20. Stepanek, L., Kasal, P., Mestak, J.: Evaluation of facial attractiveness for purposes of plastic surgery using machine-learning methods and image analysis. In: 20th IEEE International Conference on e-Health Networking, Applications and Services, Healthcom 2018, Ostrava, Czech Republic, 17–20 September 2018, pp. 1–6 (2018). https://doi.org/10.1109/HealthCom.2018.8531195
21. R Core Team: R: A Language and Environment for Statistical Computing. R Foundation for Statistical Computing, Vienna, Austria (2013). http://www.R-project.org/. ISBN 3-900051-07-0
22. King, D.E.: Dlib-ml: a machine learning toolkit. J. Mach. Learn. Res. **10**, 1755–1758 (2009)
23. Chambers, J.: Statistical Models in S. Chapman & Hall/CRC, Boca Raton (1992). ISBN 041283040X
24. Friedman, N., Geiger, D., Goldszmidt, M.: Bayesian network classifiers. Mach. Learn. **29**, 131–163 (1997). ISSN 0885-6125
25. Breiman, L.: Classification and Regression Trees. Chapman & Hall, New York (1993). ISBN 0412048418
26. McCulloch, W.S., Pitts, W.: A logical calculus of the ideas immanent in nervous activity. Bull. Math. Biophys. **5**, 115–133 (1943)
27. Stallkamp, J., Schlipsing, M., Salmen, J., Igel, C.: The German traffic sign recognition benchmark: a multiclass classification competition. In: The 2011 International Joint Conference on Neural Networks, pp. 1453–1460 (2011). https://doi.org/10.1109/ijcnn.2011.6033395
28. Ramaswamy, S., Tamayo, P., Rifkin, R., Mukherjee, S., Yeang, C.H., Angelo, M., Ladd, C., Reich, M., Latulippe, E., Mesirov, J.P., Poggio, T., Gerald, W., Loda, M., Lander, E.S., Golub, T.R.: Multiclass cancer diagnosis using tumor gene expression signatures. Proc. Natl. Acad. Sci. **98**, 15149–15154 (2001). ISSN 0027-8424

Recognition of Antisocial Behavior in Online Discussions

Kristína Machová(✉) and Dominik Kolesár

Technical University, Letná 9, 04200 Košice, Slovakia
kristina.machova@tuke.sk

Abstract. The paper is focused on recognition of antisocial behavior in social media. User generated content in the form of discussions is essential to the success of many online platforms. While most users tend to civil accepting the social norms, others engage in antisocial behavior negatively affecting the rest of the community and its goals. Such behavior includes harassment, bullying, flaming, trolling, etc. The contribution is focused on classification of troll posts in online discussions to distinguish them from creditable posts. The work proposes a machine learning approach to build the classification model for toxic posts identification on an extensive dataset. The following machine learning methods were used: k Nearest Neighbors, Naïve Bayes Classifier, Decision Trees, Logistic regression and Support Vector Machine. These machine learning methods were used in combination with three different feature representations of texts of online discussions as a binary vector, a bag of words and the TF-IDF weighting scheme. The paper contains also the results of experiments with all learned models for toxic posts recognition.

Keywords: Antisocial behavior · Trolling posts recognition · Machine learning · Texts representation · Natural language processing

1 Introduction

Internet is a place where users communicate and express their thoughts. On social media such as Facebook, twitter, Google+, MySpace, Instagram, users can post and other users can post on. On news sites such as Yahoo News, New York Times, The Guardian, Fox News, Daily Mail, users can discuss the topic in the comment section. On QA (Question Answer) communities such as Stackoverflow or Quora, users can ask questions and answer questions of others.

While most users tend to civil accepting the social norms, others engage in antisocial behavior negatively affecting the rest of the community and its goals. Antisocial content may be of a dual nature. First, it can represent information that will affect and manipulate its recipients. Second, it is misinformation, which is caused by misunderstanding without manipulation [1].

Because of existence of an antisocial and disinformative content in online discussions, some forms of regulation of social media posts should be introduced. The community of informatics is expected to propose means for detecting these unhealthy web phenomena in an automatic way.

J. Świątek et al. (Eds.): ISAT 2019, AISC 1051, pp. 253–262, 2020.
https://doi.org/10.1007/978-3-030-30604-5_23

2 Antisocial Content

The antisocial content includes mostly bullying, harassment, flaming and trolling [2, 3]. *Bullying* is a use of force, threat or coercion to aggressively dominate, abuse or intimidate others. The behavior is habitual and a perspective of social or physical imbalance is in place, which distinguishes it from conflict. Cyberbullying has the same characteristics, but taking place on the internet and is not physical. In 2011 9% of 6–12 grades students report they have been bullied online. The person bullied usually feels totally overwhelmed which can make them feel embarrassed that they are going through such a devastating time and not knowing what support is available. *Harassment* and cyber-harassment shares the same characteristics as bullying, but the targeted person is based on gender, race, religion, color, age, disability or national origin. Sexual Harassment refers to unwanted sexual advances. *Flaming* is an insulting and hostile interaction between persons over the internet, often involving use of profanity. Many times can be just swapping insults back and forth. It frequently is a result of discussion of heated real world issues in politics, religion, philosophy, etc. *Trolling* is when a person sows discord on the internet by starting quarrels and upsetting people by posting inflammatory or off topic messages with the motif of provoking users to respond emotionally or in another way distort the flow of discussions for his own amusement. Trolling is referred to as the activity of a troll. *"Troll"* is an internet slang term that identifies individuals who intentionally attempt to incite conflict or hostility by publishing offensive, inflammatory, provocative or irrelevant posts. Their intent is to upset others and to produce a strong emotional response, mostly negative one. They use it as a bait for engaging new users in the discussion [3].

Because of these negative behaviors, popular sites like National Public Radio, The Atlantic News, The Week, Mic, Recode, Reuters, Popular Science, The Verge, USA Today were forced to shut down the comments sections. New York Times website instead of shutting it down is employing 14 moderators which serve as a gateway between comments and the site. Every comment is at first added to the queue checklist, waiting to be approved or denied. Better way is approving the comments in automatic way.

A demand for reliable automatic tools able to score and identify undesired behavior in online discussion is present. An example of such tool is present in paper [4]. It is based on disseminating propaganda. Propaganda tries to relativize reality by generating arguments that distort the truth. Sometimes this truth distortion could be generated automatically using algorithms based on similarity measures. There are various disinformation techniques. These techniques are discussed in paper [5], including consequences of their usage. The paper presents a probability approach to detection of relativized statements. An opinion sharing by product reviews is a part of online purchasing. This opinion sharing is often manipulated by fake reviews. Dematis et al. [6] presents an approach which integrates content and usage information in the fake reviews detection. The usage information is based on reviewers' behavior trails. In this way, a reviewer reputation is formed. The ability to define a reviewer reputation can help us to recognize trolls in social media. Also machine learning offers possibility to create tools able to score and identify antisocial behavior in online discussions. Our approach is based currently on using machine learning methods.

3 Used Machine Learning Methods

Machine learning is a section of artificial intelligence and its goal is to give computers the ability to 'learn', often with help of statistical algorithms. The term was established by Arthur Samuel in his paper [7]. Based on a type of learning, machine learning methods can be divided into two categories, supervised learning and unsupervised learning methods. According to [8] *supervised learning* is a task where computer is given input-output examples and attempts to learn a model, which can be used for mapping given input into unknown output values. *Unsupervised learning* is explained as a task (typically clustering) where computer learns from only input values. The following machine learning methods were used for recognition of toxic posts in online discussions.

3.1 K-nearest Neighbors

Nearest neighbor classifiers are the simplest of classifiers. Yet, it is still used and performs well. It is called a memory classifier because it requires only to remember training samples. Every new sample is given a class of the nearest neighbors in the training set. Many times it is benefiting to assign a class of the majority of the nearest *k* training samples. The distance between a new sample and other samples have to be calculated for all samples in training set. Thinking about it further, a reasonable adjustment would be to give more weight to a nearer sample in the voting process [9].

K-nearest neighbors' advantages are easy implementation, interpretation and debugging. Disadvantages are poor prediction speed and memory consumption for large datasets. K-nearest neighbors cannot cope enough with noise and so it is out performed by more sophisticated methods for complex tasks [10]. Modified versions of K-nearest neighbors are reported to perform well on tasks such as text categorization [11].

3.2 Naïve Bayes Classifier

Naive Bayes is a probabilistic classifier based on Bayes' theorem and independence assumption between features. Algorithm was widely studied in the 1950s.

Let us assume that event *A* and event *B* are independent then their conditional probability is defined according Bayes theorem (1).

$$P(A/B) = \frac{P(A)P(B/A)}{P(B)} \tag{1}$$

In practice *P(B)* can be an estimated constant calculated from the dataset. Replacing *P(B)* with a constant β^{-1}, formula (1) is then expressed as

$$P(A/B) = \beta P(A)P(B/A). \tag{2}$$

Let us assume that A represents class and B represents a feature relating to this class A. Then this equation handles only one feature. Let us extend the rule with more features. Then the conditional probability of class A on features B, C is following.

$$P(A/B, C) = \beta P(A)P(B, C/A) = \beta P(A)P(B/A)P(C/A) \tag{3}$$

That assumes that features B and C are independent of each other. Then a simplifying above expression is possible using replacement of *P(B, C/A)* to *P(B/A)P(C/A).* For n observations – features $x_1, \ldots, x_n$ the conditional probability for any class y_j can be expressed as below.

$$P(y_j/x_1, \ldots, x_n) = \beta P(y_j) \prod_{i=1}^{n} P(x_i/y_j) \tag{4}$$

This model is called naive Bayes classifier. Naive Bayes is often applied as a baseline for text classification, however its performance is reported to be outperformed by support vector machines [12].

3.3 Decision Trees

Decision tree is a model which uses a tree of decisions to predict a label for a new sample. Assuming a standard top-down approach, method starts with a full dataset in one root node. A generated question divides a node to sub-nodes each representing answers to the question. Focusing on subsets, they are generated according a class diversity. The are two most commonly used types of diversity functions, Information Entropy and Gini index [13]. Best ranked questions generate minimal disorder.

The advantages of decision tress are their intuitive interpretation and non-linear solution. Algorithm was already successfully used for part-of-speech tagging [14] categorizing text documents and parsing [15]. Natural text processing includes a large amount of words as features. Generated tree is therefore very robust.

3.4 Logistic Regression

Logistic regression is a statistical technique to estimate parameters of a logistic model. Logistic model is a model where linear combinations of independent variables are transformed using a specific type of logistic function, mostly a sigmoid function. Logistic regression was developed by David Cox [16]. An important fact comparing logistic regression with support vector machines, is that it does not maximize the margin between classes. Methods for adjusting logistic regression to converge to support vector machines solution exist [17].

3.5 Support Vector Machine

The method separates the sample space into two or more classes with the widest margin possible. The method is originally a linear classifier, however, it can relatively efficiently perform non-linear classification by using a kernel. Continuing to complete the solution, creating the widest margin between samples, it was observed that only nearest points to the separating street determines its width. They are called support vectors. The objective is to maximize the width of the street, which is known as the primal problem of support vector machines [18].

4 Approach to Toxic Posts Recognition Based on Machine Learning

4.1 Text Data Preprocessing

We have provided a data preprocessing using three different text data representations: bag of words, binary vectors and TF/IDF weighting. Data preprocessing is a process of transforming raw data form text into a form suitable as input of a machine learning algorithms. For example, in natural language text processing, before applying classification, text must be transformed into a word vector.

The most common method of handling this is a strategy known as **Bag of Words** utilizing word frequency as a feature. Words can be recognized as strings separated by a white-space, thought in practice it is more complicated. Occurrences of tokens are counted in documents. As a result, we have a fixed numerical vector.

When a bag of words representation is replaced with only ones and zeros (such that one indicates word occurrence), vector is known as a **Binary Vector** representation.

TF-IDF in other words Term Frequency subtracted with Inverse Document Frequency is a measure of how important relation is between a word and a document. TF part of this measure represents term frequency of a term in a document. IDF part of the measure (Inverse Document Frequency) has higher value, when the term from the given document is not very often used in the whole corpus of documents, so the term is not very general. Highly scored terms within a document are considered to represent topic of the document.

4.2 Obtained Datasets

Our goal is to identify troll posts. The term troll post for purposes of this work is interchangeable with the term toxic post. Labeled dataset was created by merging data from two sources, labeled Wikipedia talk comments obtained from figshare.com [19] and labeled comments obtained from Kaggle Toxic Comment Challenge at kaggle.com [20]. Wikipedia talk comments are divided into four separate data sets with different labels: personal attack, aggression, toxic and neutral. Kaggle comments include 7 labels, toxic, severe toxic, obscene, threat, insult, identity hate, and neutral. The two datasets and types of toxic labels were aggregated under one dataset and one label: toxic assigned a value 1, otherwise neutral or non-toxic comments was assigned a value 0. Together making a little less than 400,000 comments from which are approximately 40,000 toxics.

4.3 Models Building

Training and testing sets were extracted in a stratified fashion with a ratio 80:20 (%). Hold-out cross-validation strategy was implemented. Training set was used for determining the best parameters via cross-validation with grid search and testing final performance was done only once on the hold-out set. Five folds were used for cross-validation.

Given a labeled dataset of comments, they have to be preprocessed before learning. Firstly, binary vector was extracted and used as an input for machine learning methods. Models were evaluated and scores recorded. Upgrading to a count vector - bag of words representation same process was repeated expecting better results. It was assumed, that bag of words approach is more informative. Lastly Tf-Idf weighting was applied. To identify troll posts these methods were implemented: K-nearest neighbors (KNN), naive Bayes (NB), decision trees (DT), logistic regression (LR) and support vector machines (SVM) in combination with three mentioned feature representations.

5 Results of Machine Learning Models

Trying to apply k-nearest neighbor method for classification of such high volume (400,000) dataset has achieved weaker results. Basic k-nearest neighbors' algorithm requires to compute distances between a new sample and all training samples. Algorithms exist to speed classification and generally use less memory, for example BallTree and KDTree. They need the dataset present in an uncompressed format, which is unrealistic as well. A solution may be in use of approximate nearest neighbors' methods. Result tables below still include k-nearest neighbors; these are obtained with using only 1/10 of all samples in the dataset.

All results are shown in Tables 1 and 2. The obvious classic measure of effectivity of a classification model were used: a precision, a recall and a F1 measure (which is balanced average of the precision and the recall). Comparing binary vector and bag of words representations in Table 1. SVM method is the best in F1 measure and in precision measure which is sensitive on number of FP (fault positive) classifications. LR was the best in recall measure which is sensitive on FN (fault negative) classifications. The similar conclusions can be derived from Table 2 for Tf-Idf representation with one exception – model based on NB is the best in precision measure (which is sensitive on number of FP classifications).

Observing Table 1, it is obvious that NB performs better when repeated words in comments are recorded with bag of words approach instead of being ignored. Even Tf-Idf representation enhances its prediction capabilities. LR and SVM on the side seems to embrace simplicity of the binary vector. DT is reasonably performing almost equally for every input. It tells that continues thresholds generated by questions on features does not appear to influence performance for text classification. Average results in Table 2 give us information about effectivity of used machine learning methods abstractedly from input data representation. Let us consider only F1 measure, which is sensitive on both false positive as false negative, then the best model is model trained using support vector machine.

Table 1. Results of testing machine learning methods (KNN, NB, DT, LR and SVM) for binary vector and bag of words representations using precision, recall and F1 measures in percentage.

ML method	Binary vector			Bag of words		
	Precision	Recall	F1	Precision	Recall	F1
KNN	38.93	48.71	43.27	39.23	44.08	41.51
NB	63.02	53.57	57.91	**75.47**	66.06	70.45
DT	70.57	74.70	72.57	70.41	74.56	72.43
LR	71.80	**79.28**	**75.35**	42.81	**85.89**	57.14
SVM	**72.44**	78.89	**75.53**	**71.44**	77.65	**74.42**

Table 2. Results of testing methods (KNN, NB, DT, LR and SVM) for Tf-Idf weighting representations using precision, recall and F1 measures accompanied with Average values.

ML method	Tf-Idf weighting			Average		
	Precision	Recall	F1	Precision	Recall	F1
KNN	46.23	59.62	52.08	41.46	50.80	45.62
NB	**80.72**	57.28	67.01	**73.07**	58.97	65.12
DT	70.35	74.83	72.52	70.44	74.70	72.51
LR	49.98	**84.92**	62.93	54.86	**83.36**	65.14
SVM	73.82	66.66	**70.06**	**72.57**	74.40	**73.34**

6 Related Works and Discussion

Machine learning approach was used in other works. The paper [21] presents an approach for recognition of inappropriate comments to online discussion. The best results were achieved using Bi-dir. RCNN (Regional Convolutional Neural Network) and LSTM (Long Short Term Memory – recurrent neural network) models. Their results (0.79 of Accuracy) are comparable to our results. In the paper [22] a novel computational approach is introduced to address the problem called Bag of Communities (BoC). They applied BoC toward identifying abusive behavior within a major Internet community. The BoC models use the output likelihoods from the following classifiers as internal estimators: Naive Bayes, Linear Support Vector Classification and Logistic Regression. Their results in static BoC are similar to our results. Key aspects of the approach in the paper [23] are the monitoring and analysis of the most recently published comments. The paper presents an application of text classification techniques for detecting whether an aggressive action actually emerges in a discussion thread. The authors experimented with various forms of representations of input texts (unigram, bigram, trigram, PP feature set) in combination with RBF (Radial Basis Function), SVM (Support Vector Machines) and Hidden Markov Model classifier. Their results were also comparable with our results.

In this article, we have focused on identification of troll posts in other words, toxic posts, but another important research area is the identification of authors of these posts. Work [24] proposed an approach to detect such users from the sentiment of the textual content in online forums. In experiments, they used features derived from a recursive neural tensor network sentiment analysis model trained on a movie reviews data set. They achieved a final result of 78% and 69% generalized ROC for the binary troll classification. Another approach [25] is not based on the sentiment analysis model, but on using several machine learning models, particularly k-Nearest Neighbour (kNN), Naive Bayes, and C4.5 decision tree algorithms. Their tests showed that C4.5 has a better performance on troll detection with achieved Accuracy 0.89. From the comparison of these two approaches it seem to be better to use machine learning methods than the sentiment analysis.

When we talk about the machine learning, the use of deep neural networks has become very popular. This approach was used in the work [26] to track the toxic posts. In the paper, they proposed Deep Neural Network (DNN) architectures to classify the overlapping sentiments with high accuracy. Their classification framework does not require any laborious text pre-processing and its empirical validation showed good performance: for CNN up to F1 = 0.69, for Bi-GRU up to F1 = 0.7 and for Bi-LSTM up to F1 = 0.7. Our results presented in this paper are better. Also another study [27] addresses the questions of toxic online comments classification using a logistic regression model and three neural networks models – the convolutional neural network (Conv), long short-term memory (LSTM), and Conv + LSTM. The combined model Conv + LSTM is the most effective with accuracy 0.983.

In the study [28] a novel hybrid approach for detecting automated spammers is presented. The novelty of the proposed approach lies in the characterization of users based on their interactions with their followers. Nineteen different features were identified for learning three classifiers, namely, random forest, decision tree, and Bayesian network. The discrimination power of different feature categories is also analyzed in the study. The community-based features were determined to be the most effective for spam detection, whereas metadata-based features are proven to be the least effective, which is an interesting conclusion. Similarly, the achieved results in the F score are quite intriguing (the best performance using Random Forest 0.979, Decision Tree 0.943, Bayesian Network 0.942). In the future, we would like to focus on similar hybrid approaches.

7 Conclusions

To identify troll posts, five classifiers were implemented. Each was trained with three selected feature representations. KNN classifier failed to be implemented for full volume of the dataset and was trained only on part of total samples. Results show there is not a feature representation generally performing best for all classifiers. NB tends to do well with bag of words representation. KNN show highest score with Tf-Idf scoring. DT shows consistent scores through all representations. SVM and LR achieved the highest scores, particularly 75,53% and 75.35% using binary representation. Generally, the SVM model had the best results according F1-measure.

The best model was selected and used to build an application with a graphical user interface allowing the user to insert text of a post and see how each model classifies his input. For future, we would like include some specific methods of machine learning with the potential to improve accuracy. Among them are the use of ensemble learning methods, deep learning neural networks and use of unsupervised learning algorithms.

In following research, we tend to improve the effectivity of sentiment analysis using information about an authority or trolling of the given reviewer while his/her posts are analyzed from the point of sentiment or opinion polarity [29]. We would like to develop a hybrid approach able to analyze toxic texts in relation to attributes of their authors – spammers or trolls.

Acknowledgements. The work presented in this paper was supported by the Slovak Research and Development Agency under the contract No. APVV-017-0267 "Automated Recognition of Antisocial Behavior in Online Communities" and the contract No. APVV-015-0731.

References

1. Kumar, S., West, Ř., Leskovec, J.: Disinformation on the web. In: Proceedings of the 25th International on Word Wide Web – WWW16. Association for Computing Machinery, pp. 591–602. ACM, Montreal (2016)
2. Lessne, X., Hermalkar, X.: Student Reports of Bullying and Cyber-Bullying: Results from the 2011 School Crime Supplement to the National Crime Victimization Survey, Web Tables, NCES 2013-329. http://nces.ed.gov/. Accessed 25 May 2019
3. March, E.: 'Don't feed the trolls' really is good advice – here's the evidence. https://theconversation.com/dont-feed-the-trolls-really-is-good-advice-heres-the-evidence-63657. Accessed 25 May 2019
4. Wang, S.E., Garcia-Molina, H.: Disinformation Techniques for Entity resolution. In: Proceedings of the 22nd ACM International Conference on Information and Knowledge Management, New York, USA, pp. 715–720 (2013)
5. Řimnáč, M.: Detection of a disinformation content – case study Novičok in CR. In: Proceedings of the Conference Data a znalosti & WIKT 2018, Brno, Vysoké učení technické, pp. 65–69 (2018)
6. Dematis, I., Karapistoli, E., Vakali, A.: Fake review detection via exploitation os spam indicators and reviewer behaviour characteristics. In: Proceedings of the 44th International Conference on Current Trends in Theory and Practice of Computer Science beyond Frontiers - SOFSEM 2018. Lecture Notes in Computer Science, Krems an der Donau, pp. 1–14. Springer, Heidelberg (2018)
7. Samuel, A.L.: Some studies in machine learning using the game of checkers. IBM J. Res. Dev. **3**(3), 210–229 (1959)
8. Russell, S.J., Norvig, P.: Artificial Intelligence. A Modern Approach, 3rd edn. Prentice Hall, Pearson Education, New Jersey (2010). ISBN-13 978-0-13-604259-4
9. Tan, S.: Neighbor-weighted K-nearest neighbor for unbalanced text corpus. Expert Syst. Appl. **28**(4), 667–671 (2005)
10. Cunningham, P., Delany, S.J.: k-Neighbour classifiers. Technical report, pp. 1–17, Dublin (2007)
11. Jiang, S., et al.: An improved k-nearest neighbor algorithm for text categorization. Expert Syst. Appl. **30**(1), 1503–1509 (2012)
12. Pang, B., Lee, L., Vaithyanathan, S.: Thumbs up? Sentiment classification using machine learning techniques. In: Proceedings of the Conference on Empirical Methods in Natural Language Processing (EMNLP), Philadelphia, pp. 79–86 (2002)

13. Kingsford, C., Salzberg, S.L.: What are decision trees? Nat. Biotechnol. **26**(1), 1011–1013 (2008)
14. Orphanos, G., et al.: Decision Trees and NLP: A Case Study in POS Tagging. Academia, pp. 1–7 (1999)
15. Magerman, D.M., et al.: Statistical decision-tree models for parsing. In: Proceeding ACL 1995 Proceedings of the 33rd Annual Meeting on Association for Computational Linguistics, pp. 276–283 (1995)
16. Cox, D.R.: The regression analysis of binary sequences. J. R. Stat. Soc. **20**(2), 215–242 (1958)
17. Zhang, J., et al.: Modified logistic regression: an approximation to SVM and its applications in large-scale text categorization. In: Proceedings of the Twentieth International Conference on Machine Learning, Washington DC, pp. 888–895 (2003)
18. Ben-Hur, A., et al.: Support vector clustering. J. Mach. Learn. Res. **2**(2), 125–137 (2001)
19. Wulczyn, E., Thain, N., Dixon, L.: Ex machina: personal attacks seen at scale. In: Proceedings of the International World Wide Web Conference (WWW 2017), Perth, Australia, 3–7 April 2017, pp. 1391–1399 (2017)
20. Hosted by Kaggle.com, Toxic Comment Classification Challenge. https://www.kaggle.com/c/jigsaw-toxic-comment-classification-challenge. Accessed 26 May 2019
21. Švec, A., Pikuliak, M., Šimko, M., Bieliková, M.: Improving moderation of online discussions via interpretable neural models. FIIT Slovak University of Technology, Bratislava, Slovakia, pp. 1–6 (2018)
22. Chandrasekharan, E., Samory, M., Srinivasan, A., Gilbert, E.: The bag of communities: identifying abusive behavior online with preexisting internet data. In: Proceedings of the CHI Conference on Human Factors in Computing Systems, pp. 3175–3187. ACM (2017)
23. Ventirozos, F.K., Varlamis, I., Tsatsaronis, G.: Detecting aggressive behavior in discussion threads using text mining. In: CICLing 2017. LNCS, vol. 10762, pp. 420–431. Springer, Cham (2018)
24. Seah, C.W., et al.: Troll detection by domain-adapting sentiment analysis. In: Proceedings of the 18th International Conference on Information Fusion Washington, DC, pp. 792–799 (2015). 978-0-9824-4386-6/15/$31.00 ©2015 IEEE
25. Mutlu, B., et al.: Identifying trolls and determining terror awareness level in social networks using a scalable framework. In: Proceedings of IEEE International Conference on Big Data, pp. 1792–1798 (2016). 978-1-4673-9005-7/16/$31.00 ©2016 IEEE
26. Saeed, H.H., Shahzad, K., Kamiran, F.: Overlapping toxic sentiment classification using deep neural architectures. In: Proceedings of IEEE International Conference on Data Mining Workshops (ICDMW), pp. 1361–1366 (2018). 2375-9259/18/$31.00 ©2018 IEEE
27. Saif, M.A., et al.: classification of online toxic comments using the logistic regression and neural networks models. In: Proceedings of the 44th International Conference on Applications of Mathematics in Engineering and Economics, pp. 1–5. AIP Publishing (2018). 978-0-7354-1774-8/$30.00
28. Fazil, M., Abulaish, M.: A hybrid approach for detecting automated spammers in Twitter. IEEE Trans. Inf. Forensics Secur. **13**, 1556–6013 (2018)
29. Mikula, M., Machová, K.: Combined approach for sentiment analysis in Slovak using a dictionary annotated by particle swarm optimization. In: Acta Elektrotechnica et Informatica, vol. 18, no. 2, pp. 27–34 (2018). ISSN 1335–8243

Evaluation of the Classifiers in Multiparameter and Imbalanced Data Sets

Piotrowska Ewelina(✉)

Opole University of Technology, Prószkowska 76 Street, 45-758 Opole, Poland
e.piotrowska@po.edu.pl

Abstract. The paper discusses the basic problems resulting from the classification of imbalanced data, which are additionally described by a large number of parameters. The paper also presents various optimization methods, including the use of a synthetic indicator that is the product of specificity and the power of sensitivity, which was proposed by the author.

Keywords: Imbalanced data · Optimization methods · Classification

1 Introduction

The continuous development of measurement techniques affects the collection of more and more data. These data become the basis for building expert systems in which classification is an important task. Unfortunately, many problems are characterized by an imbalanced data distribution. Two types of classes can be distinguished in the classification of such data: a minority class and a majority class. The minority class is characterized by a significantly smaller number of cases, usually not exceeding 10% of the size of the collection. Examples of data imbalance include the detection of Oil Spills in Satellite Radar Image [9], the detection of fraud phone calls [4] or monitoring the damage to the helicopter gearbox [8]. This problem also occurs in medicine in the analysis of data from screening tests [2, 3].

2 Analysis of Multiparameter Measurements

Most learning algorithms assume balancing of classes. Therefore, the classification based on unbalanced data sets causes difficulties in the learning process and lowers the predictive accuracy. The low quality of classifications may also result from wrong conditioning of minority class data, such as: too few objects, overlap of majority class objects with minority class or ambiguity of boundary objects [3, 5, 6, 11].

A growing number of parameters in the measurements is an additional problem. It leads to a more accurate description, but unfortunately it has a negative impact on the possibilities of interpretation. The more parameters are observed, the larger the size of the space of consideration is. It is also an important calculation problem that results from the complexity of memory and time of implemented algorithms.

J. Świątek et al. (Eds.): ISAT 2019, AISC 1051, pp. 263–273, 2020.
https://doi.org/10.1007/978-3-030-30604-5_24

Microscopic diagnostics is an example of research where imbalanced data is generated. Its aim is not to describe the anatomical location of the disease, but to determine the nature of the observed changes. For this purpose, microscopic specimens are analysed and their morphometric features are calculated. Each morphometric parameter is a numerical description of the morphological object (Fig. 1). It can refer to such properties as the number of objects, their size, shape, optical properties, texture or topology. The choice of parameters depends on the type of genetic material being tested.

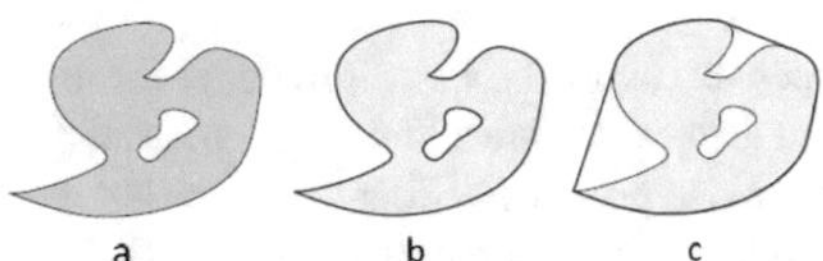

Fig. 1. An example of morphometric parameters: (a) surface area, (b) circumference, (c) convex circumference.

The calculation of cell morphometric parameters is used in cancer diagnostics. As a result of the analysis of microscopic specimens, a set of parameters that can contain several hundred elements is determined. The analysis of bladder cells is a good example [10]. As a result of the analysis, over 200 features were identified. The data set contained approximately 23,000 cases. Cancer cells were only 3% of the set.

3 Classification Measures in Imbalanced Data Sets

Proper preparation of the data set, selection of the classification algorithm and assessment of the classifier's effectiveness are important issues in the analysis of multiparameter measurements. The classifier construction process is a complex task. In the paper [10] an iterative process of determining sets of features is proposed. First, a set of training and validation data is established. In the next step the reduction of the feature space begins. Then, based on the training data on the reduced number of features, the supervised learning is carried out. Based on the validation set, the classification is made and the quality criteria for the classification are calculated. If the calculated quality of classification does not meet expectations, the set of features or the level of imbalance changes are modified. The whole process is repeated [10].

The performance measure of the classifier is its ability to correctly predict or separate classes. The contingency table, also known as the confusion matrix (Fig. 2), is a basic tool for evaluating classifiers.

Accuracy is one of the measures of the classification quality. It describes the percentage of objects classified correctly. The sensitivity and specificity coefficients are a complement to the accuracy coefficient [1]. Sensitivity (recall, hit rate, true positive rate) is a measure of the classifier's ability to predict the i-th class correctly. Specificity (true negative rate) is an estimation of the probability of not belonging to an i-th class, provided that the object does not belong to the i-th class.

		Predicted class	
		N_i	P_i
Actual class	N_i	TN_i	FP_i
	P_i	FN_i	TP_i

Fig. 2. The binary contingency table for i-th class.

Imbalanced class distribution also causes problems in the interpretation of the classification quality. In such data sets accuracy, which applies to all correct classifications, should not be a main parameter. An example of such a situation is shown in Fig. 3. In a set containing 100 objects, 95 are assigned to the negative class and 5 to the positive class. Although the classifier does not classify correctly any object from the positive class, its accuracy is up to 95%.

		Predicted class	
		0	1
Actual class	0	95	0
	1	5	0

Fig. 3. An example of the classification matrix

In the analysis of imbalanced data, the classification measure should take into account the significance of the minority class by maximizing the number of correct classifications in the minority class (TP) and minimizing the number of incorrect classifications in the majority class (FP).

The selection of the optimal classifier, which is characterized by high sensitivity with high specificity, is one of the tasks that should be solved. High sensitivity ensures the detection of cancer cells, while high specificity - the detection of healthy cells. Because in medical issues it is more important to indicate the occurrence of the disease, the classifier should be as sensitive as possible. An additional difficulty is that the tested criteria of sensitivity and specificity are contradictory. An increase in the value of one of the criteria causes a decrease in the other [7].

Figure 4 shows the relationship between sensitivity and specificity for various classifiers. The points correspond to the classification based on a reduced set of features determined by the method of searching for relative reducts [10]. Five classification algorithms were used in the analysis. The first classifier was based on the rough set theory (RS). The second classifier chosen for the analysis was Naive Bayes Classifier (NB). NB is an example of a probabilistic classifier, using statistical knowledge about the training set. Due to the assumption of statistical independence of variables, the NB classifier was particularly suitable for a large number of dimensions of the input variable space. The data were classified also with the methods of Linear Discriminant Analysis (LDA) and Square Discriminant Analysis (QDA). Their aim was to find such functions in the space of features that best distinguished the classes present in the data sets. The classification using decision trees (DT) was also used to assess the sets of features. Classification rules corresponding to branches of a tree led to a clear interpretation of the results of analyses.

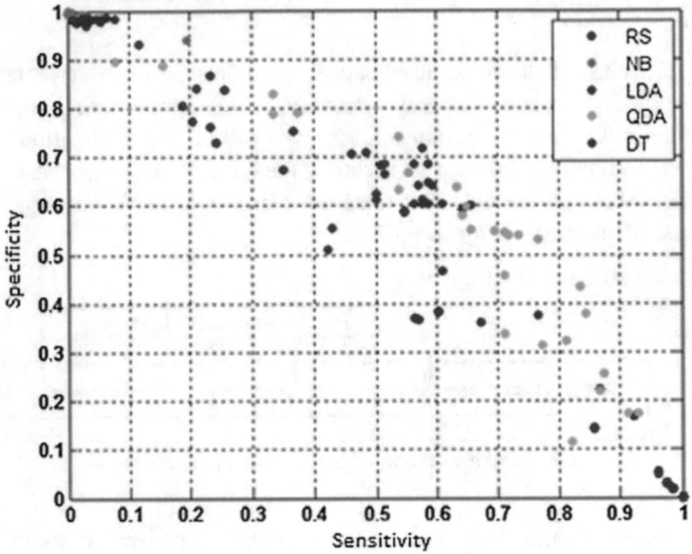

Fig. 4. Characteristics of sensitivity and specificity for selected classification methods.

4 Evaluation of Classifiers

The described problem of selecting the best classifier is an example of multicriteria optimization. Let x be the 4-dimensional vector of the parameters:

$$x = [TN, FP, FN, TP]^T, \tag{1}$$

which is evaluated by the 2-dimensional criteria vector $f(x)$ (objective function)

$$f(x) = [f_1(x) f_2(x)]^T, \tag{2}$$

where

$$f_1(x) = sen = \frac{TP}{TP + FN}, \tag{3}$$

$$f_2(x) = spe = \frac{TN}{TN + FP}, \tag{4}$$

are profit functions. Then the problem of multicriteria optimization can be defined as a task of multicriterial maximization of the objective function:

$$\max_{x} f(x). \tag{5}$$

Multicriterial optimization methods can be divided into classical methods (e.g.: the method of weighted profits, the method of inequalities) and ranking methods (e.g.: a ranking method according to Pareto-optimality, a ranking method according to the global level of optimality (GOL)).

4.1 Classical Methods

The Weighted Profit Method. The weighted profit method is an example of the scalarization approach in which the multidimensional problem is reduced to a one-dimensional task. The coordinates of the criteria vector are aggregated into a single objective function $g(x)$:

$$g(x) = w \cdot f(x), \tag{6}$$

where

$$w = [w_1 w_2], \tag{7}$$

is a standardized line weight vector so that

$$w_i \in [01], i = 1, 2. \tag{8}$$

Most often the following dependence is assumed:

$$w_1 + w_2 = 1. \tag{9}$$

The weighted profit method is simple to use but requires the selection of a suitable weight vector. Figure 5 shows the graphic interpretation of the discussed method for the RS classifier. The characteristics include the range of weight changes for a sensitivity measure from 0.5 to 1.0. According to this method, the most optimal classifier is the one with the highest sensitivity, even when the level of specificity is close to zero. It means that almost all cells in the analysed data set have been classified as cancerous.

The Method of Inequality Constraints In the method of inequality constraints, a vector of parameters is searched for which satisfies the inequality

$$f_1(x) \geq M_i . i = 1, 2, \tag{10}$$

where M_i represents a numerical limitation on the value of i-th profit function.

The graphical interpretation for the analysed measures of the RS classifier is shown in Fig. 6. Initially, the sensitivity measure was described by $sen \geq 0.7$ and the measure of specificity by $spe \geq 0.5$ (Fig. 6a). It can be seen that for such selected limit values it is not possible to find the optimal solution. In the analysed method, one should gradually lower the level of specificity until a sufficient solution is obtained. The method is simple to implement but requires appropriate determination of boundary values. The boundary can be changed iteratively to set a threshold of less significance. In Fig. 6b the specificity threshold was reduced to 0.2.

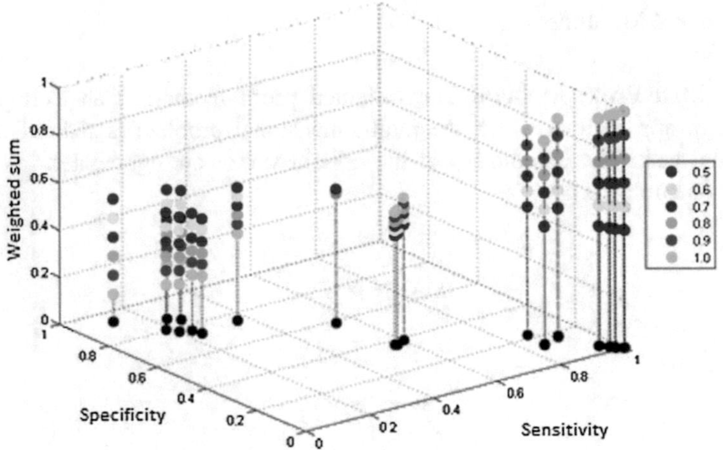

Fig. 5. The weighted profit method for the RS classifier.

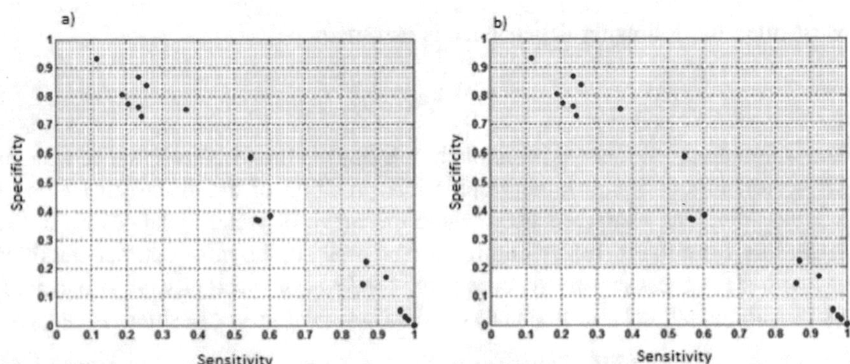

Fig. 6 Characteristics of an inequality constraint method for the RS classifier (a) $sen \geq 0.7$, $spe \geq 0.5$ (b) $sen \geq 0.7$, $spe \geq 0.2$.

4.2 The Ranking Methods

In the ranking methods, solutions are evaluated by classifying them as dominated or non-dominated solutions. If x* is a weakly dominated solution, it is impossible to increase simultaneously all criteria by choosing another acceptable solution. If x* is a strongly non-dominated solution (Pareto-optimal), then improving one of the criteria must be associated with a simultaneous deterioration of at least one of the other criteria. A set of solutions (Pareto-optimal) (Fig. 4), which should be estimated using additional criteria, is usually the result of the analysis.

The Pareto-optimality Rank. One of the possible criteria is to use the Pareto-optimality rank corresponding to the degree of domination. It allows us to map target functions into a one-dimensional space. Members of the non-dominated collection receive the highest rank. The graphical interpretation of the method is shown in Fig. 8a. Selected points correspond to classifiers characterized by high sensitivity with low specificity, high specificity at low sensitivity or classifiers with averaged values. It prevents the optimization of classifiers for maximum sensitivity.

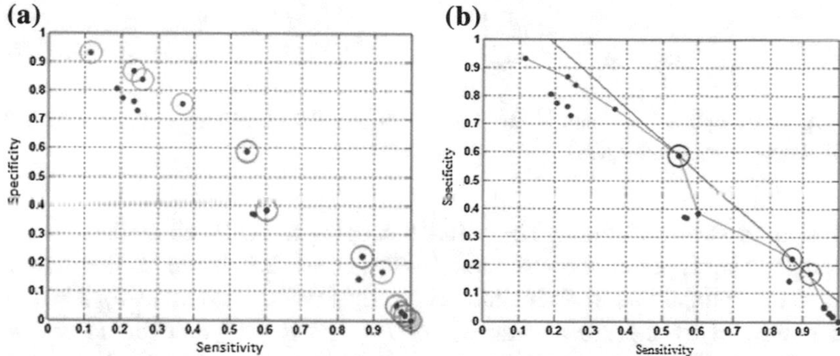

Fig. 7. (a) A set of optimal variants in the Pareto for the RS classifier. (b) Selection of the point from the Pareto set using the weighted sum method for the RS classifier.

The Global Optimality Level. The application of the global optimality level (GOL) is an improvement in the above-mentioned method. The method determines the highest value for each of the individual profit functions among all N solutions (or only P-optimal):

$$\Lambda_{i=1,2} f_{i_{max}} = \max_{j=1,2,\ldots,N} \{f_i(x_j)\}. \tag{11}$$

Then, a global level of optimality is assigned to each solution according to:

$$\eta(x_j) = \min_{i=1,2} \frac{f_i(x_j)}{f_{i_{max}}}. \tag{12}$$

The method allows for a significant minimization of the problem of ambiguous solutions. The characteristics of this method are presented in Fig. 8b. The method facilitates the selection of a classifier but favours indirect solutions.

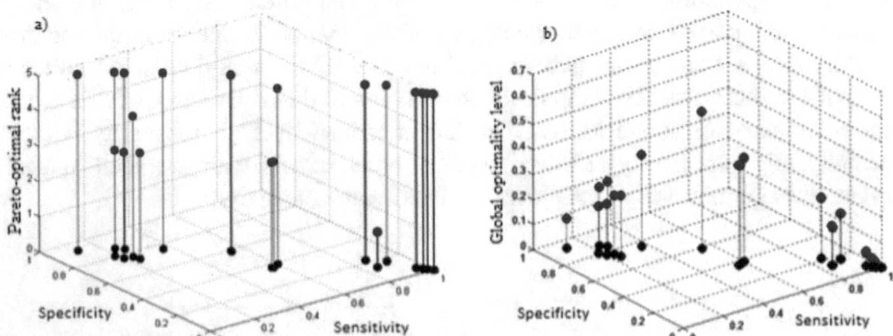

Fig. 8 A set of optimal variants in the Pareto for the RS classifier (a) Pareto-optimal rank (b) global optimality level (GOL).

The Weighted Profit Method. The points belonging to the Pareto collection can also be determined by applying the weighted profit method. It is based on the search for a tangent to the optimal Pareto set inclined at an angle determined by w_i for $i = 1, 2$. The vector w formed by the coefficients w_i for $i = 1, 2$ is perpendicular to the wanted tangent. The graphical interpretation of this method is shown in Fig. 7b. In the analysed set the vector of weights is w = [1.13 1]. The points common to the edge of the set and the tangent are the optimal solution. A serious disadvantage of this method is the inability to select any point from the Pareto collection.

4.3 The Methods Based on Classification Measures

The methods discussed above require additional calculations or estimation of values for weights. This additionally affects the uncertainty of the selected classifier. Therefore, it is important to find an objective function calculated directly from the classification measures.

The use of functions where a product of sensitivity and specificity is used in the calculations is the simplest example:

$$F_{Pro} = sen \cdot spe \tag{13}$$

Figure 10a presents the application of the F_{Pro} function in the selection of the RS classifier. It can be seen that this method, like the GOL criterion, distinguishes classifiers whose sensitivity and specificity values are similar to each other.

Another metrics used to assess the quality of classifications, which is often proposed in the literature, is the G-mean measure [11] described by:

$$G - mean = \sqrt{sen \cdot spe}. \tag{14}$$

The G-mean indicator will reach a high value when both the sen and spe indices are close to 1. The smaller G-mean value will appear when even one of the indicators has a small value.

The F-value [10] measure is another measure of the classification quality assessment in the unbalanced data analysis. It is described by:

$$F - value = \frac{(1+\beta^2) \cdot sen \cdot ppv}{\beta^2 \cdot sen + ppv} \tag{15}$$

where ppv is the precision. This measure describes the relationship between the three values: TP, FP and FN (false negative). The β coefficient corresponds to the relative importance of ppv in relation to sen and it most often takes the value β = 1.

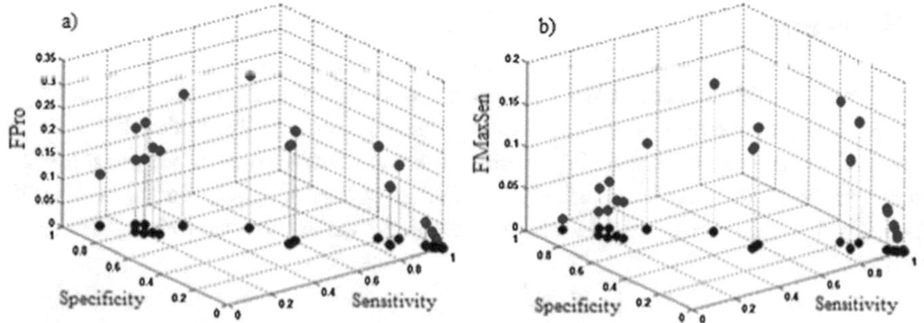

Fig. 9. A set of optimal variants using synthetic indicators (a) FPro (b) FMaxSen.

In the imbalanced data sets it is necessary to maximize the detection of a minority class. Therefore, modification of the product was proposed by applying the F_{MaxSen} function as the product of specificity and power of sensitivity [10]:

$$F_{MaxSen} = sen^2 \cdot spe \tag{16}$$

The characteristics of the F_{MaxSen} function are shown in Fig. 10. The second power of the sensitivity coefficient emphasizes its validity. It allows to search for a classifier characterized by maximum specificity at maximum sensitivity. The target function defined in this way was used to select the RS classifier on reduced sets of features. The characteristics are shown in Fig. 9b. The use of this function allowed the selection of

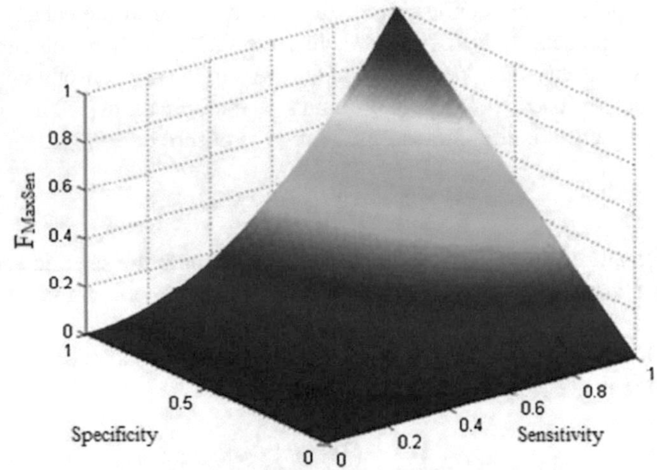

Fig. 10. Characteristics of the objective function for the decision system.

classifiers characterized by the highest possible sensitivity value, as well as classifiers with average values.

5 Summary

Analysis of multidimensional data which are also imbalanced is very complex task. The set of features should be reduced and the classification methods should be adapted. However, the developed methods are not universal. That is why there are many indicators available to assess the classifier. The proposed F_{MaxSen} index is important especially in tasks with imbalanced distribution, where the minority class is true positive.

References

1. Armitage, P.: Metody statystyczne w badaniach medycznych. Państwowy Zakład Wydawnictw Lekarskich, Warszawa (1978)
2. Bator, M.: Automatyczna detekcja zmian nowotworowych w obrazach mammograficznych z wykorzystaniem dopasowania wzorców i wybranych narzędzi sztucznej inteligencji. Instytut Podstawowych Problemów Techniki PAN. Praca doktorska pod kierunkiem prof. dr hab.inż. Mariusz Jacek Nieniewski (2008)
3. Chawla, N.: Data mining for imbalanced datasets: an overview. In: Maimon, O. Rokach, L. (eds.) Data Mining and Knowledge Discovery Handbook, Part 6, pp. 875–886 (2010)
4. Fawcett, T., Provost, F.: Adaptive fraud detection. J. Data Min. Knowl. Disc. **1**(3), 291–316 (1997)
5. Fernández, A., García, S., Herrera, F.: Addressing the classification with imbalanced data: open problems and new challenges on class distribution. In: Hybrid Artificial Intelligent Systems. LNCS, vol. 6678, pp. 1–10 (2011)

6. García, V., Sánchez, J.S., Mollineda, R.A.: Exploring the performance of resampling strategies for the class imbalance problem. In: Trends in Applied Intelligent Systems. LNCS, vol. 6096, pp. 2010, pp.541–549 (2010)
7. Górecki, H.: Optymalizacja i sterowanie systemów dynamicznych", Uczelniane Wydawnictwa Naukowo-Dydaktyczne Akademii Górniczo-Hutniczej w Krakowie, Kraków (2006)
8. Japkowicz, N., Myers, C., Gluck, M.: A novelty detection approach to classification. In: Proceedings of the Fourteenth Joint Conference on Artificial Intelligence, pp. 518–523 (1995)
9. Kubat, M., Holte, R.C., Matwin, S.: Machine Learning for the detection of oil spills in satellite radar images. J. Mach. Learn. (Special Issue on Applications of Machine Learning and the Knowledge Discovery Process) **30**(2–3), 195–215 (1998)
10. Piotrowska, E: Analysis of morphometric features for computer aided medical diagnosis. PhD thesis, Opole University of Technology (2012)
11. Stefanowski, J., Wilk, S.: Combining rough sets and rule based classifiers for handling imbalanced data. In: Czaja L. (ed.) Proceedings of Concurrency, Specification and Programming, CS&P 2005 Conference, vol. 2, pp. 497–508 (2005)

Image Processing

Time of Flight Camera Calibration and Obstacle Detection Application for an Autonomous Vehicle

Leanne Miller, Antonio Ros García, Pedro J. Navarro Lorente(✉),
Carlos Fernández Andrés, and Raúl Borraz Morón

Technical University of Cartagena, Cartagena, Spain
{leanne.miller, antonio.ros, pedroj.navarro, carlos.fernandez, raul.borraz}@upct.es

Abstract. In autonomous driving, the ability to correctly detect and calculate the distance to objects is a fundamental task of the perception system. The objective of this paper is to present a robust object detection system using a Time of Flight (ToF) camera. The system is designed for covering the blind spot areas of vehicles and to be integrated with an autonomous vehicle perception system. ToF cameras generate images with depth information in real time as well as providing grayscale images. The accuracy of the measurements obtained depends greatly on the correct calibration of the camera, therefore in this paper, a calibration process for the Sentis 3D-M420 ToF camera is presented. For the object detection system that has been developed, different descriptors and classifiers have been analyzed. To evaluate the system tests were performed in real situations using a series of images obtained with the Sentis 3D-M420 camera during autonomous driving tests.

Keywords: Time of flight · LiDAR technology · Object detection · KNN classifier · Calibration

1 Introduction

Image processing systems are currently used in a wide range of applications, one of these being autonomous driving. The perception system of an autonomous vehicle is one of the most important systems as the safety of the vehicle depends on the correct interpretation of its surroundings. The most common technologies used onboard autonomous vehicles for perception are: LiDAR (2D and 3D), RADARs, cameras (Visible (VIS), infrared (IR), ToF) or ultrasonic sensors [1]. In computer vision for autonomous driving, obtaining distance information as well as detecting objects and reconstructing the environment is a fundamental task. LiDAR sensors tend to have a higher precision than computer vision systems [2]. 3D LiDARs are used for object detection and avoidance, 3D mapping and Adaptive Cruise Control [3]. However, due to the high cost of 3D LiDAR sensors, computer vision cameras and ToF cameras are often used.

J. Świątek et al. (Eds.): ISAT 2019, AISC 1051, pp. 277–286, 2020.
https://doi.org/10.1007/978-3-030-30604-5_25

It is also important to distinguish between vision and perception. Any camera is capable of seeing the world in the same way as humans do, however cameras alone are unable to interpret the images. For human beings, the main perception method is sight, so much so, that when driving, we use rear view and side view mirrors to extend the field of view. However, these still fail to give a 360º view and there is always a blind spot.

The present work consisted in the development of an object detection system which covers the blind spot area of a vehicle using a ToF camera. The system presented is then tested in real life situations for curb detection. Using this type of camera has the advantage of being able to calculate the distance from the surrounding objects to the vehicle. A significant amount of data can be obtained even though the resolution of the image is lower than that provided by computer vision cameras [4]. Computer vision cameras can also be used but to calculate the distance to the objects, complicated stereo matching techniques must be applied [5].

ToF cameras provide real time distance estimates for each pixel using the ToF principle. ToF systems consist of optical transmitters and receivers [6]. Due to the fact that this technology is recent, for most applications, so far researchers relied on the calibration provided by the manufacturer [7]. This paper aims to provide a calibration method for a ToF camera, the Sentis 3D-M420, for use on an autonomous vehicle for close range object detection.

2 Time of Flight Technology

ToF cameras measure the time that the light signals take to reach the object and return reflected [8]. ToF cameras provide a depth image in which each pixel instead of indicating color as a conventional camera would, give the distance to the corresponding point in the environment. The light emitter sends a near infrared (NIR) light signal and the distance to the object is calculated by and measuring the time the signal takes to return to the receiver [9]. Two types of ToF technology currently exist, the first measures the time directly, the second however, calculates the phase difference by correlating the optical signal received by the camera with the reference signal which is electrical. Figure 1 shows an example of the ToF signals where the phase difference between the signal emitted and the signal received can be observed.

With the speed of light a known constant ($c = 3 \times 10^8$ m/s), the distance can be calculated directly using (1), where d is the distance to the object, $\Delta\varphi$ the phase difference between the emitted signal and the received signal and f_{mod} the frequency modulation of the signal emitted.

$$d = c\,\Delta\varphi/(4\pi f_{mod}) \tag{1}$$

The second procedure is known as “continuous wave” technology. This process is carried out directly by the camera, and consists in the correlation of four signals displaced 90º: C0(0º), C1(90º), C2(180º) and C3(270º). These signals are measured in periods of T = 1 / f_{mod} and with these values it is possible to calculate the distance, phase difference and amplitude using expressions (2) and (3):

$$\Delta\varphi = arctan(C1 - C3/C0 - C2) \tag{2}$$

$$a = \sqrt{(C1 - C3)^2 + (C0 - C2)^2}/2 \tag{3}$$

In addition to the depth image obtained, the camera also provides a grayscale image obtained from the amplitude signal received. It is important to take into account that the ToF camera provides distances, not 3D coordinates, however these can be obtained mathematically.

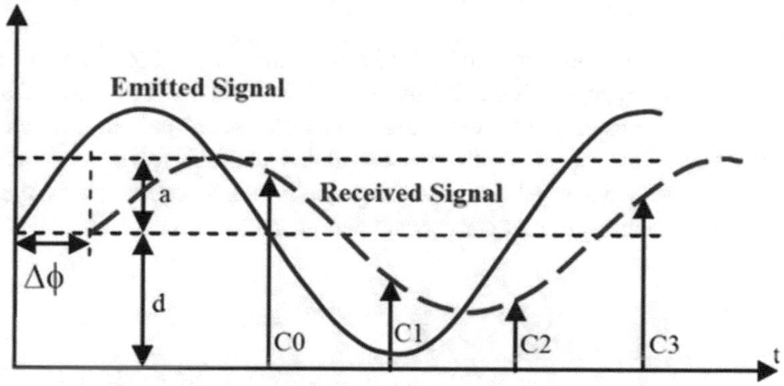

Fig. 1. Signals emitted and received by a ToF camera [4]

The Sentis 3D-M420 Camera

The Sentis 3D-M420 is a ToF camera based on different modules which satisfy a variety of demands and applications [10]. The module used in this work is the TIM-UP-19K-S3 USB 2.0. Some characteristics of this module are:

- 24 850 mm high power LEDs.
- Up to 160 frames per second.
- An angle of view of up to 110º.
- 120 × 160 pixel resolution.

The Sentis 3D-M420 ToF camera uses beams of amplitude modulated light and obtain distance information by measuring the phase difference between the reference signal and the reflected signal. Next to the lens, the camera has matrices of LEDs which emit light signals close to the infrared range known as NIR light.

3 Calibration Process

The greatest advantage of ToF cameras is the ability to extract depth information in order to determine geometrical characteristics and planes. In addition, the correct calibration allows the depth image to be transformed into a 3D point cloud. The disadvantage is the fluctuation in the distance measurement precision, due to interference from external factors, for example, sunlight, the reflectivity of the surfaces, etc. In order to reduce these errors, the camera must be calibrated correctly.

Before performing the calibration process, the desired lens must be mounted on the camera module and the focus adjusted. To adjust the focus, the camera needs to be placed facing a chess board pattern at a distance of at least 1.5 m [11]. The lens then has to be turned clockwise or counter clockwise until a sharp amplitude image is obtained. Once the corrected point cloud has been obtained, the lens calibration file is updated.

The following tests were carried out with the same chess board pattern that was used previously to adjust the focus of the lens (Fig. 2). To avoid errors from interference, different groups of pixels belonging to the chess board pattern were chosen and then in 30 images, the mean depth was calculated and the distance obtained. The calibration process is designed for obtaining depth measurements with the highest possible precision, setting the camera parameters to constant values.

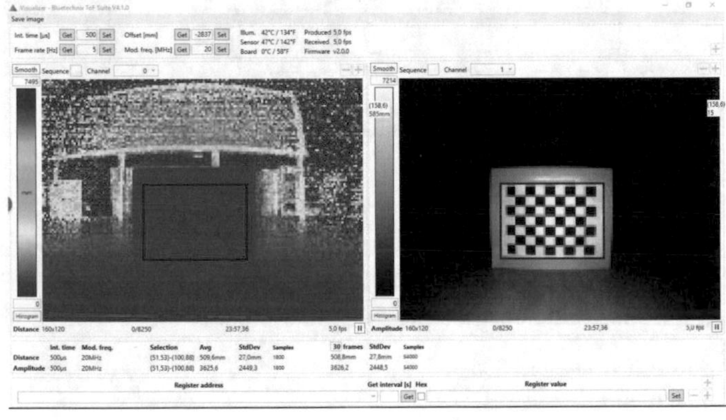

Fig. 2. Calibration using the chess board pattern

3.1 Tests with Default Parameters

The camera is initially calibrated for a field of view of 110°. The first tests were carried out with this configuration, with the following parameters: Integration time: 500 µs, Frames/s: 5, Offset: −2837, Modulation frequency: 20 MHz.

The test consists of obtaining depth measurements at different distances: 500 mm, 1000 m, 1500 mm and 2000 mm, with the default parameters.

At a distance of one meter, the distance error was very low (Table 1), however at 2 m, the error was 12.59% (0.251 m) which is not acceptable if the camera is used for close range obstacle detection.

Table 1. Tests with default parameters for a 110° lens

Distance (mm)	Measured distance (mm)	Integration times (μs)	Modulation frequency (MHz)	Error (%)
500	512.3	500	20	2.46
1000	996.7	500	20	0.33
1500	1402.4	500	20	6.51
2000	1748.3	500	20	12.59

3.2 Integration Time Variation

The integration time defines the time that the camera sensor requires in order to receive the light signal reflected by the surrounding objects. If the integration time is low, the distance to objects that are further away will not be as accurate, due to the fact that for large distances the light signal will not have time to return and activate the camera sensor. However, if the integration time is too great, the camera reduces its acquisition velocity and the depth maps may reflect incorrect and oversaturated distances [12].

In the first test, the error obtained was high and there was a lot of noise in the readings. This can be reduced by increasing the integration time. The following tests consisted of varying the integration time for a distance of 1.5 m.

In this test, the errors obtained were of the same magnitude (1.89–7.57%). When the integration time was increased the errors were reduced significantly. The most notable improvement was observed with an integration time greater than 5000 μs (less than 5.79% error).

3.3 Frequency Modulation Variation

The modulation frequency changes the physical appearance of the mathematical cosine model of the light signal. This parameter determines the maximum distance to an object that the camera is capable of detecting. This value is known as the "No uncertainty/ ambiguity range".

This test consisted in changing the modulation frequency and the errors obtained fluctuated greatly with values between 0.81 and 9.45%. The manufacturer recommends a frequency modulation of 20 MHz, therefore, this parameter will be set to the default value.

3.4 Offset Variation

The offset is used to compensate the measurements and is a value added to the measurement to obtain greater precision. This test was performed with a reference distance of 500 m and the offset value is adjusted to minimize the error for this distance.

The errors obtained with this configuration increased for a distance of more than 1500 mm but were no greater than 3.3% which is very acceptable.

3.5 Calibration Process

After carrying out the previous tests, a calibration method has been developed for the Sentis 3D-M420 ToF camera. The process consists of the following stages:

1. Lens adjustment: The 110º lens is adjusted to the correct focus, at which a chess board pattern can be clearly distinguished at a distance of one meter.
2. Integration time adjustment: To achieve this, a balance between the amplitude image that is not saturated and a depth image with the lowest amount of noise possible.
3. Offset Adjustment: Once the integration time has been set, the offset is adjusted to a reference distance.

For our use of the ToF camera for close range object detection on an autonomous vehicle, the following final calibration parameters were used: Integration time: 1000 μs, Frequency modulation: 20 MHz, Offset: −2874 mm (with a reference distance of 500 mm). The results obtained are shown in Table 2.

The distance measured and the real distance can be adjusted by means of a minimum squares regression. The resulting linear equation that best fit the results is (4):

$$y = 0.989x + 20.493 \tag{4}$$

The correlation coefficient obtained with this fit was $R^2 = 0.9984$, therefore this linear regression is adequate for adjusting the distance measurements obtained.

Table 2. Final calibration for use on an autonomous vehicle

Distance (mm)	Measured distance (mm)	Integration times (μs)	Modulation frequency (MHz)	Offset (mm)	Error (%)
500	500.2	1000	20	−2874	0.04
750	762.1	1000	20	−2874	1.61
1000	1016.8	1000	20	−2874	1.68
1250	1256.8	1000	20	−2874	0.54
1500	1479.5	1000	20	−2874	1.37
1750	1692.3	1000	20	−2874	3.30
2000	1945.7	1000	20	−2874	2.72
2500	2515	1000	20	−2874	0.60
3000	3043.8	1000	20	−2874	1.46

4 Application for Curb Detection

Following the calibration, a perception system has been designed that is able to detect curbs and calculate the distance between the vehicle and the curb. The ToF camera was installed on the CIC test vehicle.

In object classification and pattern recognition algorithms, the first step is feature extraction. To perform this task, the histogram or orientated gradients (HOG) [13] and local binary [14] methods are used.

The first process consists of obtaining the necessary features to train the classifier to detect a curb. This is done by defining a rectangular window size from which the features will be extracted. Window sizes of 10×40 and 20×40 pixels are chosen as curbs are narrow and elongated in shape.

A metric analysis was obtained from the training and determined which of the window sizes gave the best results. After the samples were extracted from the depth and amplitude images and the features were obtained, the features vector is formed. To make sure that the right amount of information is contained in the features vector, the classifier performance is analyzed.

The size of the features vector depends on the cell size. For a cell size of 2×2 the features vector has a length of 6156, whereas with a cell size of 8×8, the length of the vector is just 144. A balance must be found between achieving a good performance with the classifier and obtaining a too greater amount of data.

4.1 Classifier Training Process

For training the classifier 660 samples were obtained from 150 images. Of these samples, 210 contained a curb and 450 did not. The KNN [15] and SVN [16] classifiers were used and the cross validation leave one out method was then used to test their performance. The configuration of the classifiers is shown in Table 3.

To obtain the highest performance with the classifier, different combinations have been tested using different filters to reduce the noise in the image. The HOG and LBP descriptors were tested and the descriptor that provided the best curb definition with a features vector was chosen. Different window sizes were also tested, to find which one provided sufficient information but without redundant pixels. Using the same parameters, the HOG features provided an improvement of 26,44% compared to the results obtained from the same tests using the LBP features.

The process that was followed consisted of the following tasks, in order: (1) Feature selection, (2) Window size selection, (3) Image selection, (4) Image filtering.

Table 3. Configuration of the KNN and SVM classifiers

K (Neighbourhood):	4
Standardization:	True
Window Size:	20×40
Filter:	Median
Descriptor:	HOG
Feature Number:	2592
Performance:	98.333%

The KNN classifier obtained the best performance with 97.78% (Table 4) using the leave one out technique. The filter used on the sample was a median filter, applying the HOG descriptor with 2592 features and a cell size of 4 × 4.

The accuracy is calculated by means of a ROC curve, obtaining an area below the curve of 0.998. This means that given two samples, one containing a curb and one without, the classifier will identify them correctly.

Table 4. Best performance obtained with the Classifiers

	KNN	SVM
Performance	97.78%	95.33%
Error	2.22%	4.67%

4.2 Distance Calculation

Trigonometry was applied to the depth information provided by the camera to calculate the distance from the vehicle to the curb. First the curb candidates perpendicular to the camera were taken from the central area of the image. Then to classify the candidates, the median filter and HOG features vector were used. Finally, once the curb was located in the image, the distance from the curb to the vehicle was obtained using Pythagoras and the distance from the camera to the curb (Fig. 3) [4]. The camera measurements were corrected using the linear equation obtained from the calibration.

Fig. 3. Calculation of the curb to the vehicle distance

4.3 Tests Performed

To test the system in real situations, 30 new images were used. The testing process consisted of the following stages:

1. To reduce the noise, a median filter was applied to the distance and amplitude images.

2. Candidates were taken from the images to later be classified. A 20×40 window was used to scan the images to determine if a curb was present.
3. In each window the features were identified and then classified. To increase the computational efficiency, two parts of the images were automatically discarded: The lower area containing the edge of the vehicle and the side edges where an excess of noise is always found.

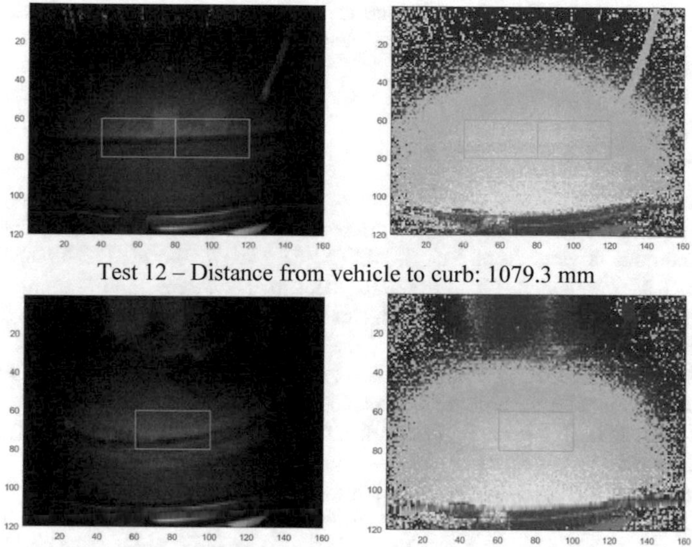

Test 12 – Distance from vehicle to curb: 1079.3 mm

Test 22 – Distance from vehicle to curb: 1060.4 mm

Fig. 4. Results obtained with the curb detection system

Out of 30 tests performs, the curbs were correctly identified in 24 of the images and in the other 6 classification errors were obtained. Therefore the performance of the system is 80% (Fig. 4).

5 Results and Conclusions

Computer vision systems based on ToF cameras are likely to play an important part in the future development of autonomous vehicles due to factors such as size, weight and affordability in comparison with other sensors such as 3D LiDAR.

In this paper an experimental calibration method has been presented, in which by adjusting the integration time, frequency modulation and offset parameters of the Sentis 3D-M420 ToF camera, distance measurements can be obtained with minimal error. This analysis has demonstrated that by modifying the default values provided by the manufacturer for these parameters, the error in distance measurements can be reduced significantly.

Once correctly calibrated. The camera has been used for the design and implementation of a curb detection system for an autonomous vehicle. Although in this work the system has been used to detect curbs, the system could also be trained to detect pedestrians or other objects in close range of a vehicle. The system was tested using images obtained in real traffic situations and a performance of 80% was achieved.

Acknowledgements. This work was partially supported by ViSelTR (ref. TIN2012-39279), DGT (ref. SPIP2017-02286) and UPCA13-2E-1929 Spanish Government projects, and the "Research Programme for Groups of Scientific Excellence in the Region of Murcia" of the Seneca Foundation (Agency for Science and Technology in the Region of Murcia-19895/GERM/15).

References

1. Borraz, R., Navarro, P., Fernández, C., Alcover, P.: Cloud incubator car: a reliable platform for autonomous driving. Appl. Sci. **8**, 303 (2018). https://doi.org/10.3390/app8020303
2. Navarro, P.J., Fernández, C., Borraz, R., Alonso, D.: A machine learning approach to pedestrian detection for autonomous vehicles using high - definition 3D range data. Sensors 1–13 (2015)
3. Rosique, F., Navarro, P.J., Fernández, C., Padilla, A.: A systematic review of perception system and simulators for autonomous vehicles research. Sensors (2019)
4. Ros, A., Miller, L., Navarro, P., Fernández, C.: Obstacle Detection using a Time of Flight Range Camera. IEEE. (2018)
5. Scharstein, D., Szeliski, R.: A taxonomy and evaluation of dense two-frame stereo correspondence algorithms. Int. J. Comput. Vis. **472**, 7–42 (2002)
6. Ringbeck, T., Hagebeuker, B.: A 3D time of flight camera for object detection (2007)
7. Zhu, J., Wang, L., Yang, R., Davis, J.: Fusion of Time-of-Flight Depth and Stereo for High Accuracy Depth Maps (2008)
8. Foix, S., Alenya, G., Torras, C.: Lock-in Time-of-Flight (ToF) cameras: a survey. IEEE Sens. J. **11**, 1917–1926 (2011). https://doi.org/10.1109/JSEN.2010.2101060
9. Li, L.: Time-of-Flight Camera – An Introduction. (2014)
10. ToF 3D Cameras - Bluetechnix
11. Modification of TIM Lenses, http://datasheets.bluetechnix.at/goto/TIM/HOWTO_Modification_of_TIM_Lenses.pdf
12. Gil, P., Kisler, T., García, G.J., Jara, C.A., Corrales, J.A.: ToF Camera calibration: an automatic setting of its integration time and an experimental analysis of its modulation frequency. RIAI - Rev. Iberoam. Autom. e Inform. Ind. **10**, 453–464 (2013)
13. Dalai, N., Triggs, B., Rhone-Alps, I., Montbonnot, F.: Histograms of oriented gradients for human detection. In: IEEE Computing Society Conference on Computer Vision and Pattern Recognition. CVPR 2005, vol. 1, pp. 886–893 (2005). https://doi.org/10.1109/CVPR.2005.177
14. Ojala, T., Pietikäinen, M., Mäenpää, T.: Multiresolution gray-scale and rotation invariant texture classification with local binary patterns. IEEE Trans. Pattern Anal. Mach. Intell. **24**, 971–987 (2002). https://doi.org/10.1109/TPAMI.2002.1017623
15. Weinberger, K., Blitzer, J., Saul, L.: Distance metric learning for large margin nearest neighbor classification. Adv. Neural. Inf. Process. Syst. **18**, 1473 (2006)
16. Cortes, C., Vapnik, V.: Support-vector networks. Mach. Learn. **20**, 273–297 (1995)

The Implementation of a Convolutional Neural Network for the Detection of the Transmission Towers Using Satellite Imagery

Paweł Michalski[1], Bogdan Ruszczak[1(✉)], and Pedro Javier Navarro Lorente[2]

[1] Opole University of Technology, Prószkowska 76, 45-758 Opole, Poland
{p.michalski,b.ruszczak}@po.edu.pl
[2] Universidad Politécnica de Cartagena, Plaza del Hospital, 1, 30202 Cartagena, Spain
pedroj.navarro@upct.es

Abstract. This paper presents the implementation of a supervised machine learning method for the detection of transmission towers. For this purpose, the authors have relied upon satellite images of an object taken at four magnification levels. The data processing was carried out by using a real-time object detection convolutional neural network. The data where the position and a label of the transmission tower were marked manually has been divided into a training, validation and test sets, which were used for later method evaluation. The analysis, that is provided further in the article, covers a network operating within a few alternative training scenarios and within a data which consists of various image magnification levels. It is extended with the discussion on the optimal parameters setting and their influence on the outcome classification. After evaluating several different network configurations using 4944 labelled satellite images, the highest performance obtained using the test set was of an accuracy of 0.9676, a precision of 0.9522, and recall of 0.9361.

Keywords: Deep learning · Convolutional neural networks · Satellite imagery · Image processing · Overhead high-voltage power lines

1 Introduction

Overhead high-voltage power lines (OHVL) are a crucial element of an electric power system. The lines stretch on significant distances, hence both the time and cost related to their maintenance are high. Nowadays, the most common way to diagnose them and to carry out usual inspections of their condition is by using data collected during helicopter overhead flights. Recently it has become more common for the data to be collected during flights performed by unmanned aerial vehicles (UAVs) [1, 2]. The following devices are used for data acquisition purposes: standard cameras providing images within the spectrum of visible light [3], spectral cameras, as well as systems operating in LIDAR technology, which register spatial data as points [4, 5].

As has been shown inter alia in [6, 7], in situations of wide-spread faults in electrical grids, it is very important to assess the extent of a fault and to act in a timely

J. Świątek et al. (Eds.): ISAT 2019, AISC 1051, pp. 287–299, 2020.
https://doi.org/10.1007/978-3-030-30604-5_26

manner when making decisions with regards to repairs once a diagnosis of the situation is available. In some situations, the use of UAVs may be difficult and carrying out a flight along the elements of an OHVL may be dangerous, moreover, it can incur high costs (especially in the case when manned vehicles are to be used). A viable alternative to the above solutions is to acquire information about a condition of an overhead power line through satellite imaging. This approach certainly has its limitations due to the resolution with which particular power line elements can be registered, as it is necessary for an image being acquired to be of suitable quality. The research presented below considers using such imaging for the purposes of carrying out an analysis of a condition of OHVL. Also, it covers an issue of software implementation for automatic detection of power lines transmission towers on acquired images. Acquisition of satellite data provides information on the current condition of overhead power lines but it could also be a supplementary information source for the emergency systems related to extreme events [8].

2 The Satellite Imagery Acquisition

Carrying out an analysis of imagery and electrical energy objects detection based on an acquisition of satellite images requires due care in acquiring records of a suitable resolution. In order to acquire images of a quality which enables further analysis, the ground sample distance (GSD) criterion (used by satellite system operators) may be applied. According to it, satellite imaging systems are divided into groups as follows: (less than 1 m) very high resolution, (1 to 5 m) high resolution, (5 to 30 m) medium resolution and (more than 30 m) low resolution [9].

The satellites from the very high resolution group covered in the paper [10], are as follows: IKONOS-2 (GSD 0.82 m), EROS B (0.7 m), KOMPSAT-2 (1.0 m), KOMPSAT-3 (0.7 m), Resurs-DK 1, WorldView-1 and -2 (0.46 m), GeoEye-1 (0.46 m), Cartosat-2 series (0.82 m), Pleiades 1 and 2 (0.5 m) and SkySat-1 and -2 (0.9 m). Special attention should be paid to the satellites from the Sentinel-2 group, which also register images in 13 spectral channels (in the range 443–2190 nm) and are most commonly used in studies mapping the vegetation condition [11, 12].

Unlike helicopters or UAVs, satellite systems do not require good weather conditions for a safe flight to be carried out and there is no limitation as to a distance that they may travel. However, this approach is weakened by the cloudy conditions and may render registration of objects impossible. It is further undermined by the fact that ad hoc availability of satellite systems is limited. In case of commercial satellites, the revisit time, i.e. the time elapsed between two successive observations of the same ground point, at present takes about 1 day [13].

3 Overhead Power Lines Diagnostics

The image material registered during a flight has to be subsequently processed. The aim of this processing has two objectives: the first one is to create a classification of data, which should provide information as to whether a particular object of interest is present on an image and the second one is to carry out localization which should determine its precise location on that image. Also, the depicted object should be later analyzed in order to detect any damages or faults. This process is usually carried out by qualified diagnosticians. At present, the issue of unsupervised fault detection for OHVL is relatively intractable as there is an insufficient database of samples which can be used to create an effective classifier of specific faults in particular elements. Moreover, the issue of classification and localization of power lines elements is a subject of several research studies. The relevant works involve image processing methods, which are further outlined in this paper. Main objectives of studies conducted in this area of research are as follows:

- diagnostics of the condition of an overhead power line corridor [14],
- detection of conductors [15, 16],
- detection of transmission towers [17],
- detection of various transmission tower elements or additional equipment of power lines (i.e.: power line insulators, dumpers, connectors) [3, 18].

4 Implementation of Deep Learning Methods for the Object Detection

Various methods of machine learning are used in order to carry out a process of automatic classification and localization of objects in large data sets [19–21]. At present, the most commonly implemented tool for this purpose are convolutional neural networks (CNNs). They are valued amongst researchers for their high efficiency [22–24]. This type of neural networks was developed in 2012. Its development was possible due to two factors: a significant increase in computing power (mainly a result of parallelization in computing techniques which use graphics processing units) and making several large sets of training data widely available (ImageNet 2009 was the first and popular dataset of labelled images) [25].

Development of several implementations of CNN has been summarized in the paper [26]. Therein, it was highlighted that subsequent structures of the CNN are increasingly deeper and more complex. However, it seems apparent that each of these networks has a similar structure. The selection of the effective network architecture is essential, it requires also optimization of its structure, which is time-consuming due to their complexity and the fact that each layer learns separately.

One of the popular CNN implementations, which deserves special attention, is the network architecture based on the YOLO (You Only Look Once) algorithm [27]. Therein, the process of detecting objects was rephrased as a single regression problem. The YOLO algorithm allows for the whole input image to be processed. As a result, it enables the information contained within particular classes to be saved. This was not possible for other CNNs where traditional methods using a sliding window were used. The structure of the network was inspired by the structure of GoogleLeNet network [28]. It consisted of merely 24 convolutional layers and 2 fully-connected layers. In the study [29], the authors succeeded in diminishing the imperfections of the previous version of YOLO such as frequent localization errors. In YOLOv2 system, the input image resolution has been changed from 448 x 448 pixels (in the first YOLO version) to 416 × 416 pixels in YOLOv2. This approach enables a particular dissection of an analyzed image. This means that the researched object, which is often located in the center of the frame, can be related to a separate and undivided image sample. The application of a 32 multiplier factor for downsampling enables subsequent fragmentation of the input image to get single bounding boxes. The input image is split into 13 regions. For each of the regions, the algorithm generates 5 bounding boxes containing the described object. After that, the algorithm creates a ranking and chooses the one with the best score. YOLOv2 is based on a new classification model called Darknet [30]. It consists of 19 convolutional layers and 5 pooling layers. The speed of large image sets processing is the main advantage of the YOLOv2 system. It allows for the visual data to be analyzed in real time. The tempo of manipulation of consecutive registered images is particularly important in case of processing satellite images as very often they depict large areas and they came in large quantities.

The network training stage is important for later networks functioning and has a key impact on its performance. CNNs belong to a group of methods called supervised learning algorithms. In the training phase, a set of purposely prepared training samples has to be introduced. The samples contain objects that are to be subsequently identified. As a result, the process is intended to enable correct objects detection and classification.

The training process in YOLOv2 uses a stochastic gradient descent method. It involves also the use of various resolutions of the input frame. However, the multiplier of 32 should keep within the range between the minimal possible resolution (320 × 320) and the maximum one (608 × 608). The change of resolution is random and occurs every 10 cycles of network training.

5 The Subject of Study and Data Preparation

In the given study, CNN was implemented with the purpose of detecting transmission towers on satellite images. This required a set of training data to be prepared and labelled in order to train a new class – 'transmission tower' using the fine-tuned classifier. The satellite data has been acquired through Google Maps API system. It was divided, according to the magnification level, into folders 'zoom 18', 'zoom 19', 'zoom 20', and 'zoom 21'. Hand-made labels were an important element of the prepared dataset. The labels referred to a data which contained information about the location and the size of an exact part of an image, where a particular object which was to be analyzed was present.

For the training purposes and further verification of the network, the following sets: of satellite images showing transmission towers, as well as a set of reference images without transmission towers on them, were prepared. A training subset, which was split randomly, contained 618 images in each folder (total number of images used during the training stage was 2472 for all magnification levels). Within the training subset, each image contained transmission tower. There were 309 images in the evaluation set and the test set for each magnification level.

The whole dataset which was used to train, evaluate and test the described method contained 4,944 satellite images (1,236 for each magnification level). Figure 1 presents example images of objects within the analyzed class.

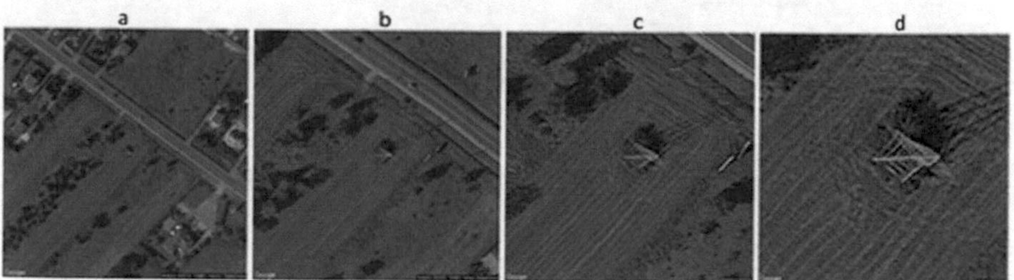

Fig. 1. Magnification level on satellite images respectively: a–18, b–19, c–20, d–21

In order to make the developed classification more versatile, the given images represent various types of transmission towers (110 kV, 220 kV and 400 kV) that were built in different periods of time. This was intended to allow for the most adaptable classifier to be developed. For this reason, it was necessary to use satellite images from a few regions. They show transmission lines located in Dolnośląskie Voivodeship, Śląskie Voivodeship, Małopolskie Voivodeship and Wielkopolskie Voivodeship. The collected images were taken within the magnification range between 18 and 21. Therein, 1 pixel represents an areas of the following length: 60 cm (zoom 18), 30 cm (19), 15 cm (20) and 7.5 cm [31]. The size of the images acquired for the purpose of the analysis was 640 × 640 pixels.

For the purpose of the study, an implementation of YOLOv2 network was used in TensorFlow environment, called Darkflow. During the training process a pre-trained network was used. Its final layer named region, that is designed to classify objects, was modified for the purposes of the research as presented on Fig. 2.

The aim of the classifier work is to detect the object and to predict its localization in the processed image. In the presented implementation, a confidence score parameter is assigned in order to assess the prediction. This parameter is computed using two components. The first one refers to an assessment (or probability) of the index object being located in the investigated region: PR (Object). The second component refers to an assessment of similarity between the indicated region and the correct location of the analyzed object: IoU (Intersection over Union, also called Jaccard index). It can be described with the following formula:

$$IoU = (GT \cap Prediction)/(GT \cup Prediction) \quad (1)$$

This parameter enables the evaluation of the detector accuracy. It is assigned as a result of comparing the area of the indicated object (prediction) and previously inputted information as to where the object is located (ground truth – GT). The relation between them has been shown in Fig. 3 where an example of a coefficient result is used.

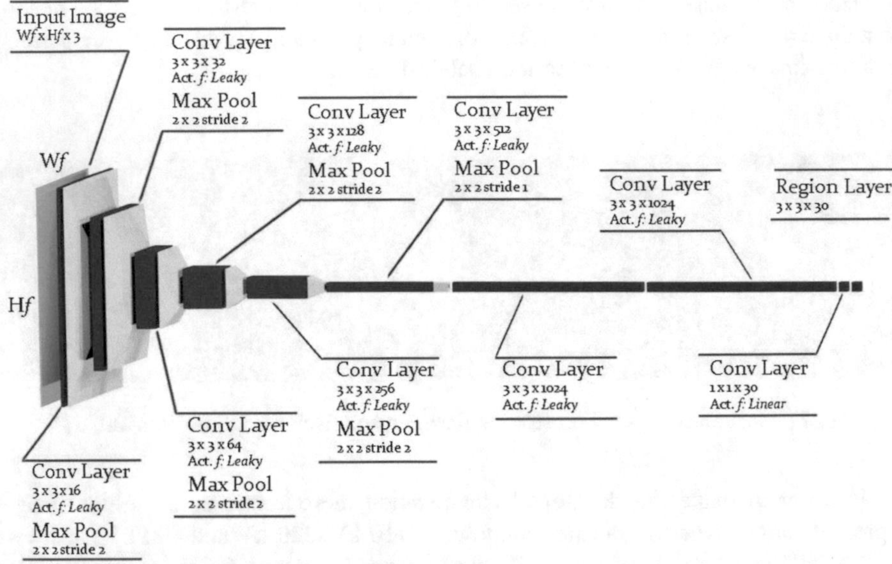

Fig. 2. Structure of used YoloV2 architecture with modified region layer. Modifications of W_f and H_f input parameters are discussed in Sect. 6.

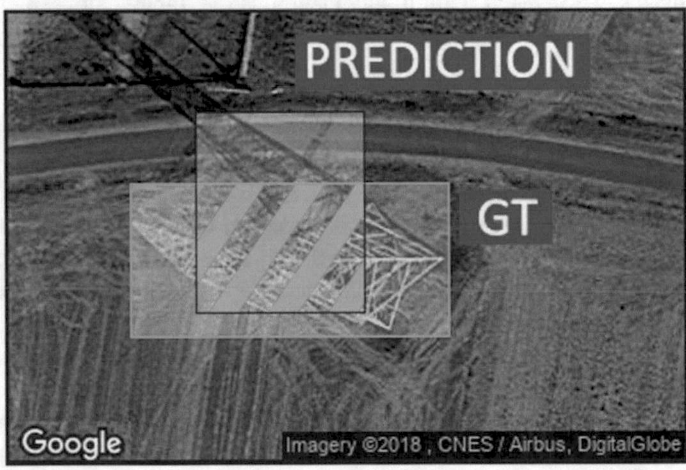

Fig. 3. An example of an analyzed image with an emphasis on the input information (GT) and the outcome (PREDICTION)

6 The Results Overview

The main objective of the analysis was to ascertain whether a CNN can be used for the detection of transmission towers on satellite images. In order to carry out a functioning assessment of the tool, some performance measures were assigned to the whole set of images: Accuracy, Recall, Precision, False Positive Rate and Negative predictive Value. These were based on examination of four possible outcomes (Fig. 4): detecting an object (true positive), incorrect identification of an object (false positive), lack of identification of an object (false negative) or a situation where an object was not identified within the reference set (true negative).

Fig. 4. Selected examples of all four possible identification results.

In the first phase of the analysis, the network was trained on a set of images acquired at a single magnification level of a satellite image. This was subsequently verified against test data for the same magnification level. The results obtained for the following four respective cycles are presented in Table 1. The best scores, accuracy = 0. 9612 and precision = 0.9514, were achieved when images at magnification level = 19 were used. It is worth noting, that the lowest accuracy was achieved for the images taken at the highest zoom where the analyzed objects are most precisely depicted. However, this approach does not allow for sufficient generalization. High precision (1.0) and low FPR indicate a lack of false identification of objects in this case. Nevertheless, this is correlated with a low number of positive identifications.

The table covers also the score result generated by the network itself with every iteration (therefore the mean was selected to generalize the whole set of scores). As it turns out that in case of images taken with lower magnification level the network detection was done with higher probability (median score for 18 magnification level was 0.8853 versus 0.2989 for the 21 magnification).

Table 1. Evaluation of the transmission towers detection efficiency based on training and validating images with the same magnification level

Imagery magnification of training set	Accuracy	Precision	Recall	Negative pred. value	Median score
18	0. 9029	0.9641	0.8703	0.8310	0.8853
19	0. 9612	0.9832	0.9514	0.9308	0.6869
20	0. 8706	1.0000	0.7849	0.7546	0.4518
21	0. 7087	1.0000	0.5161	0.5775	0.2989

At the next stage of the analysis, an attempt was made to identify the optimal combination of training and test sets. During the training process a combine set of all respective image magnification sets was used. At the same time, the network, which was trained on that data, was examined against respective test sets. The results for this stage are summarized in Table 2.

It appears that as an outcome of this process, higher accuracy of classification of the trained pattern was achieved for nearly all variants, except for zoom 18. The level of accuracy = 0.9325 for magnification = 20 was noted, while precision and recall were also higher than 0.9.

Table 2. Assessment of efficiency in detecting transmission towers based on training images from combined sets of different magnification levels.

Training set	Validating set	Accuracy	Precision	Re-call	Negative pred. value	Median score
All	18	0.8203	0.7469	0.9383	0.9279	0.9953
All	19	0.9097	0.8525	0.9850	0.9833	0.9906
All	20	0.9325	0.9102	0.9548	0.9555	0.9942
All	21	0.7892	0.7730	0.8034	0.8059	0.9156
All	All	0.8629	0.8186	0.9202	0.9156	0.9953

Further, the accuracy of the same classifier was examined against the test images taken from combined sets of all magnifications. However, in this case, the achieved accuracy and precision turned out to be lower than the best score.

It is worth noting, that when the test data for zoom 21 was tested against the classifier trained with a combined set of images, it allowed for much better results to be achieved (Accuracy21-21 = 0.6350 and AccuracyAll-21 = 0.7892). Network training based on a larger dataset (which contained 4 times more images and various magnifications) turned out to be more efficient than training and testing carried out on satellite images acquired at the same magnification level. This approach may enable a more versatile tool to be developed. Such a tool would be geared up to classify objects on satellite images at various magnification levels.

The next stage involved an examination of the threshold parameter setting and its impact on the achieved results. The threshold is used to assess which of the classifier results are considered correct. A calculation was carried out for groups of results for consecutive rates of the threshold from 0 up to 1 with step 0.01. The more restrictive was the threshold (i.e. set to a high level) the lesser number of results were marked as TP. On the other hand too low a threshold level may result in a large number of falsely indicated predictions. The results of this study were presented in a ROC curve (Fig. 5).

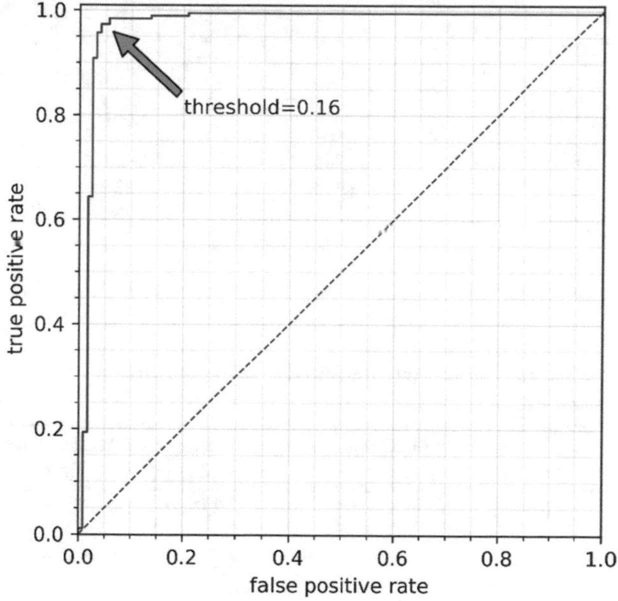

Fig. 5. ROC curve for the threshold examination.

The optimal setting for the threshold parameter calculated using the evaluation set was 0.16. At this level, a higher accuracy = 0. 9676 is achieved as compared to beforehand, while the rates of precision = 0. 973 and recall = 0.974 remain high.

A further experiment was designed to ascertain the optimal size of a frame which is to be used in the process of testing the performance of the network functioning independently to the size of the frame used before on the training stage. Figure 6 provides a comparison of the obtained results reflecting various frame sizes.

The calculations were carried out for a network which was trained on images acquired at all magnification levels. Afterwards, the network was tested against images acquired only at magnification 19. One can clearly see that the higher precision was achieved for smaller frame sizes. Also, the higher this parameter was, the greater the improvement in its accuracy was obtained.

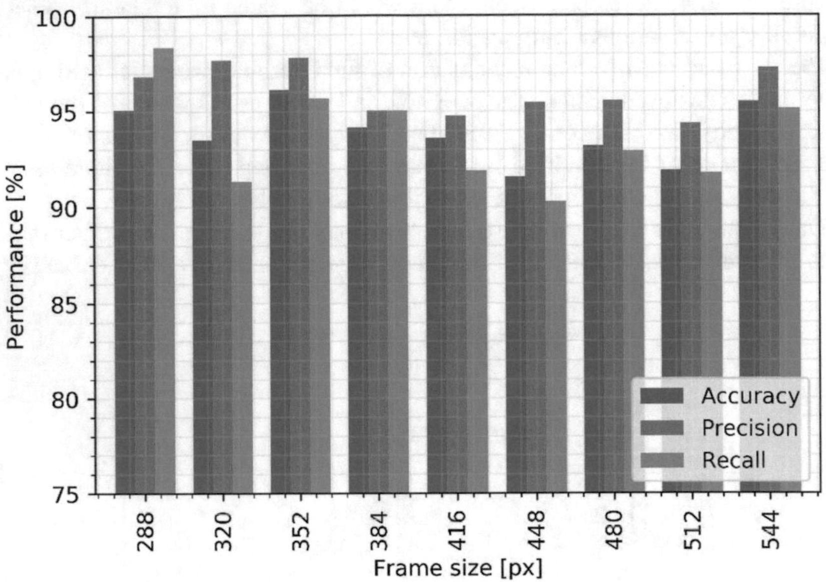

Fig. 6. A performance metrics comparison various frame sizes set during the training process.

Table 3. Performance results for the threshold = 0.16 and frame size: 352 × 352 pixels checked using the test set.

Performance metric	Results evaluated using the whole test set	Results evaluated using set of magnification 19 only
Accuracy	0.9462	0.9676
Precision	0.9522	0.9730
Recall	0.9361	0.9741
Fall-out	0.0443	0.0403
Negative predictive value	0.9407	0.9597
IoU (median)	0.6600	0.6900
Score (median)	0.6272	0.7046

The highest values of accuracy and precision and the second highest value of recall were obtained for the frame size 352 × 352 pixels. The above parameter setting is different from its default value, which is 416 × 416 pixels. A final classification accuracy of transmission towers class trained for reference was achieved at a level of 0.9676. This is a relatively high score was achieved as a result of all tested enhancements. First, the threshold of reading the results was adjusted; secondly, a modification to the frame size during the verification process was made; and thirdly, the process was carried out on all combined sets of images. Table 3 presents a complete set of data with regards to the efficiency of a network working at the above settings.

7 Conclusions

Satellite imagery may be a good source of vision data. As shown in this paper, it may be used as a training set to create a classifier for transmission towers with implementation of convolutional neural network.

In the presented study, the highest performance for an implemented CNN (with accuracy = 0. 9676, precision = 0. 973, and recall = 0. 974) was noted when the training data contained images of various magnifications and thc test set was comprised of images taken at magnification 19.

Modifications to some of the parameters of the network may result in a performance increase. On examination it was noted that precision improved when the size of the input image was changed at the network verification stage. The best result was achieved when this parameter was changed from the default value to 352 × 352 pixels.

Optimal threshold settings were also ascertained for analytical purposes and for the confidence score parameter. The best setting was 0.16. This helped to enhance the performance of a classification process even further.

The application of the satellite imagery could be a valuable source of information and allow for fast acquisition of images of quite large areas. This may be particularly useful in case of extensive faults of the power network. Further studies plan to include inter alia the following issues:

- preparation of new classes of OHVL and other networks of industrial objects,
- attempts to create datasets for damaged elements of power lines infrastructure (e.g. as a result of extreme events) and further preparation of a CNN to enable classification of such faults,
- implementing alternative convolutional neural networks architectures and comparing their performance against the dataset presented in this paper,
- object detection performance analysis for a trained CNN when other sets of satellite images are used (e.g. The Inria Aerial Image Labelling DATASET [32]).

References

1. Eck, C., Zahn, K., Heer, P., Imbach, B.: Vision-based guidance algorithms for UAV power line inspection, Lucerne University of Applied Sciences and Arts, Benedikt Imbach Aeroscout GmbH (2012)
2. Colomina, I., Molina, P.: Unmanned aerial systems for photogrammetry and remote sensing: a review. ISPRS J. Photogramm. Remote Sens. **92**, 79–97 (2014). https://doi.org/10.1016/j.isprsjprs.2014.02.013

3. Tomaszewski, M., Krawiec, M.: Detection of linear objects based on computer vision and Hough transform, Przegląd Elektrotechniczny, (Electrical Review), vol. 88/10b (2012)
4. Zhang, Y., Yuan, X., Fang, Y., Chen, S.: UAV low altitude photogrammetry for power line inspection. ISPRS Int. J. Geo-Inf. **6** (2017). https://doi.org/10.3390/ijgi6010014
5. Ahmad, J., Malik, A.S., Xia, L., Ashikin, N.: Vegetation encroachment monitoring for transmission lines right-of-ways: a survey. Electr. Power Syst. Res. **95**, 339–352 (2013). https://doi.org/10.1016/j.epsr.2012.07.015
6. Dzierżanowski, L., Ruszczak, B., Tomaszewski, M.: Frequency of power line failures in life cycle, electrodynamics and mechatronic systems. In: IEEE International Symposium on Electrodynamic and Mechatronic Systems, SELM 2013, pp. 55–56 (2013)
7. Tomaszewski, M., Bartodziej, G.: Prevention of effects of overhead lines failures caused by ice and snow adhesion and accretion. Cold Reg. Sci. Technol. **65**(2), 211–218 (2011). https://doi.org/10.1016/j.coldregions.2010.08.002
8. Ruszczak, B., Tomaszewski, M.: Extreme value analysis of wet snow loads on power lines. IEEE Trans. Power Syst. **30**(1), 457–462 (2015)
9. Dowman, I.J., Jacobsen, K., Konecny, G., Sandau, R.: High resolution optical satellite imagery. Whittles Publishing, Dunbeath (2012)
10. Matikainen, L., Lehtomäki, M., Ahokas, E., Hyyppä, J., Karjalainen, M., Jaakkola, A., Kukko, A., Heinonen, T.: Remote sensing methods for power line corridor surveys. ISPRS J. Photogramm. Remote Sens. **119**, 10–31 (2016). https://doi.org/10.1016/j.isprsjprs.2016.04.011
11. Moeller, M.S.: Monitoring powerline corridors with stereo satellite imagery. In: MAPPS/ASPRS Conference, San Antonio, Texas, pp. 1–6 (2006)
12. Pouliot, D., Latifovic, R., Pasher, J., Duffe, J.: Landsat super-resolution enhancement using convolution neural networks and sentinel-2 for training. Remote Sens. **10**(3), 394 (2018). https://doi.org/10.3390/rs10030394
13. I. Worldview-, WorldView-4 Features & Benefits, (n.d.) 4–5. https://dg-cms-uploads-production.s3.amazonaws.com/uploads/document/file/196/DG2017_WorldView-4_DS.pdf
14. Ahmad, J., Saeed, A.: A novel method for vegetation encroachment monitoring of transmission lines using a single 2D camera, pp. 19–440 (2015). https://doi.org/10.1007/s10044-014-0391-9
15. Yan, G.J., Li, C.Y., Zhou, G.Q., Zhang, W.M., Li, X.W.: Automatic extraction of power lines from aerial images. IEEE Geosci. Remote Sens. Lett. **4**, 387–391 (2007). https://doi.org/10.1109/lgrs.2007.895714
16. Zhou, G., Yuan, J., Yen, I.-L., Bastani F.: Robust real-time UAV based power line detection and tracking. In: 2016 IEEE International Conference on Image Process, pp. 744–748 (2016). https://doi.org/10.1109/icip.2016.7532456
17. Tan, T., Ruan, Q., Wang, S., Ma, H., Di, K.: Advances in Image and Graphics Technologies (2018). https://doi.org/10.1007/978-981-10-7389-2
18. Tomaszewski, M., Osuchowski, J., Debita, L.: Effect of spatial filtering on object detection with the surf algorithm. In: Biomedical Engineering and Neuroscience, Advances in Intelligent Systems and Computing, vol. 720, p. 121–140. Springer (2018). https://doi.org/10.1007/978-3-319-75025-5_12
19. Sampedro, C., Martinez, C., Chauhan, A., Campoy, P.: A supervised approach to electric tower detection and classification for power line inspection. In: Proceedings of the International Joint Conference on Neural Networks, pp. 1970–1977 (2014). https://doi.org/10.1109/ijcnn.2014.6889836
20. Ko, K.E., Sim, K.B.: Deep convolutional framework for abnormal behavior detection in a smart surveillance system. Eng. Appl. Artif. Intell. **67**, 226–234 (2018). https://doi.org/10.1016/j.engappai.2017.10.001

21. Ferreira, A., Giraldi, G.: Convolutional Neural Network approaches to granite tiles classification. Expert Syst. Appl. **84**, 1–11 (2017). https://doi.org/10.1016/j.eswa.2017.04.053
22. Nguyen, V.N., Jenssen, R., Roverso, D.: Automatic autonomous vision-based power line inspection: a review of current status and the potential role of deep learning. Int. J. Electr. Power Energy Syst. **99**, 107–120 (2018). https://doi.org/10.1016/j.ijepes.2017.12.016
23. Gallego, A.J., Pertusa, A., Gil, P.: Automatic ship classification from optical aerial images with convolutional neural networks. Remote Sens. **10**(4), 511 (2018). https://doi.org/10.3390/rs10040511
24. Koga, Y., Miyazaki, H., Shibasaki, R.: A CNN-based method of vehicle detection from aerial images using hard example mining. Remote Sens. **10**(1), 124 (2018). https://doi.org/10.3390/rs10010124
25. Li, K.: ImageNet : a large-scale hierarchical image database ImageNet : a large-scale hierarchical image database (2009). https://doi.org/10.1109/cvpr.2009.5206848
26. Michalski, P., Ruszczak, B., Tomaszewski, M.: Convolutional neural networks implementations for computer vision. In: Biomedical Engineering and Neuroscience, Advances in Intelligent Systems and Computing, vol. 720, pp. 98–110. Springer (2018). https://doi.org/10.1007/978-3-319-75025-5_10
27. Redmon, J., Divvala, S., Girshick, R., Farhadi, A.: You only look once: unified, real-time object detection. In: Proceedings of the IEEE Conference on Computer Vision and Pattern Recognition, pp. 779–788 (2016)
28. Szegedy, C., Liu, W., Jia, Y., Sermanet, P., Reed, S., Anguelov, D., Erhan, D., Vanhoucke, V., Rabinovich, A.: Going deeper with convolutions, arXiv:1409.4842 (2014). https://doi.org/10.1109/cvpr.2015.7298594
29. Redmon, J., Farhadi, A.: YOLO9000: better, faster, stronger, conference on computer vision and pattern recognition (2016). https://doi.org/10.1109/cvpr.2017.690
30. Redmon, J.: Darknet: Open source neural networks in c (2013–2016) 5. http://pjreddie.com/darknet/
31. Setting Zoom Levels of Google Image, MicroImages, Inc. TNTgis - Advanced Software for Geospatial Analysis (2014)
32. Maggiori, E., Tarabalka, Y., Charpiat, G., Alliez, P.: Can semantic labeling methods generalize to any city? The Inria Aerial Image Labeling Benchmark. In: IEEE International Geoscience and Remote Sensing Symposium (2017)

The Image Classification of Workstation Instructions Using Convolutional Neural Networks

Daniel Halikowski[1(✉)] and Justyna Patalas-Maliszewska[2(✉)]

[1] Institute of Technical Science, University of Applied Science in Nysa, ul. Armii Krajowej 7, 48-300 Nysa, Poland
daniel.halikowski@pwsz.nysa.pl

[2] Faculty of Mechanical Engineering, Institute of Computer Science and Production Management, University of Zielona Góra, ul. Licealna 9, 65-417 Zielona Góra, Poland
j.patalas@iizp.uz.zgora.pl

Abstract. In recent years, the progress of computerised vision technology has resulted in the recognition of objects in images and video sequences regaining a high level of attention. Convolution neural networks have become one of the basic tools used for recognising patterns and tasks when classifying objects placed on images. The article presents the architecture and the principles of the operation of convolutional networks. The most popular methods for classifying objects, using SVM and the Softmax function, are presented. A model for classifying objects, using convolutional neural networks, is also proposed; this enables the employee training process to be automated in the workplaces of enterprises. An analysis of the case study, previously defined, was also carried out.

Keywords: Deep learning · Convolutional networks · Feature classification · Workstation instruction

1 Introduction

Due to the continuous turn-over of employees in enterprises, management boards expect a solution which, upon implementation, will help to improve training process. In particular, it is necessary to support the process of training employees in workstation positions in the field of the automatic building of workstation instructions, in which images, or animated instructions, can be used. For this purpose, it is necessary to build a database of images depicting operations which have been carried out correctly and then to use CNN to recognise: (1) the action performed, (2) the tool used, (3) modus operandi, (4) the element of the machine serviced. In other words, the manager of, for example, the production department of an enterprise, in accepting a new employee, organises training for him to carry out tasks that would require additional costs were an outside expert to be employed or if time had to be spent delegating an experienced employee. By using the automatic process of generating workplace instructions with the use of CNN, a company can reduce both the costs and the time associated with training employees.

J. Świątek et al. (Eds.): ISAT 2019, AISC 1051, pp. 300–309, 2020.
https://doi.org/10.1007/978-3-030-30604-5_27

Our proposed approach focusses on developing of work instructions based on the extraction and comparison of the features of objects using convolutional neural networks and it includes the following elements: (1) Defining training images from the video sequence of real performed work; (2) Defining the set of graphic instructions of workstation; (3) Acquiring training image features while using convolutional neural networks (CNN); (4) Defining an ontology; (5) Comparing training image features with the features of graphic instructions while using the ontology, (6) Defining the classes of objects appearing on the analysed images, (7) Determining the sequence of workstation instructions. The problem in this paper is how to automatically obtain a workstation instruction assignment in order to improve the effectiveness of the employee training process in companies.

In the second chapter of the present article, the subject literature has been reviewed with a view to image classification using a convolutional neural network and also a model for classification of workstation instruction images, based on the extraction and comparison of the features of objects has been proposed. The third section presents the developed ontology and an example of the implementation of the proposed model based on work activity, namely servicing a given boiler element. Finally, the general action detection scheme, based on video sequence analysis and also conclusions are presented.

2 Image Classification Using a Convolutional Neural Network

Deep learning is one of the foundations of cognitive computing. Deep learning is closely related to machine learning [1], in which the model learns to perform classification tasks directly from analysed data. Deep learning is related to deep neural networks (DNN - deep neural networks), recurrent neural networks (RNN - recurrent neural networks [13]) and convolutional neural networks [9, 11] (CNN - convolutional neural networks). DNN networks are characterised by a high degree of complexity and are generally determined as neural networks in which there are multiple, *that is, more than two*, layers. If there is feedback on the network, we term such networks RNN. The structure of this type of network also has many hidden layers, between which there are peer-to-peer connections. The weights of these connections are often initiated using pre-supervised or unsupervised learning which is based on previously processed data. Convolutional neural networks (CNN) are one of the most popular Deep Learning techniques. The architecture of this type of network is multi-layered and constitutes a set of interconnected layers that can learn classification while extracting features of the objects under examination.

In the case of image analysis, the neural network uses convolutional layers to study spatial or temporal neighbourhoods in order to construct new representations of features. A CNN network can have tens or hundreds of layers, each learning to detect different image characteristics. For each training set, transformations are used, the result of which is the input data for the next layer. The final effect of the network's operation is a set of features which facilitate the unambiguous identification of objects on the image [1].

CNN networks allow the determination of neuron weights for individual network layers; this results in the fact that individual layers will represent common, learning pattern features and on this basis, the representations of features of complex objects may be formed [2] (see Fig. 1) in successive layers of the network [2] and low-level features such as edges, lines and corners and higher-level features, such as structures, objects and shapes.

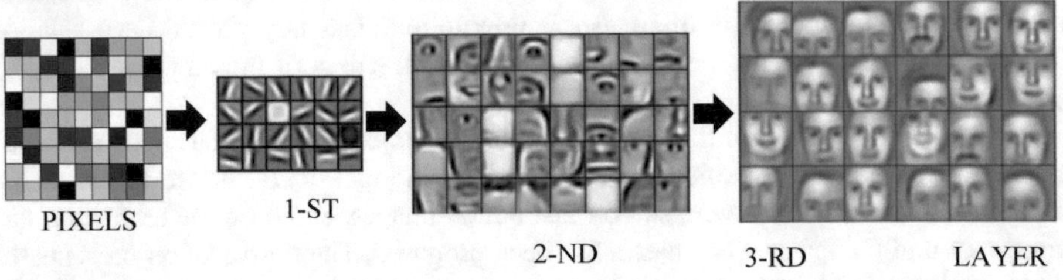

Fig. 1. Representations of features of low and high level objects (Source: https://stats.stackexchange.com/questions/114385/what-is-the-difference-between-convolutional-neural-networks-restricted-boltzm)

When using network learning methods with back error propagation, the neuron weights of individual network layers will be properly initiated and the information already processed in previous calculations will be used to tune the network.

The operation of the CNN network includes a separate part of the image and leads to the detection of edges and colours along with parts of the objects; this results, ultimately, in the recognition of the entire object. Teaching the network to recognise the object requires a large amount of learning data with collections of over 2000 pictures [1]. Each set of training images has a label assigning the object in the image to a specific class [11]. CNN image classification consists in processing the input image and finding features that assign the object, or objects, located in the image, to the appropriate category.

The architecture of the CNN network is a system of connected processing units, or neurons, whose output data is input data for subsequent processing units [1]. Each of the neurons has a weight W = $(w_1, w_2, \ldots, w_n)$ where $n \epsilon$N, input vector with the same dimension X = $(x_1, x_2, \ldots, x_n)$ where $n \epsilon$N and threshold or bias. Processing units are grouped into layers, creating subsequent layers (e.g., convolutional). In order to teach the CNN network, each input image will be forwarded through a series of convolutional layers with filters (*Kernals*), pooling layers, and fully connected layers [1].

After passing through all the layers, the probability of the class of the extracted object, based on the feature vector, can be calculated [1, 10]. This can be done by omitting the last layer of the CNN network and classifying it, using the SVM classifier (see Fig. 2) [1, 4, 5, 7–10, 12, 13, 16–18], or by creating a classification layer based on the Softmax function [1, 3, 9, 10, 12, 15] (e.g. PatreoNet, AlezNet, CaffeNet, VGG, OverFeatS, OverFeatL, GoogLeNet).

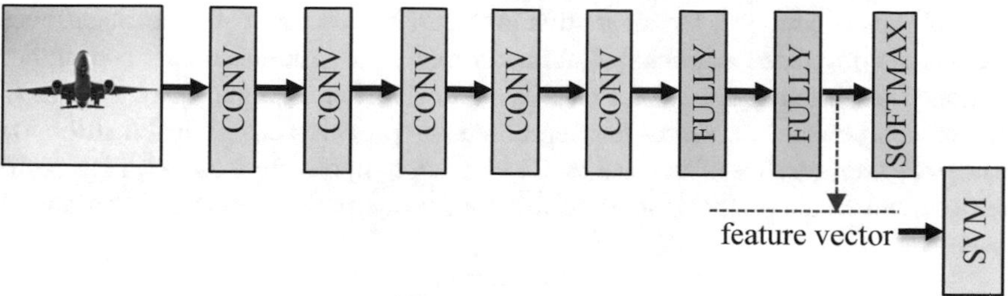

Fig. 2. CNN as feature extractor

In the case of classification by SVM, the last layer of the CNN network is omitted [9] and the pre-trained network is used as an extractor of features that will be processed further [1].

The SVM, that is, the supporting vector technique, is designed for classification and regression tasks. In the case of linearly separable data, the SVM method allows a hyperplane to be established that divides the data set into two classes with a maximum separation margin. For linearly, non-separable data, a hyperplane that classifies objects with the minimum of error can be defined [19]. When creating a hyperplane, one should also be mindful of the maximum margin of trust, or *margin of separation*, this being the distance of the hyperplane from the nearest data.

SVM has been formulated as a binary classification method. By having both the training data and the corresponding labels (m_n, l_n), $n = 1, \ldots, N$, $m_n \in R^D$, $l_n \in \{-1, +1\}$ then operating the method is based on the following formula.

$$\min_{w} \frac{1}{2} v^T v + C \sum\nolimits_{n=1}^{N} \max\left(1 - v^T m_n l_n, 0\right) \tag{1}$$

where $v^T v$ is the *Manhattan norm* (L1), C is the *penalty parameter - also termed the capacity* [9, 21] which may be an arbitrary value or a selected value, using hyper-parameter tuning), v is the vector of the coefficients; the index n numbers the N learning cases. In the above equation, bias can also be included by increasing the elements of the m_n vector through scalar 1.

Class prediction is based on the formula [15]:

$$\arg\max_l (v^T m) l \tag{2}$$

Having output k (predicted class) from SVM as:

$$a_k(m) = v^T m \tag{3}$$

The predicted class is determined by the formula:

$$\arg\max a_{k_k}(m) \tag{4}$$

As already mentioned, the alternative to the SVM classifier is the Softmax function which is a normalised exponential function that is a generalisation of a sigmoidal function. This function is used for classifying objects belonging to many classes (>2). The function generates a vector that represents the probable categorical distribution of the object's membership of the classes. The equation, upon which is based the Softmax function - and which predicts the probability of the object class- appears as follows [15].

$$p_i = \frac{\exp(a_i)}{\sum_j^{10} \exp(a_j)} \tag{5}$$

where $p_i \epsilon <0, 1>$

The prediction follows the classification:

$$\hat{i} = \arg\max_i p_i \tag{6}$$

Based on the research literature presented, the approach to instructions at the workstation and image classification, using convolutional neural networks, is formulated (Fig. 3)

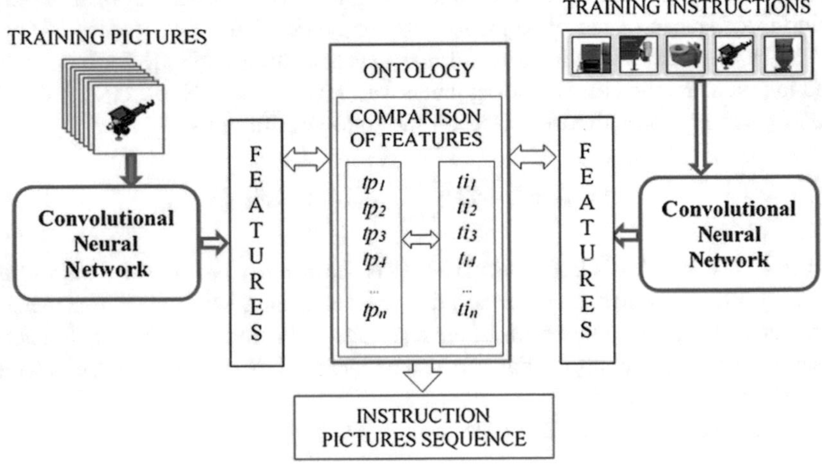

where:
tp_n - a training pictures vector element, $n \epsilon N$,
ti_n - a training instruction vector element, $n \epsilon N$

Fig. 3. An approach to the classification of workstation instruction images, based on the extraction and comparison of the features of objects

The process of image recognition and analysis is closely related to the process of video analysis, understood as a sequence of individual images. The probability of belonging to a class is calculated on the basis of the features extracted from individual frames.

The proposed approach aims to classify the features of objects, located on the processed image, in order to generate workstation instructions for the maintenance department. The use of a ontology allows to compare the features of training pictures and training instructions, which will allow to determine the sequence of workstation instructions.

In the first stage of our proposed approach the training images from the video sequence are obtained, which represent the individual stages of the performed works. Their selection is based on defined conditions for individual steps in the video sequence. In the second stage, the set of graphic instructions of workstation is defined. Next, for each image features using the convolutional neural networks (CNN) are generated (stage 3). Then, they are compared with the features of graphic instructions previously received using the defined ontology (stage 4 and stage 5). The result of the comparison allows to determine the classes of objects appearing on the analysed image (stage 6), which will allow the system to generate the appropriate graphic instructions (stage 7).

The implementation of the model will allow a scenario for service activities to be put in place at the workstation on the basis of the data collected.

3 A Case Study

The identification and classification of objects is carried out during analysis of the training material. Where it is possible to compare the features of such objects with the set of features of the model objects associated with the training process being undertaken, it is possible to analyse activities carried out, in terms of the accuracy of the training process. Operation of the computer system supporting the training process which generates the set of graphic instructions, aims to process the video sequence, in order to extract objects appearing in the material analysed; these are then compared with the set of features previously created which are related to the training process. As a result, the system will be able to analyse activities currently being performed and will be able to decide on the accuracy of the training process.

Implementation of the training process should take place according to a specific scenario of activities. The proposed model assumes that it will generate data for the algorithm, so that a scenario of conduct, for a specific workstation, will be automatically created. However, in order for the training process to proceed properly, it is necessary not only to teach the system to recognise objects, but also to remember the order in which they occur. This requires preparation and analysis of the model training material on the basis of which it will be possible to determine the activity currently being performed. The CNN prepared network should, however, be able to recognise objects contained in the processed material, and therefore must be instructed as to the features which should be looked for when analysing the image being processed. For this purpose, a set of graphic instructions should be created which will serve as a reference point for the features distinguished during the processing of the reference material. It also requires a hierarchical set of classes to be created that map objects both in the graphic instructions and in the reference material. With a system prepared in such a way, objects can be classified easily; this will allow the correlation of the classes of analysed objects of the video sequence processed, along with the set of graphic instructions, to be quickly and easily determined. This, in turn, will be the basis for decisions on the activity currently being performed. The hierarchical set of classes defined, which is the basis for ontology, is presented in Fig. 4.

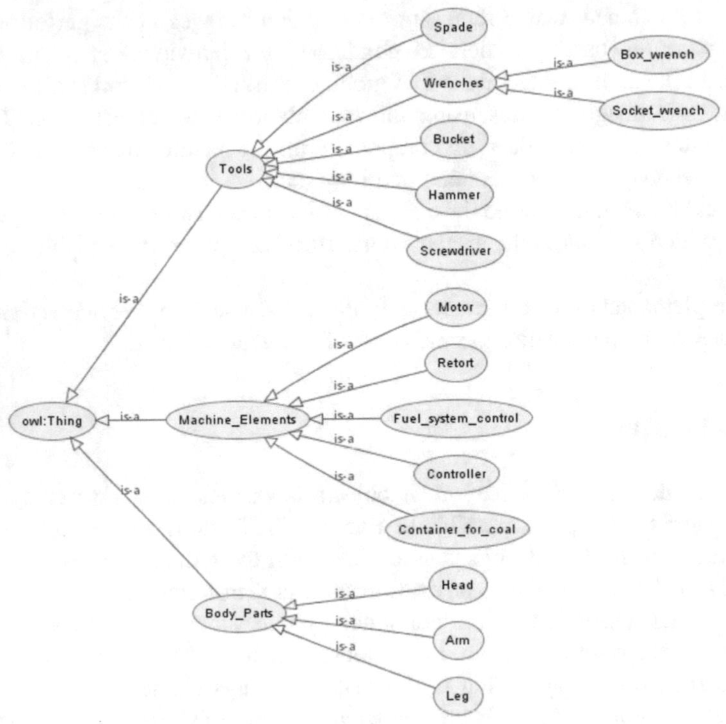

Fig. 4. Hierarchical set of classes (Protégé)

In having a defined group of classes to be the basis upon which the system can recognise and classify objects in the image, an algorithm should be created to detect activities currently being performed. The case under consideration concerns the service analysis related to the control of a solid fuel furnace retort. The reference material on which an employee performs a given activity, contains objects, based upon which, the system will decide as to the accuracy of undertaking a given activity in the future. Control of the retort servicing process requires the employee to appear on the video material himself, alongside the furnace elements and the correct tools for the job of accurately servicing a given boiler element.

In analysing the reference material in terms of the beginning and end of the service activity undertaken, it can be assumed that with the employee appearing on the registered image, the service activity has been initiated (status 1, previously status 0). The ensuing absence of the employee from the team can be taken as the end of the service activity (status 2). In this case, however, information obtained from the material registered between the two states should also be taken into account. Data on the objects recorded on the video sequence, that is, the order and the time of the occurrence of events, will be the basis for verifying the correctness of the service activity. Based on this data, the system will be able to control ongoing work, that is, it will be able to analyse registered objects either individually or in groups; it will be able to control the order in which an object or objects appear(s) and will also be able to control the appropriate stages of the work, that is, the time at which an object occurs in the frame.

An example of video sequence analysis, with the classification of objects of an event, is shown in Fig. 5.

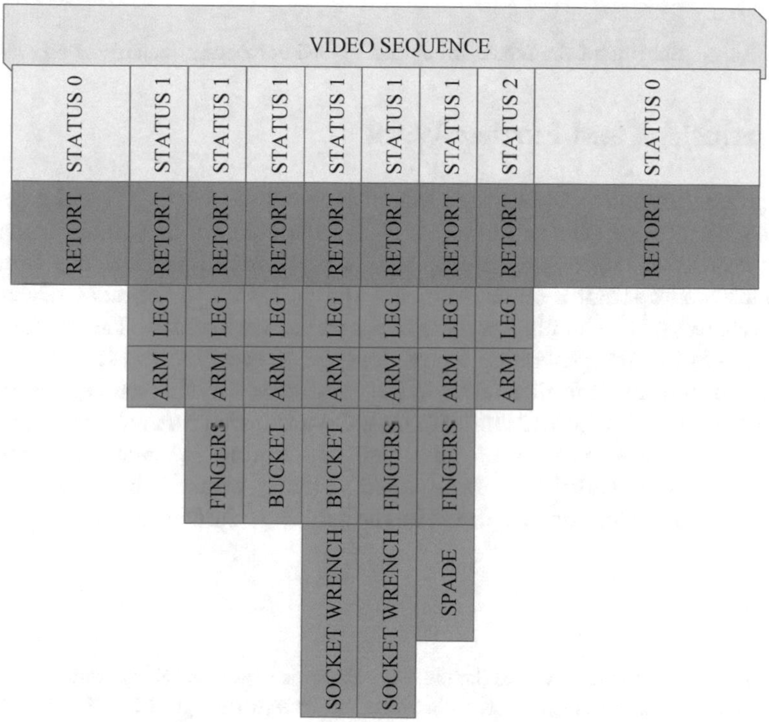

VIDEO SEQUENCE
STATUS 0 RETORT
STATUS 1 RETORT LEG ARM
STATUS 1 RETORT LEG ARM FINGERS
STATUS 1 RETORT LEG ARM BUCKET
STATUS 1 RETORT LEG ARM BUCKET SOCKET WRENCH
STATUS 1 RETORT LEG ARM FINGERS SOCKET WRENCH
STATUS 1 RETORT LEG ARM FINGERS SPADE
STATUS 2 RETORT LEG ARM
STATUS 0 RETORT

Fig. 5. An example of video sequence analysis, along with classification of the objects in the image.

The ontology created for the needs of the system is to properly represent the knowledge acquired by the system, which must be interpretable by people and computer systems as unambiguously as possible. In this particular case, it should be that set of knowledge obtained at the stage of the image processing of graphic instructions and model training material. The general action detection scheme, based on video sequence analysis, is shown in Fig. 6.

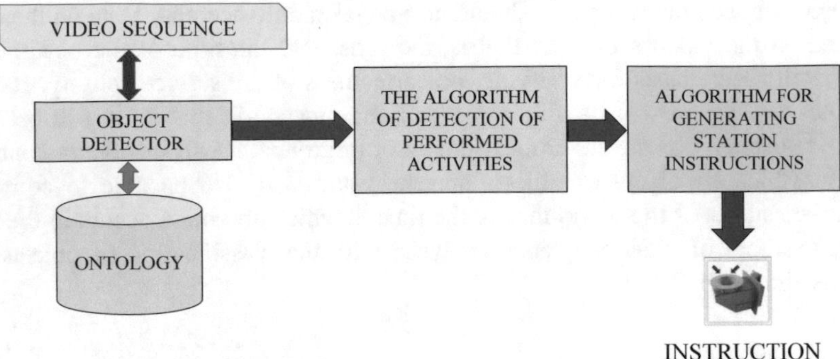

Fig. 6. Scheme of the model for extracting and comparing features of objects

4 Conclusions and Further Work

The approach presented, allows knowledge which has been acquired in the process of extracting features of objects registered during the creation of training material and graphic instructions for a given workplace, to be classified and formally stored. The formulated model is not a universal model and is limited to the areas related to the implementation of the training process at a specific workstation. The model is not a complete solution for generating instructions for a specific workstation. It will be necessary to implement further stages in the creation of a model, among which will be the formulation of an algorithm for automatically generating instructions in the form of animated film, the formulation of an algorithm allowing irregularities in tasks performed at a given workstation to be detected along with the implementation of a computer system; this, however, must be the subject of further research.

References

1. Nogueira, K., Penatti, O.A., dos Santos, J.A.: Towards better exploiting convolutional neural networks for remote sensing scene classification. Pattern Recogn. **61**, 539–556 (2017)
2. Khan, A., Sohail, A., Zahoora, U., Qureshi, A.S.: A survey of the recent architectures of deep convolutional neural networks. arXiv preprint arXiv:1901.06032 (2019)
3. Bengio, Y.: Learning deep architectures for AI. Found. Trends® Mach. Learn. **2**(1), 1–127 (2009)
4. Alexe, B., Deselaers, T., Ferrari, V.: Measuring the objectness of image windows. IEEE Trans. Pattern Anal. Mach. Intell. **34**(11), 2189–2202 (2012)
5. Harzallah, H., Jurie, F., Schmid, C.: Combining efficient object localization and image classification. In: IEEE 12th International Conference on Computer Vision, pp. 237–244. IEEE (2009)
6. Viola, P., Jones, M.J.: Robust real-time face detection. Int. J. Comput. Vision **57**(2), 137–154 (2004)
7. Kumar, M.D., Babaie, M., Zhu, S., Kalra, S., Tizhoosh, H.R.: A comparative study of CNN, BoVW and LBP for classification of histopathological images. In: IEEE Symposium Series on Computational Intelligence (SSCI), pp. 1–7. IEEE (2017)

8. Sicre, R., Awal, A.M., Furon, T.: Identity documents classification as an image classification problem. In: International Conference on Image Analysis and Processing, pp. 602–613. Springer, Cham (2017)
9. Agarap, A.F.: An architecture combining convolutional neural network (CNN) and support vector machine (SVM) for image classification. arXiv preprint arXiv:1712.03541 (2017)
10. Seo, Y., Shin, K.S.: Hierarchical convolutional neural networks for fashion image classification. Expert Syst. Appl. **116**, 328–339 (2019)
11. Yan, Z., Zhang, H., Piramuthu, R., Jagadeesh, V., DeCoste, D., Di, W., Yu, Y: HD-CNN: hierarchical deep convolutional neural networks for large scale visual recognition. In: Proceedings of the IEEE International Conference on Computer Vision, pp. 2740–2748 (2015)
12. Yu, H.J., Son, C.H.: Apple leaf disease identification through region-of-interest-aware deep convolutional neural network. arXiv preprint arXiv:1903.10356 (2019)
13. Han, H., Li, Y., Zhu, X.: Convolutional neural network learning for generic data classification. Inf. Sci. **477**, 448–465 (2019)
14. Ponti, M.A., Ribeiro, L.S.F., Nazare, T.S., Bui, T., Collomosse, J.: Everything you wanted to know about deep learning for computer vision but were afraid to ask. In: 30th SIBGRAPI Conference on Graphics, Patterns and Images Tutorials (SIBGRAPI-T), pp. 17–41. IEEE (2017)
15. Tang, Y.: Deep learning using linear support vector machines. arXiv preprint arXiv:1306.0239 (2013)
16. Bhuvaneswari, R., Subban, R.: Novel object detection and recognition system based on points of interest selection and SVM classification. Cogn. Syst. Res. **52**, 985–994 (2018)
17. Cadoni, M., Lagorio, A., Grosso, E.: Incremental models based on features persistence for object recognition. Pattern Recogn. Lett. **122**, 38–44 (2019)
18. Sharma, V.K., Mahapatra, K.K.: Visual object tracking based on sequential learning of SVM parameter. Digit. Signal Proc. **79**, 102–115 (2018)
19. Borah, P., Gupta, D.: Review: support vector machines in pattern recognition. Int. J. Eng. Technol. **9**(3S), 43–48 (2017)
20. Nagi, J., Di Caro, G.A., Giusti, A., Nagi, F., Gambardella, L.M.: Convolutional neural support vector machines: hybrid visual pattern classifiers for multi-robot systems. In: 11th International Conference on Machine Learning and Applications, vol. 1, pp. 27–32. IEEE (2012)
21. Statsoft Electronic Statistic Textbook. https://www.statsoft.pl/textbook/stathome_stat.html?https%3A%2F%2Fwww.statsoft.pl%2Ftextbook%2Fstsvm.html
22. PyImage Search. https://www.pyimagesearch.com/2016/09/05/multi-class-svm-loss/
23. Cross Validated. https://stats.stackexchange.com/questions/114385/what-is-the-difference-between-convolutional-neural-networks-restricted-boltzma

Automatic Tissue Type Classification in Large-Scale Microscopic Images Using Zernike Moments

Aneta Górniak(✉) and Ewa Skubalska-Rafajłowicz

Wrocław University of Science and Technology, Wrocław, Poland
{aneta.gorniak, ewa.skubalska-rafajlowicz}@pwr.edu.pl

Abstract. In this paper, we propose an approach in identification of histological sections of human tissue in large-scale microscopic images on the basis of sample tissue fragments from the image. The method uses pattern recognition properties of Zernike moments in the form of image descriptors consisting of sequences of Zernike moments. The goal is to acquire a robust and precise method that allows for identification of the original source of the tissue fragments in the microscopic images. The approach relies on machine learning to perform the final identification of the tissue subject from the constructed image descriptors. The method is verified by a series of experiments on a set of microscopic slides of histological sections. The results and their analysis are presented in the conclusion of the paper.

Keywords: Image processing · Microscopic image · Histological sections · Tissue images · Large-scale images · Classification · Zernike moments

1 Introduction

Biomedical image processing is a broad field of study. It finds application in many disciples ranging from diagnosis and planning to ending treatment or even monitoring disease progression [5]. The term 'medical image' itself covers a wide variety of subjects. Images can differ in the terms of methods of acquisition e.g. MR imaging systems, CT scanning or a microscope survey, or the source material e.g. a brain MRI scan, an abdominal ultrasound image or a microscopic image of pleura of the lungs [14].

Microscopic images of histological sections form one of the categories of medical images that can carry information necessary to diagnosis or the study of pathology. They can be characterized by elaborate structures both on the microscopic and macroscopic levels, from the morphology of the object in the image to the phenotype of its building cells [8].

The complex nature of histological sections and the large data format of microscopic images constitute a very difficult classification problem. For one, there is a massive amount of data to process which in turn requires a lot of computational resources and often complex and customized methods of analysis. The content of images is something that needs to be taken into account. The internal structure of histological samples can be strongly diversified. Multitudes of small cells make up

J. Świątek et al. (Eds.): ISAT 2019, AISC 1051, pp. 310–319, 2020.
https://doi.org/10.1007/978-3-030-30604-5_28

bigger formations that can vary depending both on the region of origin in the image and the object in the image. Reduction or simplification of the data in such cases may lead to the loss or distortion of the original information. There is also the possibility that the image itself may be already distorted or missing fragments. Such cases may introduce another level of complexity to the problem. Especially, when the entirety of the object cannot be used in a viable manner or the distorted fragments cause the risk of misinterpretation of the data.

Microscopic images are a research subject in many problems that focus on analysis [10, 11], classification [13], recognition [2, 4], and many more. The large scale of images often requires automatic processing of such amount of data [12].

Zernike moments make very versatile information carriers. Due to their orthogonality, their invariants can be calculated independently of high orders without the need to recalculate low order invariants. Because of that Zernike moments do not carry redundant data and the information does not get replicated in consecutive moments. Using right processing methods they can be made invariant to rotation, translation, and scaling. These inherent properties make Zernike moments suitable for image description and consequently useful in description-related problems like object classification [1, 3] or image recognition [7].

In this paper, we presents a method of microscopic image classification for histological sections of biological tissue. This approach uses only parts of all available data to solve the posed classification problem. We reduce the amount of information that needs to be processed by sampling small portions of the image. To lessen the impact of such major loss of information, we use multiple samples from random regions of the image to create a partial profile of the subject. This way the classification algorithm does not rely on a singular fragment of the image to make the classification decision. Using multiple samples lessens the risk of confusing one type of tissue with the other and even when some of the samples get misclassified the remaining ones may lead to correct conclusion. In this method, we use accumulated Zernike moments (previously tested in [3] and [4]) to construct image descriptors for the samples that later serve as the input data for our classification algorithm.

This paper is divided into following parts. Section 2 contains the definition of Zernike polynomials and Zernike moments, and the formulas that are used in discrete image processing. The following Sect. 3 presents the concept of the image descriptor built with accumulated Zernike moments. Section 4 describes the proposed approach to microscopic image classification using multiple sampled data. Section 5 covers performed experiments with discussion of results. Finally, Sect. 6 contains brief summary and conclusion.

2 Zernike Moments for Discrete Images

Zernike polynomials are a finite set of complex radial polynomials that are orthogonal in a continuous fashion over the interior of the unit disk in the polar coordinate space [9].

Let us denote the Zernike polynomial $V_{nm}(x,y)$ of order n and repetition m as

$$V_{nm}(x,y) = V_{nm}(\rho,\theta) = R_{nm}(\rho)\exp(jm\theta) \qquad \rho \leq 1, \tag{1}$$

where R_{nm} is the real part, the radial polynomial orthogonal in the unit circle and $\exp(jm\theta)$ is the complex part; ρ and θ are polar coordinates, where ρ is the length of vector from origin to (x,y) pixel, θ is an angle between vector ρ and x-axis in a counter-clockwise direction. In this equation n is a positive integer and m is a positive (and negative) integer subjected to constraints

$$m \in \{0, \pm 1, \ldots, \pm|n| \,|\, n - |m| \text{is even}\}, \tag{2}$$

The radial polynomial R_{nm} is defined as

$$R_{nm}(\rho) = \sum\nolimits_{s=0}^{(n-|m|)/2} (-1)^s \frac{(n-s)!}{s!\left(\frac{n+|m|}{2} - s\right)!\left(\frac{n-|m|}{2} - s\right)!} \rho^{n-2s}, \tag{3}$$

where $R_{nm}(\rho) = R_{n(-m)}(\rho)$.

If we assume that $\mathrm{f}(\rho,\theta)$ is the image intensity function stretched over the unit circle, then the two-dimensional complex Zernike moment of order n and repetition m is defined as

$$A_{nm} = \frac{n+1}{\pi} \int_0^{2\pi} \int_0^1 f(\rho,\theta)[V_{nm}(\rho,\theta)]^* d\rho d\theta, \quad \rho \leq 1 \tag{4}$$

where $[V_{nm}(x,y)]^*$ is the complex conjugate of Zernike polynomial $V_{nm}(x,y)$ that follows $[V_{nm}(x,y)]^* = V_{n(-m)}(x,y)$.

A discrete two-dimensional Zernike moment of a computer digital image $\mathrm{f}(x,y)$ is defined as [9]

$$A_{nm} = \frac{n+1}{\pi} \sum\nolimits_x \sum\nolimits_y f(x,y)[V_{nm}(x,y)]^*. \tag{5}$$

We assume that the point of origin of $\mathrm{f}(x,y)$ lays in the center of the image and that the area boundary of $\mathrm{f}(x,y)$ is removed from the center symmetrically in every direction θ along the x- and y-axis. It is necessary to convert the discrete intensity function $\mathrm{f}(x,y)$ from the Cartesian coordinate system into the polar coordinate system in numerical calculations.

3 Accumulated Zernike Moments in Image Description

In this section, we establish the image descriptor that uses sequences of accumulated Zernike moments. This method was previously mentioned in [3] and [4]. Because of the properties of Zernike moments in digital image processing, it is assumed that images containing similar data have similar descriptors. This property can be used to solve such problems as identification, recognition or classification.

The image descriptor is a sequence of accumulated Zernike moments that can be presented as a vector of the size $n_{max} \times 1$, where n_{max} is the defined maximum order of the calculated Zernike moments. Each component of the vector contains the absolute accumulated value of all Zernike moments of order n, where $n = 1, \ldots, n_{max}$. The formula for the image descriptor D is

$$D = [Z_1, Z_2, \ldots, Z_{n_{max}}]^T, \tag{6}$$

where components Z_i are accumulated values of Zernike moment of the order i

$$Z_i = \left| \sum\nolimits_j A_{ij} \right|. \tag{7}$$

To calculate the singular Zernike moment the formula from (5) was used.

One of the important properties of Zernike moments is that each moment carries a unique piece of information pertaining to the image. The subsequent moments do not stack up data from the preceding moments of the lower order. This property allows Zernike moments to be calculated and used independently.

By accumulating Zernike moments we reduce the amount of information to process without duplicating the data itself or losing too much of the data quality. This approach retains part of the properties of Zernike moments and allows for faster data use, but it renders the reconstruction of the image from the accumulated data impossible.

4 Proposed Method of Image Classification

The proposed method utilizes sequences of accumulated Zernike moments as image descriptors (as mentioned in Sect. 3). They are used as the input data for the classification algorithms. Since the concept uses multiples samples for identification, the classification algorithms needs to be run for each data sample. Results from each classification are summed into one output vector that is processed by the softmax function. The final decision is made depending on the value of the acquired distribution.

The purpose of using multiple samples from the same source to make the final decision on the identification of the subject is to strengthen the decision-making process and to lessen the risk of misclassification that comes from testing on singular samples. Biological tissue is a complex structure which makeup vary not only between different types of tissue, but also within itself. Using only part of the tissue for identification may lead to wrong classification results. Increasing the number of samples that potentially may come from different regions of the tissue allows for a more informed decision by building a semi-profile of the subject in question. Moreover, this approach does not need to take into account the entirety of the tissue, but only small samples of it. It is very useful when parts of the tissue in the image are distorted, discolored or missing, thus making the classification process much more difficult or even impossible to make.

The implementation of the proposed method can be broken down into following steps:

1. Calculate sequences of accumulated Zernike moments for the input images.
2. Run classification algorithm for each Zernike sequence.
3. Sum the outputs from each classification into one output vector.
4. Run softmax function for the output vector.
5. Make the final classification decision.

In the first step, the sequence of accumulated Zernike moments that serves as the image descriptor needs to be calculated for each sample image that is being used in this classification. For the sake of the load and the speed of numeric calculations of Zernike moments, sample images are resized to the size of 120 × 120 px. For this experiment we set the maximum value of the order of the Zernike moment to $n_{max} = 40$. The moment A_{nm} of order n and repetition m is calculated according to the definition from (5). The image descriptor D is calculated according to the formula from (6) and (7).

In the second step, we run classification algorithms with the calculated Zernike sequences as the input data. The output of the classifier is a vector that contains the probabilities of the entry belonging to each of the known classes. We can denote this output as y and

$$y = [y_1, y_2, \ldots, y_K], \tag{8}$$

where K is the number of classes in this experiment, and each y_i is the probability value of the input belonging to the class i.

Classification algorithm uses build-in classification tools provided by *Mathematica* software. In this experiment, we apply three types of classifiers from this toolbox: Neural Networks (NN), Support Vector Machine (SVM) and Logistic Regression (LR).

In the next step, we combine the results of classification for multiple samples into one output vector $\bar{y}$ and

$$\bar{y} = [\bar{y}_1, \bar{y}_2, \ldots, \bar{y}_K], \tag{9}$$

where K is the number of available classes. Let $\bar{y}_i$ be the output value that is the sum of probability values for the class i from all sample images in classification, then

$$\bar{y}_i = \sum\nolimits_j^M y_{ij}, \tag{10}$$

where y_{ij} is the output of the j-th classification for the class i and M is the number of outputs used to calculate the accumulated result. The value of M equates to the number of samples used for this classification decision.

The softmax function S(·) from the fourth step is used to get the final value of the output for the classification instance that is being solved. Its form is as follows

$$S(\bar{y}_i) = \frac{\exp(\bar{y}_i)}{\sum_j \exp(\bar{y}_j)} \tag{11}$$

where $\bar{y}_i$ is the classifier's accumulated output for the class i. The output of the softmax function is the distribution of probability of classification to each class that sums to 1 for the current batch of input data [6].

The final conclusion is derived from the last output $\bar{y}$ which combines the classification results from multiples samples, filtered through the softmax function. Its purpose is to increase the chance of correct classification by increasing the pool of data that influences the decision. In the case of complex patterns, a category that histological sections fall in, even with some of the samples classified wrong, the method can still make the correct classification.

The decision criterion for the classification results is

$$\max_i S(\bar{y}_i) \tag{12}$$

The samples belong to class i that has the highest probability of classification within the output vector $\bar{y}$.

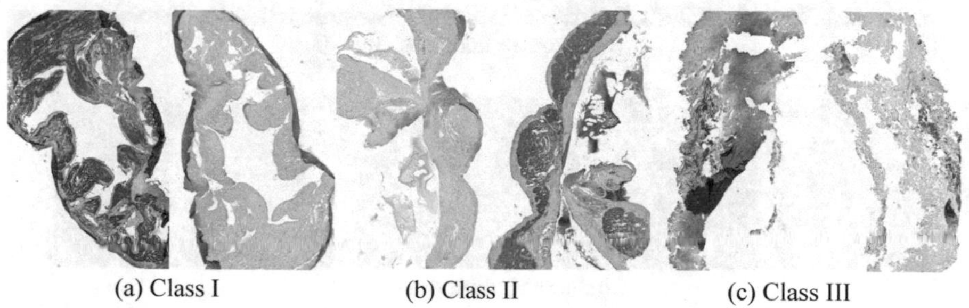

(a) Class I (b) Class II (c) Class III

Fig. 1. A representative sample of two images from each class of the microscopic image series used in the experiments.

5 Experiments and Discussion

In this experiment we test the proposed method of classification of histological sections in large-scale microscopic images. The considered approach covers classification on the basis of randomly sampled fragments of the tissue from the slides. The goal is to identify the subject in the image using only partial data from the image. It is important to note that the samples used in classification do not necessarily need to come from the same region of the image. However, it is advisable for them to contain a piece of the tissue. This approach allows for the classification of subjects that have missing fragments of their bodies or are partially distorted in some way.

The data set used in the experiment consists of microscopic images of human tissue. The source of data are image series of microscopic slides of histological sections (seen in Fig. 1). All sample images are obtained from the image series. The samples are cut out from random regions across multiple subjects of the same series of slides to provide a varied representation of the same tissue type. A representative group of data is shown in Fig. 2. As it can be seen in the examples, even if the images come from the same series their contents vary: from the structure of histological content, shape of the object to the intensity of the image. This diversity makes it difficult to identify the origin source of the tissue with standard methods.

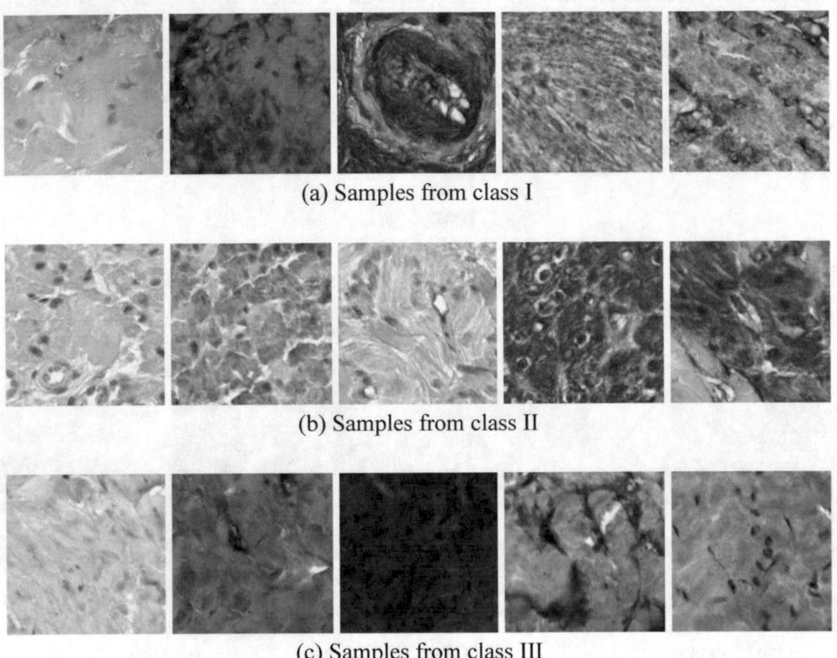

(a) Samples from class I

(b) Samples from class II

(c) Samples from class III

Fig. 2. Examples of image samples for each of the microscopic image series.

The source images of histological sections are stored in big data format ($\sim$1.5 GB). The size of the TIFF image file ranges from 15 000 px to 23 000 px in height and width, making the direct processing of these images a very resource- and time-consuming operation. Therefore, we resort to sampling of these images. Obtained sample images are saved in grayscale format, where the pixel value ranges from 0 (black) to 1 (white). The images are not normalized in any way. The size of the sample image is 400×400 px, which roughly covers from 0,08% to 0,11% of the source image.

We use three image series in this experiment that we assign to three classes. The classes are numbered in Roman numerals from I to III. The number of samples in each class is 800. The size of the training set is 75% of all samples. Testing was performed on the remaining 25% of the samples.

The experiment consists of multiple test runs with customizable parameters. These parameters are the type of the classifier and the number of sample images to use as the input data for the classification algorithm. We use three types of classifiers: logistic regression (LR), neural network (NN) and support vector machine (SVM). The number of input images starts at 1 and ends up at 10. There are 30 combinations of possible test runs: LR with 1, 2, 3, 4 to 10 input samples, NN with 1, 2, 3, 4 to 10 input samples, and SVM with 1, 2, 3, 4 to 10 input samples. The sample images are picked randomly for each test run. Therefore, the input data may represent one region of the image or come from completely different parts of the tissue. The order in which the samples are picked does not influence the final outcome since it is an accumulated value.

The goal of this experiment is to ascertain the number of input images that needs to be used to obtain a high probability of correct classification for the presented image series. The results are shown in Tables 1, 2 and 3.

Table 1. Classification results for each of the classes using LR classifier with varying number of input samples.

Cl.	Number of input samples used in classification									
	#1	#2	#3	#4	#5	#6	#7	#8	#9	#10
I	0.4200	0.5250	0.5587	0.6000	0.6375	0.6300	0.6387	0.7200	0.6975	0.7875
II	0.6088	0.7025	0.7725	0.8100	0.8125	0.8625	0.8488	0.9000	0.8888	0.9375
III	0.7462	0.8300	0.9187	0.9200	0.9563	0.9750	0.9800	1.0000	0.9788	1.0000

Table 2. Classification results for each of the classes using SVM classifier with varying number of input samples.

Cl.	Number of input samples used in classification									
	#1	#2	#3	#4	#5	#6	#7	#8	#9	#10
I	0.5587	0.5900	0.6600	0.7250	0.7250	0.7650	0.7700	0.8500	0.8100	0.9250
II	0.6312	0.7300	0.7950	0.8250	0.8625	0.9000	0.9012	0.9300	0.9225	0.9375
III	0.7113	0.8150	0.8775	0.9100	0.9313	0.9450	0.9625	0.9900	0.9788	1.0000

Table 3. Classification results for each of the classes using NN classifier with varying number of input samples.

Cl.	Number of input samples used in classification									
	#1	#2	#3	#4	#5	#6	#7	#8	#9	#10
I	0.5250	0.6325	0.6563	0.7100	0.7375	0.7725	0.7700	0.8100	0.7988	0.8375
II	0.5625	0.6625	0.6862	0.7400	0.7625	0.8100	0.8050	0.8200	0.8438	0.8750
III	0.7738	0.8475	0.9000	0.9300	0.9375	0.9675	0.9800	0.9800	0.9675	0.9750

In this experiment, we try to establish how many samples are necessary to obtain a correct classification decision for each of the classes. As seen in Tables 1, 2 and 3, the probability of correct classification rises with every additional sample. The initial classification results for a single sample range from 42% to 77% depending on the class and the classification method that is used. The lowest score is achieved with LR classifier for class I and the highest one with NN classifier for class III. Adding more samples to the decision-making process increases the probability of correct classification as we assumed in the beginning of the paper. The increase is linear and showcases that we can improve the effectiveness of the classification algorithm by using more samples as the input data. With the maximum number of samples, that is 10, the classification achieves the range from 79% to over 90%. The worst score of 79% is obtained for LR classifier and class I, and the best score of 100% for SVM classifier with class III.

The end results depend heavily on the applied classifier. LR starts with the 42–72% range of correct classifications and ends up with the 79–100% range (Table 1). For SVM it is the 56–71% range in the beginning that ends up with the 92–100% range (Table 2). And for NN it is 53–77% starting range and the 84–100% ending range for the correct classification (Table 3). As the results also show the other deciding factor are the image series. Using Fig. 2 as the reference, we can see that the class I is the most diverse out of all classes and the final results reflect that. The range of correct classification for this class is 79–93%. The most uniform class appears to be class III, where the range is 98–100%. Class II falls in the middle of the batch since its samples vary, but are not as diverse as in class I. The classification results for this class are in the range of 88–94%.

Depending on the number of samples that is used and the size of the source image we have utilized from about 0,8% to 1,1% of the original image to solve this classification problem.

6 Conclusion

In this paper we presented an interesting approach to large-scale image classification for histological images. In this method we used sequences of accumulated Zernike moments to construct image descriptors. This way we greatly reduced the number of input data for the classification algorithm (from the size 400×400 px of the sample image to the vector of 40 elements). We took advantage of the pattern recognition properties of Zernike moments and applied them to the problem of image classification. In order not to use the entirety of the stored data we resorted to sampling. We used multiple samples from the source image to build a profile of the image that later served as the input value for the classification algorithm. To test this approach, we performed a series of experiments with different types of classifiers (from *Mathematica* toolbox) and for different number of input samples.

In the end, the approach proved to be successful. As the results show, we achieved the correct classification threshold that starts in the lower 90% for most of the test cases. This method promises to be easily scalable as both the number of samples and the size of the image can be modified. There is also no restriction to the size of the source image, therefore it can easily be applied to more complex classification problems with larger image data. Because of the inclusion of Zernike moments in the method, the subject of the classification can potentially cover disciplines other than microscopic imaging that also require a strong pattern recognition feature and involve a varied image patterns like aerial glocalization or even face recognition.

Acknowledgements. The authors express their thanks to Dr. Agnieszka Malińska and professor Maciej Zabel from Poznań University of Medical Science, Poland for the microscopy section images.

References

1. Athilakshm, R., Wahi, A.: Improving object classification using Zernike moment, radial Cheybyshev moment based on square transform features: a comparative study. World Appl. Sci. J. **32**(7), 1226–1234 (2014). https://doi.org/10.5829/idosi.wasj.2014.32.07.21861
2. Fernandez, D.C., Bhargava, R., Hewitt, S.M., Levin, I.W.: Infrared spectroscopic imaging for histopathologic recognition. Nat. Biotechnol. **23**(4), 469 (2005)
3. Górniak, A., Skubalska-Rafajowicz, E.: Object Classification using sequences of Zernike moments. In: Saeed, K., Homenda, W., Chaki, R. (eds.) Computer Information Systems and Industrial Management 2017. Lecture Notes in Computer Science, vol. 10244, pp. 99–109 (2017). https://doi.org/10.1007/978-3-319-59105-6_9
4. Górniak, A., Skubalska-Rafajłowicz, E.: Tissue recognition on microscopic images of histological sections using sequences of Zernike moments. In: IFIP International Conference on Computer Information Systems and Industrial Management, pp. 16–26. Springer, Cham (2018)
5. Hill, D.L.G., Batchelor, P.G., Holden, M., Hawkes, D.J.: Medical image registration. Phys. Med. Biol. **46**(3), R1 (2001). S0031-9155(01)96876-9
6. James, G., Witten, D., Hastie, T., Tibshirani, R.: An Introduction to Statistical Learning. Springer, New York (2013)
7. Khotanzad, A., Hong, Y.H.: Invariant image recognition by Zernike moments. IEEE Trans. Pattern Anal. Mach. Intell. **12**(5), 489–497 (1990). https://doi.org/10.1109/34.55109
8. Ourselin, S., Roche, A., Subsol, G., Pennec, X., Ayache, N.: Reconstructing a 3D structure from serial histological sections. Image Vis. Comput. **19**(1–2), 25–31 (2001). https://doi.org/10.1016/s0262-8856(00)00052-4
9. Pawlak, M.: Image Analysis by Moments: Reconstruction and Computational Aspects. Oficyna Wydawnicza Politechniki Wrocławskiej, Wrocław (2006)
10. Saitou, T., Kiyomatsu, H., Imamura, T.: Quantitative morphometry for osteochondral tissues using second harmonic generation microscopy and image texture information. Scientific Reports **8**(1), 2826 (2018)
11. Shang, C., Daly, C., McGrath, J., Barker, J.: Analysis and classification of tissue section images using directional fractal dimension features. In: Proceedings 2000 International Conference on Image Processing (Cat. No. 00CH37101), vol. 1, pp. 164–167. IEEE (2000)
12. Vaidehi, K., Subashini, T.S.: Automatic classification and retrieval of mammographic tissue density using texture features. In: 2015 IEEE 9th International Conference on Intelligent Systems and Control (ISCO), pp. 1–6. IEEE (2015)
13. Vu, Q.D., Graham, S., Kurc, T., To, M.N.N., Shaban, M., Qaiser, T., Koohbanani, N.A., Khurram, S.A., Kalpathy-Cramer, J., Zhao, T. and Gupta, R.: Methods for segmentation and classification of digital microscopy tissue images. In: Frontiers in Bioengineering and Biotechnology, vol. 7 (2019)
14. Zitova, B. and Flusser, J.: Image registration methods: a survey. In: Image and vision computing, vol. 21(11), pp. 977-1000. Elsevier (2003). https://doi.org/10.1016/s0262-8856(03)00137-9

Enhanced Hierarchical Multi-resolution Imaging

P. Joy Prabhakaran(✉)

K S Institute of Technology, Kanakapura Road, Bangalore 560109, India
joy_prabhakaran@yahoo.com

Abstract. Same image and video content are often viewed at different resolutions and over networks having different bandwidths. For these conditions, we propose an Enhanced Hierarchical Multi-Resolution (EHMR) image model that uses spatial hierarchy. EHMR enables high quality image representation of different resolutions as a single bundle. Each level is clearly separable. Higher resolutions use information from lower resolutions. EHMR supports trade-off between quality and compression can be achieved. Higher resolution images are encoded using low resolution versions and a difference component. The model is enhanced by a set of optional tools of varying complexity that can be used to improve compression. The tools include a novel use of dictionary coding, a method to eliminate one bit while representing a difference, a two-step encoding of the difference image where the first step encodes key pixels and a method to implement rounding in a biased manner that improves compression without impacting error magnitude. Some of these tools are computation intensive and can be used when the additional compression is required.

Keywords: Multi-resolution · Image compression · Spatial hierarchy

1 Introduction

With increased volumes of image and video being viewed over different networks and on devices with varying form factors, a hierarchical image model that is optimized for this scenario is useful. This paper proposes such an image model.

Various hierarchical models have been proposed and tried. Much before the voluminous, internet-based image or video consumption happened, the JPEG [1] standard had included a hierarchical mode. The Enhanced Hierarchical Multi-Resolution (EHMR) we propose gives better compression. EHMR can be used to compress frames of a video. At the lowest resolution, we use DCT based image compression because popular codecs like H.264 [2], MPEG-4 [3] and VP9 [4], use DCT. EHMR also has a High Efficiency version EHMR-HE that gives higher compression efficiency. The higher efficiency comes at the cost of increased computation complexity.

EHMR stores images in a hierarchy of different resolutions. A Low Resolution (LR) image is an intrinsic part of the representation of the High Resolution (HR) image. The lowest resolution can be decoded without using information from any other resolutions. Compression could be lossy or lossless. An earlier work [5] of ours discussed lossless compression. The focus of this paper is lossy compression.

J. Świątek et al. (Eds.): ISAT 2019, AISC 1051, pp. 320–329, 2020.
https://doi.org/10.1007/978-3-030-30604-5_29

2 Features of the EHMR Image Model

An ideal, hierarchical multi-resolution image model should have the following features for it to be useful in the scenarios we are interested in:

1. Clearly separable data for each level.
2. Each resolution level should be independent of all higher resolutions.
3. Higher resolutions should use information from lower resolution.
4. Configurable quality levels.

The EHMR model tries to achieve these by using the following methods:

1. Sign elimination in difference.
2. Quantization of difference.
3. Dictionary coding.
4. Two-step representation of difference.
5. Biased rounding of difference.

Each of these will be discussed in the following sections where the encoding and decoding processes are described.

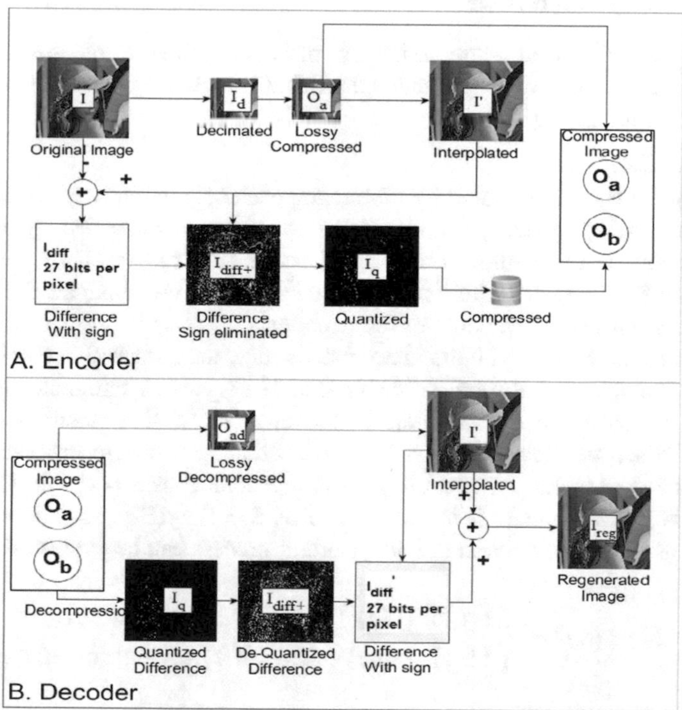

Fig. 1. The EHMR encoder and decoder.

3 The EHMR Encoder and Decoder

The EHMR image model encapsulates the same image at different resolutions. In the lossy mode, the lowest resolution in the hierarchy can be stored using JPEG or any other lossy format. We use JPEG because DCT based compression is commonly used in image and video compression.

Each of the higher resolutions are encoded using the immediate lower level and difference component. The latter is got by subtracting a HR version, generated by interpolating the LR version, and the HR version that is to be encoded. The LR version along with the difference defines the HR version.

Figure 1 shows the encoder and decoder blocks of the EHMR architecture. In this model, the image to be encoded is first decimated. The decimated image is compressed and this forms one component Oa of the EHMR representation. The decimated image is also interpolated. The difference between the interpolated image and original image is transformed as described in the next section. The processed difference, Ob, is another component of the EHMR representation. This implements two levels of hierarchy. For more levels, the LR version of one level will become the input to the encoder to generate the lower level.

3.1 Enhanced Compression

EHMR achieves enhanced compression by introducing three steps into the encoding pipeline. These are the "Difference Sign eliminated" block, the Quantized block and the Compressed block in Fig. 1.

Sign Elimination

The difference image is generated by subtracting each color component of each pixel in the HR image being encoded, from the HR version that is generated by interpolating the LR version. If each color component is of 8 bits, the result of the difference operation is a 9-bit number where one bit is for the sign. We eliminate the sign bit by using information that is available to the decoder.

If S is the source data and P the interpolated value, the error is E = P − S. If S is an 8-bit value, E can range between −255 and 255. The sign bit is eliminated by using the fact that P is known to the decoder. In the encoder, if E is positive we leave it unchanged. Else, we replace it with E + 256, which will be greater than P. During decoding, if E is less than or equal to P, S is regenerated as S = P − E. On the other hand, if E is greater than P, it is regenerated as S = P − (E − 256). For each color component at row r and column c, this encoding process can be expressed as follows:

$$I_{diff+}[r, c] = \begin{cases} I'[r, c] - I[r,c] & \text{if } I'[r, c] - I[r, c] \geq 0 \\ I'[r,c] - I[r,c] + 256 & \text{if } I'[r, c] - I[r,c] \geq 0 \end{cases}$$

where, $I[r, c]$ is a color component in the original image, $I'[r, c]$ is from the interpolated image and $I_{diff+}[r, c]$ is the resultant sign eliminated difference.

The corresponding decoding process is:

$$I_{diff}[r, c] = \begin{cases} I_{diff+}[r, c] & \text{if } I_{diff+}[r, c] \le I'[r, c] \\ I_{diff+}[r, c] - 256 & \text{if } I_{diff+}[r, c] > I'[r, c] \end{cases}$$

where, $I_{diff}[r, c]$ is the original difference value with both sign and magnitude.

Quantization
The difference image is quantized in this step. This provides a mechanism by which a quality versus compression tradeoff can be made. Additionally, a higher quantization step size results in the subsequent compression step becoming more effective. In the block diagrams, the quantized difference image is represented as I_q.

Compression
In this step, the difference image is compressed using a chosen compression technique. The difference image has many small values. This makes it a good candidate for compression. We explored different compression techniques and we found that Dictionary coding [6], using LZ77, worked best among the methods analysed. The output of this stage is represented as O_b in the block diagram.

Analysis of the Transformations in the Method
The input image goes through different transformations in the encoder and decoder. Figure 2 shows the information flow in EHMR in terms of histogram-based entropy and standard deviation. The sizes of the raw, decimated and difference images shown are computed assuming optimal entropy coding. We observe the following:

- The entropy and standard deviation of the source and decimated images are approximately equal, while these are much lower in the difference image.
- Though the output has data for 25% more pixels, it has a smaller size than the entropy coded raw image.

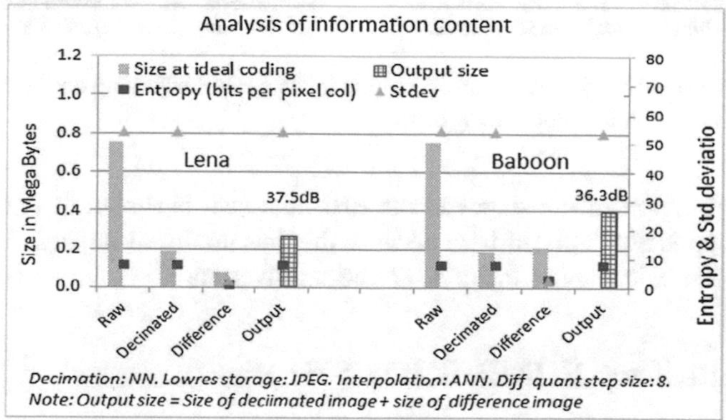

Fig. 2. Stage by stage variation in size, entropy and standard deviation.

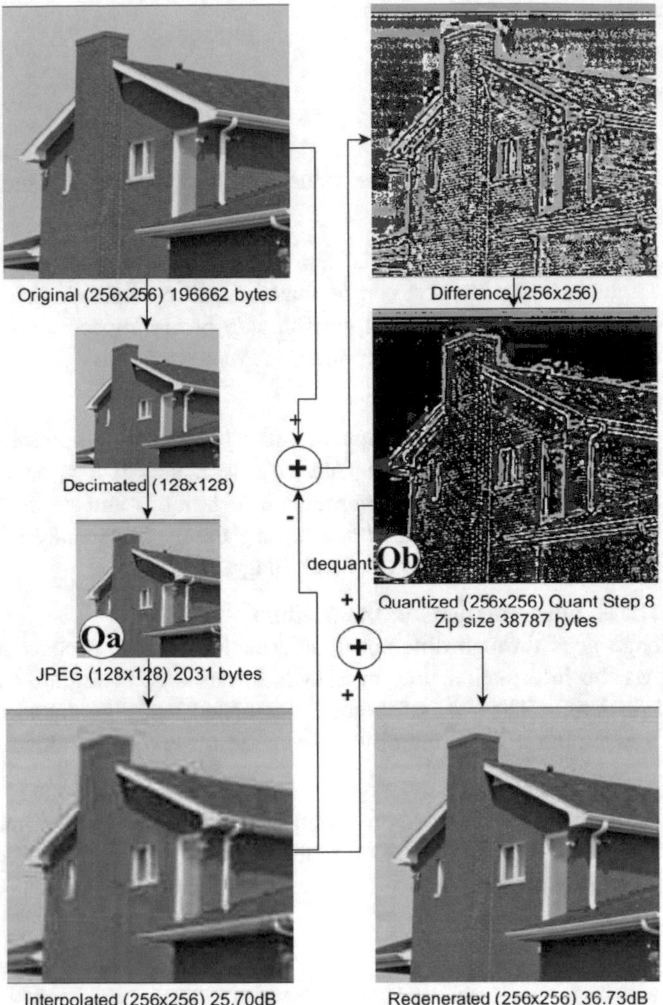

Fig. 3. A depiction of the different transformation in the EHMR encoding and decoding process. The sign elimination step is not shown here.

The visual impact of the stages in the EHMR model is shown in Fig. 3. The sign elimination step is not depicted here because that has no visual impact. An analysis of the compression efficiency is included in the results section.

4 High Efficiency EHMR (EHMR-HE)

When bandwidth is a major constraint, higher compression at the cost of additional computation may be desirable. We have developed two new algorithms that improve compression in EHMR. This is at the cost of increased complexity. When EHMR uses

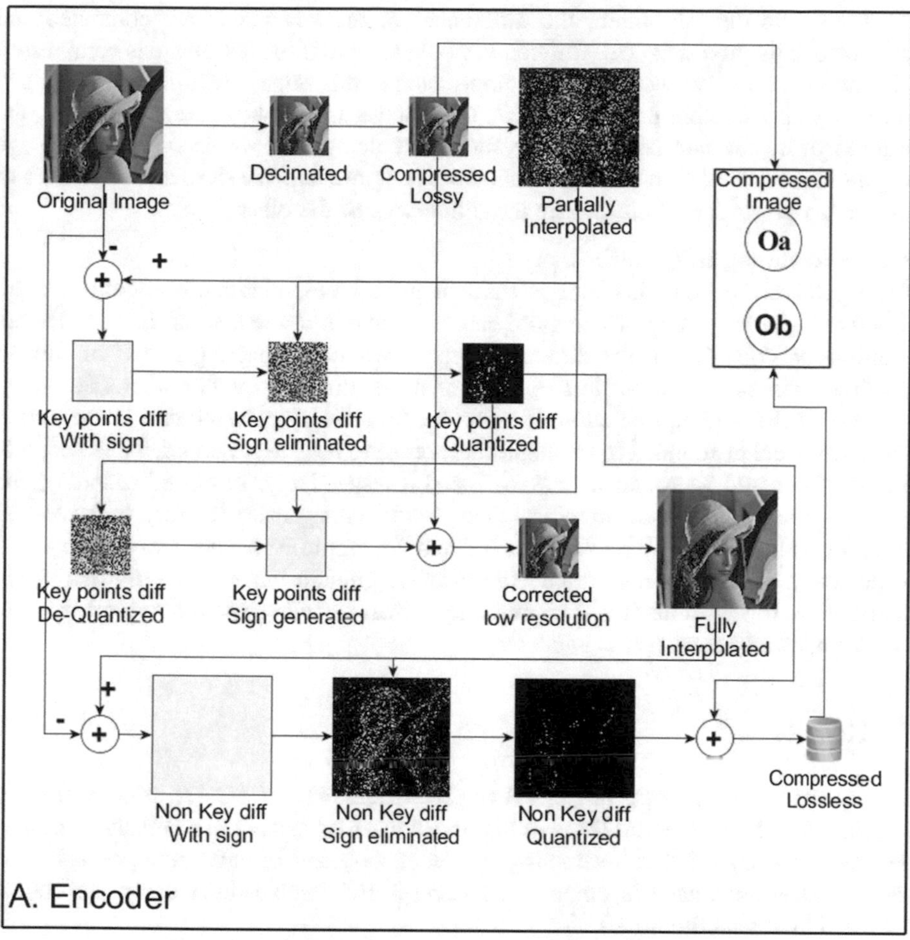

Fig. 4. The EHMR-HE architecture. Here the quantization incorporates biased rounding.

these tools, we refer to it as EHMR-HE. The block diagram of EHMR-HE is shown in Fig. 4. The new algorithms are described below.

4.1 Two-Step Difference Computation

When an image is interpolated, some locations in the interpolated image get the same value as some corresponding location in the LR image. The locations depend on scale factor and interpolation method. It is independent of the pixel's value. Based on this, pixels in the interpolated image can be classified as the follows:

Key pixels - Pixels that get their value directly from the low-resolution image.
Non-Key pixels – All pixels that are not Key pixels.

In the two-step algorithm, the differences of the key pixels are computed and quantized in the first step. This difference would be zero if the LR image is compressed with no loss, however we use lossy compression at this stage. The difference is added to the key pixels before the interpolation to compute the non-key pixels. After this, the differences for the non-key pixels are found. By doing this, we decrease the average magnitude of these differences. This enhances compression. We store the differences of all the key pixels first, followed by the differences of the other pixels.

Biased Rounding in Quantization
The hypothesis behind this tool is that, in most cases, dictionary coding is more effective if the frequency of a symbol can be increased at the cost of another. Biased quantization does this in the difference image, without impacting PSNR or sum of absolute differences. For our test set, it is seen that this helps with compression.

In normal rounding operation, if a number is mid-way between two integers, it is consistently either rounded to the higher integer or to the lower integer. Thus, 4.5, 5.5, 6.5 and 7.5 could be rounded to 5, 6, 7 and 8 respectively. In biased rounding, the floating-point numbers that are midway between two integers are rounded to the nearest even (or odd) number. Thus, 4.5, 5.5, 6.5 and 7.5 would be rounded to 4, 6, 6 and 8 respectively. This does not change the error magnitude but changes the frequency distribution. Biased rounding is identical to normal rounding when the number is not exactly midway between two integers.

5 Results

The results presented here are based on 16 images from USC [7], Kodak [8] and Hlevkin [9]. The images are listed in the first column of Table 1. The results presented here were got by using sub-sampling for decimation and bi-linear interpolation. The lowest resolution image is compressed using JPEG implemented using the codecs provided in Visual Studio 10.

5.1 EHMR with Two and Four Levels of Resolution Hierarchy

Table 1 shows the performance of EHMR when the image is encoded with two and four levels of hierarchy. Here JPEG quality is 95 and quantization step size 16.

As an example, the Lena image when encoded with four levels has a total size of 77,281 bytes and is 9.8% of its uncompressed size. This consists of four entities. The first is the image at the lowest resolution, at a scale factor of 0.125. This is a JPEG image and can be decoded independently. Since the original Lena image is of resolution 512×512, this LR image is of size 64×64. Table 1 shows that this component has a PSNR of 34.24 dB. The three other components are three dictionary-coded, difference images. The first difference image is of resolution 128×128. The 64×64 component is interpolated to size 128×128 and the first difference component is added to it to give the regenerated 128×128 version of the Lena image. The regenerated version has a PSNR of 32.2 dB. The regenerated image is interpolated and added to the next difference component to give the 256×256 version of the original, and so on. The PSNR computation for the lower resolutions is by comparing against images generated by decimating the original image to the corresponding size.

From Table 1, we see that on an average, moving from two to four levels increases the average size fraction by 2.4% and the PSNR in the HR image drops by about 5%.

Table 1. EHMR's performance with different number of resolution hierarchy-levels. The size fraction is the size of the EHMR bundle divided by the size of the uncompressed image.

	4 resolution levels					2 resolution levels		
	PSNR (dB)				Size fraction	PSNR (dB)		Size fraction
Scale ->	0.125	0.25	0.5	1		0.5	1	
lena	34.24	32.2	31.91	31.81	9.8%	34.02	33.92	7.9%
baboon	30.43	30.42	30.09	30.14	29.0%	27.54	30.64	25.7%
peppers	30.02	31.24	31.04	31.21	12.5%	34.41	33.18	9.1%
airplane	32.58	32.46	32.25	32.44	9.1%	32.13	34.91	7.5%
house	30.95	31.34	31.19	31.22	11.7%	30.37	33.27	8.9%
splash	32.07	32.19	31.82	31.93	6.5%	34.54	35.56	5.3%
jellybeans	29.52	32.77	32.81	33.02	10.3%	32.52	35.45	7.9%
car	32.48	31.67	31.42	31.54	13.6%	31.01	33.01	11.4%
sailboat	30.99	30.93	30.78	30.86	18.6%	29.22	31.83	15.7%
san_diego	33.73	30.59	30.22	30.48	21.1%	31.26	31.11	18.9%
earth	34.55	31.32	31.1	31.4	10.9%	33.25	32.88	8.5%
kodim23	33.29	33.55	33.12	33.12	5.8%	37.78	37.04	4.6%
tree	31.19	30.86	30.64	30.79	20.7%	30.47	31.99	17.3%
monarch	32.65	32.56	32.33	32.37	9.3%	35.76	35.73	6.8%
barbara	33.14	31.41	31.01	31.06	14.5%	34.2	33.06	11.6%
goldhill	35.97	31.76	31.29	31.4	10.8%	34.44	33.11	9.2%
Average	**32.36**	**31.70**	**31.44**	**31.55**	13.4%	**32.68**	**33.54**	**11.0%**

5.2 Impact of the High Efficiency Tools

In this section, we compare EHMR-HE, JPEG and JPEG2000.

PSNR vs. Compression

Figure 5 shows the average PSNR vs. compression for JPEG, JPEG2000 and EHMR-HE. We see that JPEG2000 compresses better than JPEG at all data rates. We also see that EHMR-HE is able to compress better than JPEG for data rates above 0.4 bytes per pixel. It also does marginally better than JPEG2000 at some data rates above 1 byte per pixel. Thus, EHMR-HE, a DCT based multiresolution image scheme, is able to achieve compression levels comparable to DWT based JPEG2000.

The compression level at which EHMR-HE gives better PSNR than JPEG2000 varies widely for the images in our test set. For Barbara, Kodim23, Monarch and Peppers, JPEG2000 is better for all levels of compression. On the other hand, EHMR-HE is better at certain levels of compression for the other images.

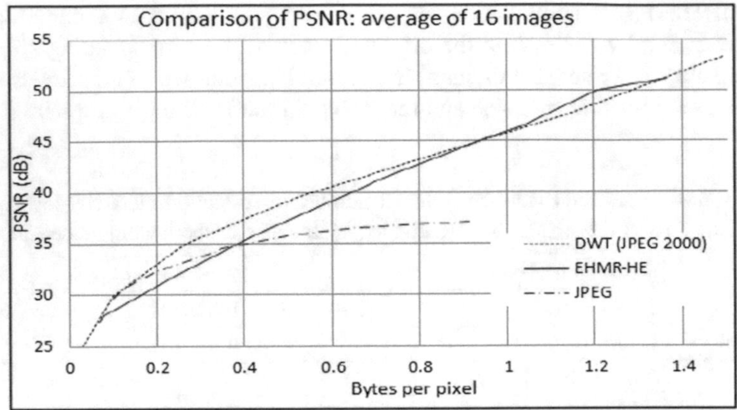

Fig. 5. PSNR vs. compression using different methods.

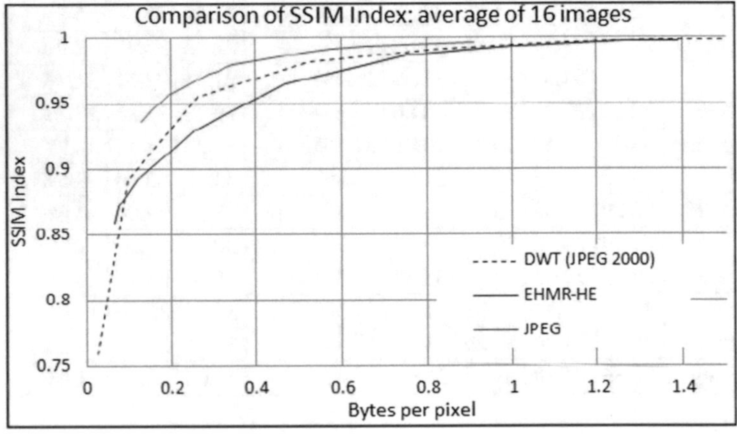

Fig. 6. SSIM index vs. compression using different methods.

SSIM index vs. Compression

Figure 6 shows the average SSIM index we get for a given compression. The results are unexpected. We see that JPEG gives a better SSIM index than JPEG2000 or EHMR-HE. However, the range of data rates achieved by JPEG is smaller than the others. At higher data rates, EHMR-HE does as well as JPEG2000.

6 Conclusion

In this paper, we proposed a hierarchical method for multiresolution image representation. We have shown that EHMR has several advantages over both traditional DCT and DWT based schemes. The model is ideally suited for various image and video consumption scenarios. EHMR separates the LR image and difference image. These can be used to generate higher resolution images. Depending on bandwidth availability, a service can have the option to provide the difference image or to not provide it. This makes it ideal for situations where bandwidth is unpredictable.

We have evaluated the impact of the transformations we carry out, in terms of certain gross parameters. The parameters were PSNR, SSIM, bytes-per-pixel, entropy, and standard deviation.

References

1. CCITT Study Group VIII and JPEG of ISO/IEC: Information Technology – Digital compression and coding of continuous-tonne still image - Requirements and Guidelines. Recommendation T.81. Technical report, ISO/IEC (1993)
2. Wiegand, T., Sullivan, G.J., Bjontegaard, G., Luthra, A.: Overview of the h.264/avc video coding standard. IEEE Trans. Circ. Syst. Video Technol. **13**(7), 560–576 (2003)
3. Richardson, I.E.G.: H.264 and MPEG-4 Video Compression: Video Coding for Next-generation Multimedia. Wiley, New York (2003)
4. Grange, A., Rivaz, P.D., Hunt, J.: VP9 Bitstream & Decoding Process Specification. Technical report, Google (2016)
5. Joy Prabhakaran, P., Poonacha, P.G: A new decimation and interpolation algorithm and an efficient lossless compression technique for images. In: 2015 Twenty First National Conference on Communications (NCC) (2015)
6. Li, Z.-N., Drew, M.S., Liu, J.: Lossless Compression Algorithms, pp. 185–224. Springer, Cham (2014)
7. USC-SIPI images. http://sipi.usc.edu/database/database.php. Accessed 31 July 2017
8. Kodak true color images. http://r0k.us/graphics/kodak/. Accessed 31 July 2017
9. Index of testimages. http://www.hlevkin.com/TestImages/. Accessed 31 July 2017

Prefiltration Analysis for Image Recognition Algorithms for the Android Mobile Platform

Kamil Buczel and Jolanta Wrzuszczak-Noga(✉)

Faculty of Computer Science and Management, Wroclaw University of Science and Technology, Wyb. Wyspiańskiego 27, 50-370 Wrocław, Poland
200420@student.pwr.edu.pl,
Jolanta.Wrzuszczak-Noga@pwr.edu.pl

Abstract. The article discusses text recognition techniques related to mobile technologies. Android application was developed for comparison of applying separate and joining of filtering methods. Three metrics based on classification of text were proposed and compared and the recognition time was analyzed.

Keywords: Image recognition · Mobile system · OCR system

1 Introduction

Text recognition and its conversion into a digital medium is one of the IT areas involving transforming images that began in the 1950s. The task of OCR (Optical Character Recognition) is the recognition of text in a scanned document so that it becomes editable. Currently, OCR techniques also involve handwriting and features of formatting [1, 3, 4, 16].

However, the problems and challenges associated with OCR techniques are changing [1, 13]. Papers [14, 15] present approach of using camera of mobile device to apply the proper filtration for image quality improve.

This article describes the process of image recognition and the influence of filtering sets on the selection of parameters of the selected classifier. The filtration algorithms were compared according to the Jaro-Winkler distance-quality score indicators, Levenshtein and the total assessment of indicators.

The contribution of this paper is to develop mobile application which bases on joining of filtering methods and test them for bills (in different form of lightness, number of folds) considering three classification metrics and time of recognition process. The aim of the work is to test of filtering methods in context of classification and time of recognition process.

2 Background

Research on the subject of text recognition began in the early 1950s. Initially, mechanical imaging was used, followed by its optical processing using rotating discs (rollers) and photocells, moving point scanners and photocell matrix [7, 8].

J. Świątek et al. (Eds.): ISAT 2019, AISC 1051, pp. 330–339, 2020.
https://doi.org/10.1007/978-3-030-30604-5_30

Another problem was the fact that in the early 1960's the OCR machines were error-prone. It was especially the case when they were dealing with poor quality prints. This was due to the wide range of fonts used or the folding of the paper surface.

In order to improve the efficiency of algorithms, vendors of OCR mechanisms pushed for the standardization of fonts, paper and ink used in printouts to be digitally processed later.

New fonts such as OCRA and OCRB were designed in the 1970s successively by the ANSI (American National Standards Institute) and ECMA (European Computer Manufacturers Association) [3]. They were quickly approved by ISO (International Standards Organization) in order to facilitate the OCR process. As a result, it was possible to obtain very good results of word recognition at high speed and at a relatively low cost.

Most OCR systems can not handle incomplete text and handwritten characters or words [2, 12, 13, 16].

Mobile Platforms

Due to the widespread use of mobile devices, the need for data collection and processing has increased even more. Current devices also provide resources that allow one to perform tasks such as digitizing text faster. The problem of the lack of standardization of fonts and distribution of presented data has returned. In addition, as images are not scanned, but captured by the camera, the process is significantly impaired by the bends or shadows that did not occur during traditional scanning.

The next chapter shows the process of text identification and filtration methods. Chapter Four contains measures of textual compliance, while the fifth one consists of the description of the experiment. The sixth describes the results of the research. The last chapter describes future developments.

3 The Process of Text Identification and Filtering Methods

Computer text recognition system involves three stages:

1. Pre-filtering - improves the quality of the source material and locates the area containing the data to be recognized,
2. Feature extraction stage - identifies and captures unique features of characters in a digital version undergoing recognition,
3. Classification stage - converts feature vectors to identify characters and words.

When using mobile devices, the contamination of the input material often takes place, due to uneven light, bends (Table 1). The use of one type of filtration does not lead to improved image clarity [9–11].

Table 1. Example of an image with uneven light and examples of filtration.

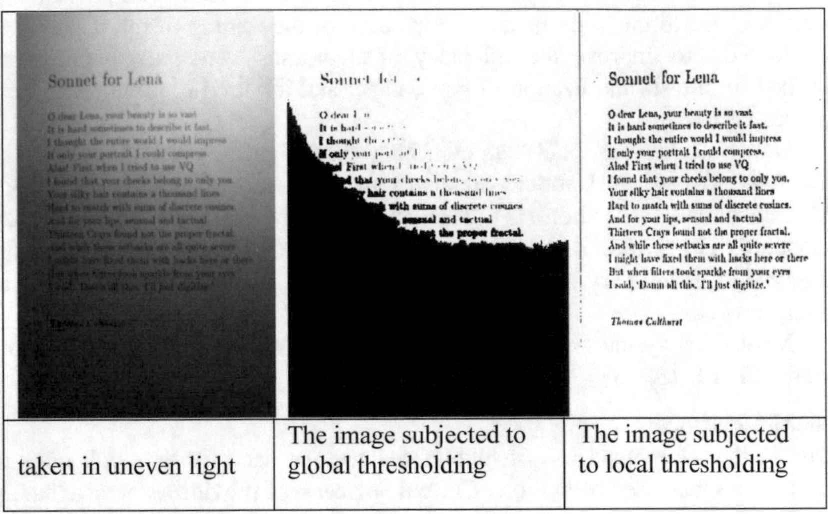

taken in uneven light	The image subjected to global thresholding	The image subjected to local thresholding

Filtration methods:

3.1 Standardization

The purpose of character standardization is to reduce the intra-class variability of shapes and character/digits colors to facilitate the extraction of objects and to improve their classification accuracy [7]. Filtration based on histogram alignment is intended to perform operations whose result in improved properties of graphics that are characterized by poor contrast. The conversion based on contrast compensation may not bring the desired effects. No further filtering in the form of binarization, for example, or using local algorithms can result in the deterioration of the classifier's performance.

3.2 Global Thresholding

Segmentation involves the division of the sample into areas meeting the homogeneity criterion. The most common, but also the simplest method of image segmentation, is thresholding. In general, it involves the determination of the T threshold value (on the scale of the analyzed sample) and comparing the value of each point of the image with the assumed value. By selecting the appropriate threshold value, it is possible to filter out the background, with interferences from the area of interest [8].

The way the global thresholding works is as follows:

1. The selection of the estimated T threshold value
2. Performing image segmentation using the T value. It will divide the image into two groups of pixels. G1 consisting of all pixels with values greater than T and G2 consisting of pixels with values less than or equal to T.

Global thresholding is characterized by speed and high efficiency for images with a simple histogram. However, its performance can bring very unwanted results in the case of photographs with, for example, uneven light.

3.3 Local Thresholding

Local thresholding, on the other hand, is based on the fragmentation into smaller areas (the so-called blocks) for which individual threshold values are calculated. The operation is based on estimating a unique threshold for each pixel depending on the grayscale pixel information of the neighboring pixels [9]. The main problem with the local thresholding is that we only consider the intensity difference, not the relationship between the pixels. This does not guarantee that the pixels are identified by the thresholds that are characterized by continuity of information. Easily, pixels which are not a part of the desired region can be included in our collection of features, and similarly single pixels in the region can be missed out (especially near the region's borders). Two static methods for determining the valuc of the local threshold are included in the article - the mean and the Gaussian methods.

4 Measures of Correctness of Text Classification

The Levenshtein and Jaro-Winkler method was used to determine similarity, as well as a summary listing of the indicators, which were referred to as the coefficient of conformity.

4.1 The Levenshtein Distance

It is a measure of the difference between two sequences (editing distance). The Levenshtein distance between two strings is defined as the minimum number of changes needed to convert one string to the other, with the allowed editing operations being:

- inserting,
- deleting,
- replacing a single character.

The Levenshtein algorithm (also called Edit-Distance) calculates the smallest number of editing operations listed above that are necessary to convert one string of characters in order to obtain another text [5]. The Levenshtein distance between the words "kitten" and "sitting" is three.

4.2 The Jaro-Winkler Distance

It is used to measure distances (differences) between two sequences of characters. The variant suggested in 1999 by William E. Winkler to the Jaro distance metric [5]. Informally, the Jaro distance between two words is the minimum number of individual transpositions required to change one word to another.

The Jaro-Winkler's string similarity analysis method is as follows (1):

$$d_w = \begin{cases} d_j & if d_j \leq weightThreshold \\ d_j + 0.1 prefixMatch(s1, s2, numChars)\left(1 - d_j\right) & if\ d_j > weightThreshold \end{cases} \quad (1)$$

where,
prefixMatch - is a length of common prefix of s1 and s2 limited to numChars
d_j – probability of Jaro distance

The Jaro-Winkler distance is described hereinafter by $\varphi_{\text{JaroWinkler}}$ and given as a percentage.

4.3 Total Distance

Contains the Jaro-Winkler and Levenshtein measures and is described as (2)

$$\varphi_{Compatibility} = \frac{\varphi_{JaroWinkler} + \frac{175 - min(\varphi_{Levenshtein}, 175)}{175}}{2} * 100\% \quad (2)$$

5 Experiment

The experiment consisted in creating a mobile application enabling the filtration and recognition of images and conducting research on various samples and different filtration methods.

5.1 Mobile Application

The application was created based on the MVP (Model-View-Presenter) architecture with the use of tools (Table 2).

Table 2. Tools used to implement the mobile application.

Name	Final version	Implementation for android
Tesseract	2017	Yes
ExperVision	2010	Yes
ABBYY FineReader	2017	Yes
Asprise OCRSDK	2015	Java
OpenALPR	2017	Yes

The application includes automatic and manual parts. The whole process was automated so that the reports containing the results and the processed images were saved in the phone's memory, marked with the appropriate names and descriptions.

The manual part of the application allows to select the type and sequence of the filters. In addition, it allows the user to change the parameters of the respective transformations. Its purpose is the visual assessment of the impact of the respective transformations on the test sample by the user.

The test device with the help of which the photographs were taken and the analyzes carried out was Honor 7.

5.2 Conducted Research

Three pictures (marked E1, E2, E3) of the same fragment of text, differing in lightness, sharpness, degree, and number of folds, were subjected to filtration and classification (Table 3). For comparison purposes, the classification of images without filtration was performed.

Input data were subjected to the following processes:

- filtration through standardization,
- filtration through global thresholding,

Table 3. Input data - samples and their histograms.

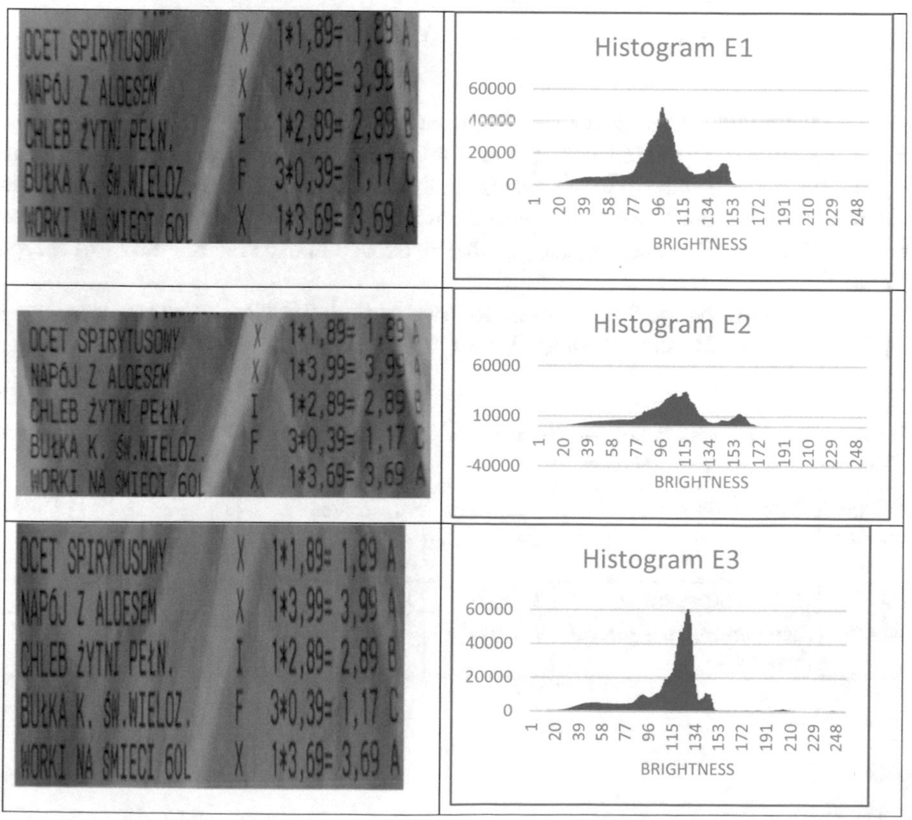

- filtration through local thresholding,
- filtration through standardization and then global thresholding,
- filtration through standardization and then local thresholding,

then, following the filtration process, all the pictures were subjected to classification.

6 Results of Experiment

The results of the experiment are presented in Tables 4, 5 and 6 and Fig. 1. Tables 4 and 5 present the results of image recognition for global thresholding and the set of standardization and global thresholding for samples E1 and E3.

Table 4. Results for the E1 sample subjected to global thresholding and for the combination of filtration with prior standardization.

Sample number	Type of filtration	$\varphi_{\text{JaroWinkler}}$ [%]	$\varphi_{\text{Levenshtein}}$ [number of transformations]	$\varphi_{\text{Compatibility}}$[%]	Time [s]
E1 Sample	Global thresholding	76.60%	29	80.01%	6,56
	Standardization + Global thresholding	79.20%	24	82.73%	4,822

With the Levenshtein distance the improvement by 5 transformations is a relatively high achievement, even considering that it translates into an increase of compatibility of just over 2.5% (Table 4). In the results for the third sample the greatest improvement was recorded. For the Levenshtein measure it is 70%, limiting the number of required transformations to 3. Value $\varphi_{\text{JaroWinkler}}$ improved by almost 17% for the third sample (Table 5).

The time of recognition process for combining filtration methods was lower comparing to one filtration method (Tables 4 and 5).

Table 5. The results for the E3 sample subjected to global thresholding and for the combination of filtration with prior standardization.

Sample number	Type of filtration	$\varphi_{\text{JaroWinkler}}$ [%]	$\varphi_{\text{Levenshtein}}$ [number of transformations]	$\varphi_{\text{Compatibility}}$[%]	Time [s]
E3 Sample	Global thresholding	82.90%	10	88.61%	1,848
	Standardization + Global thresholding	99.60%	3	98.97%	1,595

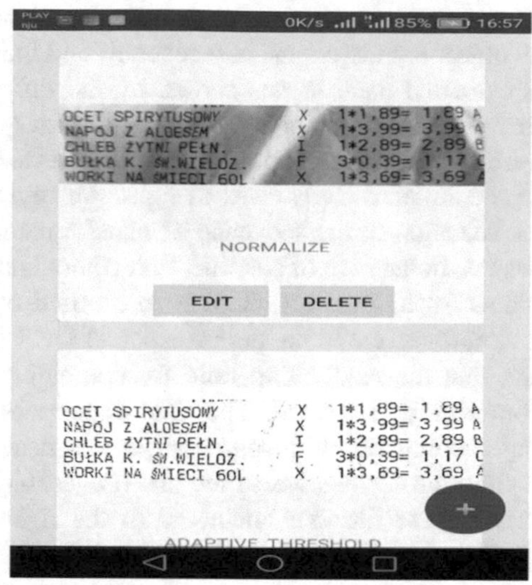

Fig. 1. Example of E1 sample subjected to standardization and local thresholding.

Table 6. Test results for samples E1, E2, E3 without filtration and subjected to standardization and local thresholding.

Sample number	Type of transformation	$\varphi_{\text{JaroWinkler}}$[%]	$\varphi_{\text{Levenshtein}}$ [number of transformations]	$\varphi_{\text{Compatibility}}$ [%]	Time [s]
E1 Sample	Without filtration	68.10%	96	56.60%	23,91
	Standardization +Local thresholding	86.97%	19	88.06%	7,71
E2 Sample	Without filtration	69.80%	89	59.45%	17,02
	Standardization +Local thresholding	85.59%	18	87.65%	9,92
E3 Sample	Without filtration	76.70%	34	78.65%	2,18
	Local thresholding	100.00%	0	100.00%	3,46

Applying the filtration increases the indicator $\varphi_{\text{JaroWinkler}}$ by about 20%. The number of transformations for sample E1 and E2 decreased approximately 4 times ($\varphi_{\text{Levenshtein}}$). Value $\varphi_{\text{Compatibility}}$ was by about c. 30% higher for the filtration method used, standardization and local thresholding or local thresholding as compared to no filtration (Table 6).

The accuracy of data extraction is crucial for the functioning of the character recognition system, as object extraction and text recognition efficiency depend largely on the quality of the extracted data. It was shown by the differences in the results achieved, in particular by means of the compatibility indicator.

The best results were obtained for the third sample, which was characterized by the highest sharpness and the smallest differences in light. An important element of the filtration process is the selection of the sequence of transformations.

For E1 and E2 samples, in the case of classification without any filterings, reaching a compliance level below 70%, local thresholding preceeded by pre-standardization resulted in the greatest improvement in the performance of OCR techniques (Table 6).

General results are, that the recognition time for combining process filtration is lower comparing to one filtration method. The time of process recognition without filtration is usually longer comparing to recognition process using filtrations methods. OCR process without filtrations brings worse metrics (Levenshtein, Jaro-Winkler and Combined index) comparing to filtration methods, so the right way of using OCR systems is to use as the first stage filtering based on standardization and as second stage thresholding techniques.

7 Directions of Future Developments

It would be sensible to carry out experiments to demonstrate the relationships between individual parameters and the working time of OCR techniques.

In addition, the range of tested filters should be increased, the iterations between the individual parameter values should be expanded, and additional methods of automatic selection of the settings should be added. The extension of the work by machine learning, additional filters and handwriting analysis would be a good foundation for wider research.

References

1. Cherietm, M., Khaarma, N., Liu, C.L., Suen, C.: Character Recognition Systems: A Guide for Students and Practitioners. Wiley, New Jersey (2007)
2. Poovizhi, P.: A Study on preprocessing techniques for the character recognition. Int. J. Open Inf. Technol. **2**(12), 21–24 (2014)
3. Schantz, H.F.: The History of OCR. Recognition Technologies Users Association, Boston (1982)
4. Christen, P.: A comparison of personal name matching: techniques and practical issues. In: Joint Computer Science Technical Report Series, pp. 1–14 (2006)
5. Jaro, M.A.: Advances in record linkage methodology as applied to the 1985 census of Tampa Florida. J. Am. Stat. Assoc. **84**(406), 414–420 (1989)
6. Kay, A.: Tesseract is a quirky command-line tool that does an outstanding job. Linux J. 24–29 (2007)
7. Russ, J.C.: The Image Processing Handbook Sixth Edition, Raleigh. Taylor and Francis Group, LLC, North Carolina (2011)

8. Dougherty, G.: Digital Image Processing for Medical Applications. Cambridge University Press, Cambridge (2009)
9. Huang, Z.-K., Chau, K.-W.: A new image thresholding method based on gaussian mixture model. Appl. Math. Comput. **205**, 899–907 (2008)
10. Chaubey, A.K.: Comparison of the local and global thresholding methods in image segmentation. World J. Res. Rev. **2**, 01–04 (2016)
11. Firdous, R., Parveen, S.: Local thresholding techniques in image binarization. Int. J. Eng. Comput. Sci. **3**, 4062–4065 (2014)
12. Isheawy, N.A., Hasan, H.: Optical Character Recognition (OCR) system. IOSR J. Comput. Eng. **17**(2), 22–26 (2015)
13. Deshpande, S., Shiram R.: Real time text detection and recognition on hand held objects to assist blind people. In: International Conference on Automatic Control and Dynamic Optimization Techniques (ICACDOT). IEEE (2016)
14. Springmann, U., Fink, F., Schulz, K.: Automatic quality evaluation and (semi-) automatic improvement of mixed models for OCR on historical documents, CoRR (2016)
15. Chao, S.L., Lin, Y.L.: Gate automation system evaluation: a case of a container number recognition system in port terminals. Marit. Bus. Rev. **2**(1), 21–35 (2017)
16. Islam, N., Islam Z., Noor N.: A survey on optical character recognition system. arXiv preprint arXiv:1710.05703 (2017)

Author Index

J. Świątek et al. (Eds.): ISAT 2019, AISC 1051, pp. 341–342, 2020.
https://doi.org/10.1007/978-3-030-30604-5